NATURAL PRODUCTS AS ANTIVIRAL AGENTS

NATURAL PRODUCTS AS ANTIVIRAL AGENTS

Edited by

Chung K. Chu

The University of Georgia
Athens, Georgia

and

Horace G. Cutler

United States Department of Agriculture
Agricultural Research Service
Athens, Georgia

SPRINGER SCIENCE+BUSINESS MEDIA, LLC

Library of Congress Cataloging-in-Publication Data

Natural products as antiviral agents / edited by Chung K. Chu and
  Horace G. Cutler.
       p.   cm.
   "Proceedings of an American Chemical Society Agricultural and Food
Chemistry Division symposium on natural products as antiviral
agents, held August 27, 1991, in New York" --Copr. p.
   Includes bibliographical references and index.
   ISBN 978-1-4613-6509-9   ISBN 978-1-4615-3414-3 (eBook)
   DOI 10.1007/978-1-4615-3414-3
   1. Antiviral agents--Congresses.   2. Natural products--Therapeutic
use--Congresses.   I. Chu, Chung K.   II. Cutler, Horace G., 1932-
. III. American Chemical Society.   Division of Agricultural and
Food Chemistry.
RM411.N42   1992
616.9'25061--dc20   92-27287
                                                              CIP

Proceedings of an American Chemical Society Agricultual and Food
Chemistry Division symposium on Natural Products as Antiviral Agents,
held August 27, 1991, in New York, New York

ISBN 978-1-4613-6509-9

© 1992 Springer Science+Business Media New York
Softcover reprint of the hardcover 1st edition 1992
Originally published by Plenum Press, New York in 1992

PREFACE

During the past fifty years, thousands of natural products have been isolated from plants, fungi, and bacteria. Apart from intense searches by pharmaceutical companies for medicinals and the concentrated effort mounted by the National Cancer Institute, many of these have not been tested in biological systems. The major reasons for this appear to be, at least, twofold. First, individual researchers looking for biologically active natural products will often isolate only small amounts of material sufficient to determine a structure and calculate the specific activity for their particular bioassay systems: insufficient funds preclude re-isolating the compound unless industrial potential is foreseen. Second, the difficulty with which original structures were proved prior to 1972. This required the isolation of relatively large quantities of a natural product and there followed extensive degradation, elemental analyses of the parent and its fragments, then synthesis, piece by piece, of the molecule. All this took time and energy. No wonder that when the structure was proved the chemist was enervated. And coupled to this was the fact that many chemists were not trained to test their materials in biological systems. In contrast, today a natural product can be isolated, its mass and molecular formula determined and, if there is some serendipity, crystals may be obtained for single crystal x-ray analysis. If conditions are near perfect, it is possible to isolate and identify a novel compound in a month. Additionally, there now exist multiple biological screening systems that are robotic and require only a few milligrams of material.

Juxtaposed to the earlier discoveries of thirty to fifty years ago there has been the advent of new diseases. One of these AIDS, has the potential for enormous destruction which some have described, perhaps hyperbolically, as becoming a second Black Plague. But, the impact of AIDS on the world could not have been understood, or predicted, twenty-five years ago. Hence, many natural products need to be re-isolated and assayed as antiviral agents in addition to those that have yet to be discovered.

We are especially grateful to the Agricultural and Food Chemistry Division of the American Chemical Society who generously sponsored the Symposium, "Natural Products as Antiviral Agents" at the 202nd Society Meeting in New York City, August, 1991. It is because of that gathering that our colleagues came together to produce this book and we thank them for their contributions. Special thanks to our research groups who carefully read all the manuscripts and made valuable suggestions.

C.K. CHU, Ph.D.
Professor of Medicinal Chemistry
Department of Medicinal Chemistry
College of Pharmacy
The University of Georgia
Athens, Georgia

HORACE G. CUTLER, Ph.D.
Research Leader,
Microbial Products Unit
Matrix Team USDA,
Agricultural Research Service
Coordinator, Natural Products
Athens, Georgia

CONTENTS

PROGRESS IN THE DEVELOPMENT OF NATURAL PRODUCTS FOR HUMAN IMMUNODEFICIENCY VIRUSES INFECTIONS

Raymond F. Schinazi

Emory University School of Medicine, Department of Pediatrics, Atlanta, Georgia 30322, and Veterans Affairs Medical Center, Decatur, Georgia 30033

INTRODUCTION

Diverse antiviral compounds are needed to meet the challenge of infections caused by viruses responsible for high morbidity and/or mortality. These viruses include human immunodeficiency viruses (HIV-1; HIV), herpesviruses, hepatitis B virus, influenza viruses, and exotic viruses. Certain antiviral compounds have been successfully derived from a variety of natural products whose structure and composition could not have been predicted by even the most imaginative of chemists. The selectivity of these antiviral agents hinges upon unique biochemical events specific for viral replication but not cellular growth. Inhibition of viral specific receptors on cells also represents an attractive target. This review addresses some of these approaches with emphasis on the potential role of natural products in treating HIV infections.

As of January 1992, more than 200,000 cases of AIDS in the United States have been reported to the Centers for Disease Control. After a latent period this infection invariably leads to death. Thus, it is imperative to develop effective, safe, antiviral drugs to suppress HIV-1 replication, prevent the progression of disease, and treat individuals with AIDS or AIDS-related complex (ARC).[1] Any step in the biosynthesis of HIV, the causative agent of AIDS, is susceptible to attack by antiviral intervention.[2,3] The most prominent of these targets is the virus-coded reverse transcriptase (RT), and the most elusive is the integrated virogene.

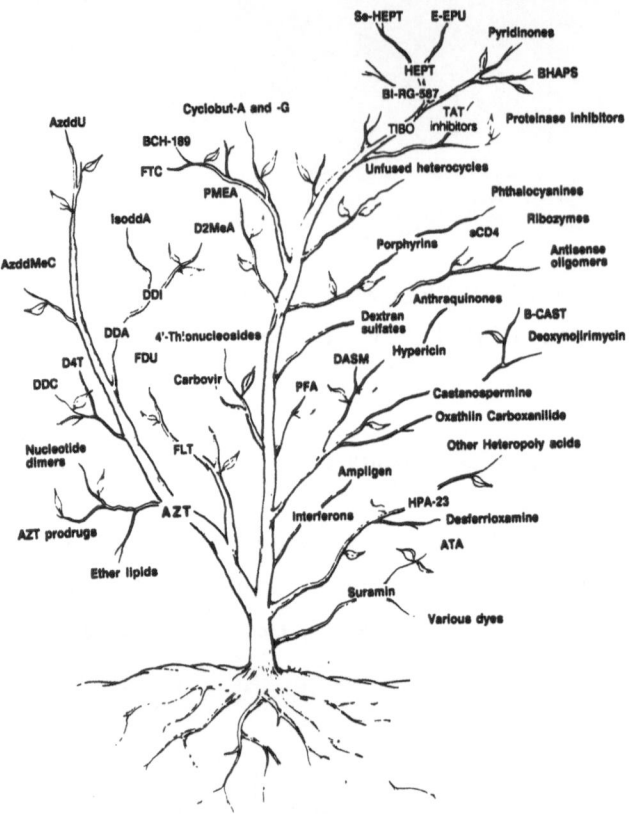

Figure 1. Evolution of antiviral compounds against HIV-1. Left side are nucleoside analogues. Right side are non-nucleoside analogues.

Multicentered placebo controlled studies demonstrating the clinical efficacy of 3'-azido-3'-deoxythymidine (AZT, zidovudine) for the treatment of AIDS have paved the way for the development of more potent and/or less toxic compounds.[4] There are great expectations that a variety of new approaches will prove to be more effective. However, genetically engineered forms of various proteins, and in particular soluble CD4, a molecule that interferes with the binding of the viral gp120 to CD4 receptors on lymphocytes, have been shown to have a narrow spectrum of activity against wild type virus and have so far failed to control HIV infection in humans.[5,6] Various CD4-antibody hybrid molecules are being evaluated.[7] More recently, the expectations of non-nucleoside RT antiviral agents have not materialized. It was found that these compounds are active only against HIV-1, and drug-resistant virus develops rapidly.[8,9]

Controlling virus infection appears to be a formidable task since we are dealing with a variety of cell types which harbor virus acutely, chronically, and latently. Modern antiviral chemotherapy will have to target these different infected cells. Reliable antiviral cell culture systems which mimic closely the situation in humans have been developed and others are still being perfected. It is necessary to find compounds or combined modalities that will suppress or inhibit the spread of HIV-1 in humans. We hypothesize that HIV infection can be controlled by safe and effective drugs that act in concert on these various types of infected cells. Studies on the mechanism by which HIV-1 establishes and maintains latency, and factors that cause reactivation of latent virus, are of critical interest. Current research on these problems will help us develop more potent and specific compounds. In addition, improved assays or animal models for predicting activity and toxicity of new agents in humans are also necessary.

This critical review focuses on some of the natural compounds that are being developed to treat HIV-1 infections. It is not meant to cover the subject exhaustively. Nucleoside analogues, oligomers, and synthetic dsRNA such as ampligen, which may be considered "natural products," will not be discussed. Of significance is that over the last 5 years the scientific community has rallied to develop a multitude of potential treatments (Figure 1). This response has now resulted in a variety of antivirals that have been approved or are close to approval for clinical trials in humans. The information on the safety and efficacy of these compounds will form the foundation for the development of additional antiviral agents for the prevention and treatment of this devastating disease.

BIONATURAL PRODUCTS

Interferons

Interferon alpha (IFN-α) is a leukocyte derived glycoprotein with antiviral and antiproliferative properties which has been extensively studied for its effect on AIDS and Kaposi's sarcoma. It has also been shown to reduce Kaposi's sarcoma lesions.[10,11] IFN-α inhibits HIV *in vitro*,[12] and in asymptomatic HIV positive individuals or early AIDS patients reduces opportunistic infections, p24 antigen, and increases circulating CD4 cells.[13,-15] However, in advanced cases of AIDS, IFN-α does not appear to be of value.[16] These studies suggest that IFN-α is probably of more benefit in patients whose immune function is still intact.

The anti-HIV activity of IFN-β or –γ as well as IFN-α has been demonstrated *in vitro* [17] in acute and chronic infections.[18] However, in another study, INF-α and -β but not -γ were shown to inhibit HIV.[19] Scanning EM observations in acutely and persistently infected cells revealed that IFN-α treatment caused accumulation of intact virus particles on the cell surface.[20] The finding that IFN-α suppresses HIV-1 replication in lymphocytes can be explained as being due to the blocking of released virus particles from infected cells. These findings were recently confirmed by Smith et al. using acutely infected C3 and Jurkat cells. In chronically infected cells, the inhibitory effect of IFN appeared to reflect decreased cell growth.[21] In contrast, Geldenman et al.[22,23] demonstrated that IFN-induced antiviral activity was more significant in monocytes than in T lymphocytes, resulting in essentially latently infected cells. The authors suggest that, in this system, IFN-α interrupts one or more early events in the virus replication cycle before the formation of proviral DNA.

Because IFN-α can be toxic at effective anti-HIV concentrations and also interferes with virus replication at a non-RT level, most current studies combine it with other antivirals, usually AZT.[24] Good synergism was shown in a study of IFN-α and AZT *in vitro*, and good *in vivo* therapeutic responses were also seen.[25] Combined IFN-α and DDC has also shown *in vitro* activity.[26] Granulocyte-macrophage colony-stimulating factor was reported to mitigate the neutropenia of combined IFN-α and AZT used in the treatment of AIDS-associated Kaposi's sarcoma.[27,28]

The major problem with using IFN in the treatment of HIV infections is that it cannot penetrate the blood-brain barrier and intravenous treatment is necessary. This may be overcome with IFN inducers or IFN preparations which are orally bioavailable.

Castanospermine (CAST), 1-deoxynojirimycin (dNM)

Several plant alkaloids are glycosidase inhibitors and have been reported to have antiviral activity.[29] These appear to interrupt glycoprotein processing by the inhibition of the endoplasmic reticulum enzyme α-glucosidase 1, which removes terminal glucose residues from precursor oligosaccharide side chains during the synthesis of mature N-linked complex glycoproteins. These compounds have been shown to block syncytium formation in CD4+ cell lines and provide protection against HIV infection in H9 cells.[30-32] Schols et al.[32a] recently reported that the activity can depend on the virus strain and cell type. *In vivo* experiments with CAST have shown moderate or no activity in Rausher leukemia virus,[33,34] but an analog, B-CAST, demonstrated antiretroviral activity as effective as AZT in inhibiting splenomegaly in Friend leukemia virus infected mice.[35]

N-(n-butyl)-deoxynojirimycin (BuDNJ) has been reported to be more potent at inhibiting HIV replication *in vitro* than CAST by reducing virus titers at much greater magnitudes.[36] Several derivatives of this compound were shown to inhibit replication of HIV. These derivatives are thought to induce hyperglycosylation of the transmembrane glycoproteins gp120/160 and gp41.[37]ˈ The biological consequences of hyperglycosylation are not known, although conformational changes in gp120 following its production in the presence of BuDNJ may contribute to the antiviral effect observed with this compound.[38] One of the limitations of these compounds may be their toxic effect on normal cell populations.[39,40] They may be better suited for combination therapy since lower doses could be utilized, and their synergism with AZT has been demonstrated.[41] Our group has been unable to confirm the anti-HIV-1 activity of CAST in human PBM cells infected with strain LAV.

DASM

Dehydroangrographolide succinic acid monoester was recently reported to selectively inhibit HIV-1 replication in H9 and PBM cells.[42] This compound is prepared from andrographolide, a major component extracted from the Chinese herb *Andrographis paniculata*. Andrographolide itself has no antiviral activity, suggesting that this compound needs to be derivatized to a more polar molecule in order to confer antiviral properties. DASM, which is composed of two different isomers, partially interfere with HIV-induced cell fusion and with binding of HIV to H9 cells. DASM failed to induce an IFN state in H9 cells. The main mechanism of action of this unusual molecule has yet to be determined. The selective antiviral activity of DASM in human PBM cells infected with HIV-1 was recently confirmed by our group (Table 1). As anticipated, DASM-E was not cross-resistant with AZT-resistant virus (Table 2).

Table 1. Effect of natural products and nucleosides on HIV-1 infection in acutely, chronically, and latently infected cell culture systems.

Compound	EC_{50} or IC_{50} in μg/ml or μM*			
	Acute PBMC Day 6[1]	Chronic H9 Day 14[1]	Latent OM10.1 Day 2[1]	^{3}H-dThd PBMC Day 2[2]
DASM-E	3.8	> 100	> 100	> 100
Compound Q (GLQ-223)	0.021	1.12	> 100	19.0
MAP-30	0.015	ND	> 100	15.5
Desferrioxamine	3.7*	> 100*	3.7*	35*
3'-Azido-3'-deoxy-thymidine (AZT)	0.006*	> 100*	> 100*	ND
3'-Fluoro-3'-deoxy-thymidine (FLT)	0.004*	79.1*	7.4*	ND

[1] RT activity was used as a measure of virus yield.
[2] Effect on cellular DNA synthesis in PHA-stimulated human PBM cells.

Table 2. Effect of DASM on a pair of AZT-sensitive and resistant HIV-1 isolates.

Compound	AZT-resistant HIV-1 EC_{50}	Slope	AZT-sensitive HIV-1 EC_{50}	Slope	Fold-increase
AZT	0.78 μM	1.12	0.0032 μM	1.92	244
DASM	25.2 μg/ml	2.99	11.82 μg/ml	1.74	2.1

Gossypol

Gossypol is a natural product extracted from the cottonseed which has been extensively studied as a potential male contraceptive and has been well reviewed by Qian and Wang.[43] Gossypol also has antiparasitic activities[44,45] and antiviral activities against enveloped viruses such as HSV-2 and parainfluenza virus type 3, but not against nonenveloped poliovirus.[46,47] Polsky et al.[48] incubated the HTLV-IIIB strain of HIV with gossypol in a cell-free medium and found that at 100 µM concentrations it prevented recovery of viable virus when subsequently incubated with H9-T cells. When isolated from cottonseed, gossypol is a racemic mixture; the enantiomeric forms are due to atropisomerism, that is, restricted rotation of the two naphthyl rings about the interlinking C-C bond. The two enantiomers have been resolved by several groups.[49] The racemic mixture (±) and the two pure enantiomers were evaluated against HIV-1 by our group in collaboration with Yale University scientists.[50] In contrast to the data of Polsky et al.[48] who evaluated racemic gossypol, we found the (-) enantiomer of gossypol to be very inhibitory (EC_{50} = 1.0-5.0 µM), while the (+) enantiomer was not active (EC_{50} = 50-100 µM), and the cytotoxicity of the (-) enantiomer to uninfected PBM cells was >100 µM. The potential significance of these observations is decreased formation of infected semen of male patients, and hence, its use prior to conception might decrease the incidence of infants born with AIDS. More than 10,000 men have been given gossypol for antifertility, and there is a very low incidence of side effects.

Hypericin, Quinones

These compounds are polycyclic quinones, occurring naturally in plants of the genus *Hypedicium* (St. John's wart). They interfere with the assembly of mature viral particles from infected cell lines and inactivate retroviruses directly.[51,52] They are known potent inhibitors of protein kinase C[53] and are dependent upon the presence of viral envelopes.[54] Antiretrovirus activity has been demonstrated in EIAV-infected cells, where virus was reduced by 99.99%.[55] Anti-HIV activity has been reported, especially on exposure to fluorescent light.[56] Inhibition of retrovirus-induced disease in mice has also been demonstrated.[51] The activity of a variety of anthraquinones related to hypericin has been extended to other viruses such as HSV-1, HSV-2, parainfluenza, and vaccinia virus.[57] Hypericin has been claimed to enter both lymphocytes and macrophages in culture, although there is no unequivocal proof of intracellular uptake.[58]

It is known that certain quinones possess interesting biological effects such as antibacterial, antiviral, and antitumor activity. These properties may be related to their ability to form free radicals or to intercalate with nucleic acids. For example, streptonigrin, an antimicrobial and antitumor antibiotic, and its analogs exhibit potent inhibitory activity on avian myeloblastosis virus RT.[59] Daunomycin and adriamycin also inhibit the avian myeloblastosis virus RT.[60] It has been reported that adriamycin inhibits HIV-1 infection and replication.[61] As mentioned above, Meruelo et al.[51] reported that hypericin and pseudohypericin, polycyclic quinone analogs, are also highly effective against a variety of nonhuman retroviruses *in vitro* and *in vivo*. The authors claim that hypericin and pseudohypericin probably interfere with viral infections and/or spread by direct inactivation of the virus or by preventing virus shedding, budding, or assembly at the cell membrane. In contrast to the report that hypericin had no effect on avian or murine RT,[51,62] we found that fresh solutions of hypericin inhibited HIV-1 RT at submicromolar concentrations. However, this inhibition was dependent on the protein concentration in the radioactive mix. Various anthraquinones substituted with hydroxyl, amino, halogen, carboxylic acid, substituted aromatic group, and sulfonate were tested to determine their activity against HIV-1 in primary human lymphocytes. Among the compounds tested, polyphenolic and/or polysulfonate substituted anthraquinones were found to possess the most potent antiviral activity. Hypericin, an anthraquinone dimer previously shown to have activity against nonhuman retroviruses also exhibited anti-HIV-1 activity in lymphocytes. Some of the active anthraquinones inhibited HIV-1 RT.[63,64] However, this enzyme inhibition was selective only for 1,2,5,8-tetrahydroanthraquinone and hypericin. Hypericin interacts nonspecifically with protein, suggesting that this interaction may dictate its inhibitory effect on the viral RT.[63]

The hypericin-containing herbal extract has been given to 26 patients and monitored by a community based group over a period of 4 months.[65] Reversible hepatotoxicity was noted in some patients. An initial Phase I study will begin as soon as sufficient compound is available. It appears that the plant material is more potent than synthetic material (Meruello, personal communication).

Alkaloids

Novel isoquinolines derived from the tropical plant *Ancistrocladus abbreviatus* were recently shown to inhibit HIV-1 and HIV-2 cytopathic effects in culture.[66] The active ingredients were found to be a novel atropisomeric pair of michellamines, with the A form exhibiting more potent activity than the B form. The EC_{50} and IC_{50} values in CEM-SS cells were about 20 and 200 µM, respectively. These compounds are quite polar and are polyhydroxylated. Extracts from *A.tectorius* which contain the monomeric alkaloids are not active.

Priel et al. reported that camptothecin, a cytotoxic alkaloid isolated from *Camptotheca acuminata*, inhibited HIV-1 in H9 cells.[66a] This compound was evaluated since it inhibits viral-associated topoisomerase I, which apparently is present in HIV-1, equine infectious anemia virus (EIAV), and Moloney murine leukemia virus. The authors report that this compound did not affect cell viability of H9 cells when cells were exposed to the drug for 1 h, the drug removed, and the viability measured for 3 days. However, cell number was affected when a similar study was performed over 7 days.

Certain alkaloids have also been reported to enhance HIV replication in culture. For example, Schweitzer et al.[66b] recently reported that morphine can stimulate virus replication in pretreated HIV-permissive human Kupffer cells. The significance of these findings with respect to heroin use among HIV-1 infected individuals is not known.

Pine Cone Extracts

Japanese investigators recently reported that aqueous extracts from the Japanese white pine tree (*Pinus parviflora*) have anti-HIV, antitumor, and antimicrobial effects.[67-69] The most potent extract fraction, PC6, inhibited the expression of HIV proteins in CEM cells, but did not affect posttranslational processing of the proteins. PC6 was also shown to inhibit both reverse and forward transcription *in vitro*. Interestingly, supernatant from CEM cells treated with PC6 induced an unknown factor with a molecular weight of 30 kD which had potent anti-HIV activity. This factor has been named HIV-inhibiting factor (HIF). It remains to be determined if highly purified HIF has activity *in vivo*. PC6 is one of the few natural products that dose-dependently inhibit acute and chronically infected cells at relatively nontoxic concentrations. PC6 was shown to inhibit HIV-1 RT by 50% at 4 µg/ml. The effect of the compound on cellular DNA polymerases was not reported.

Melanins

Synthetic melanins obtained by chemical oxidation of L-tyrosine and L-dopa inhibited acutely infected MT-2, H9, and PBM cells by 50% at concentrations of 0.15-10 µg/ml.[70] These compounds were also effective against HIV-2 and SIV, and appear to inhibit binding of gp120 to the CD4 receptor on lymphocytes. However, no effect on HIV-1 RT was noted. The spectrum of activity was specific for retroviruses since the compounds do not inhibit vesicular stomatitis virus or encephalomyocarditis in culture. The pharmacological and toxicological properties of these compounds have not been examined.

Tannins

A variety of tetragalloylquinic acids were isolated from commercial tannic acid and then characterized. The compounds were found to have weak activity against HIV in H9 cells and HIV-1 RT.[71] Unfortunately, these compounds are potent inhibitors of DNA polymerases α, and β, and modest inhibitors of γ. A variety of lignans were reported to inhibit HIV-1 in culture.[72] Extracts of *Hyssop officinalis* contained antiviral fractions that were identified to contain caffeic acid, unidentified tannins, and other active components.[73] Recently Nakashima et al.[72a] reported on the anti-HIV activity in MT-4 cells of 87 chemically defined tannins and related compounds. They demonstrated that some of these compounds inhibited HIV adsorption to the cells.

DHEA

Dehydroepiandrosterone (DHEA) is one of the most abundant naturally occurring adrenal steroids in humans. Several beneficial biological attributes of DHEA have been claimed.[74] This compound is a potent noncompetitive inhibitor of mammalian glucose-6-

phosphate dehydrogenase (G-6-PD), a rate-limiting enzyme which leads to the production of NADPH for biosynthetic reductions in cytosol and the production of ribose-5-phosphate as a precursor for nucleotide synthesis. Our studies indicated that DHEA was a modest selective inhibitor of HIV-1 replication in primary human lymphocytes.[74] In these cells, the drug appeared to be more potent than DHEA-sulfate but less potent than Epi-Br and AZT. DHEA was also a modest inhibitor of HIV-1 replication in human macrophages. Multiple-drug effect analyses were used to quantitatively determine the interaction of AZT and DHEA in human PBM cells infected with HIV-1 at a ratio of 1:1000 and 1:4000. Analyses of the cell culture RT and EIA data indicated mostly antagonistic interactions. At therapeutic levels, no apparent toxicity to uninfected cells was observed for the drugs alone or in combination.[74] The mechanism for the antiviral activity of DHEA and its interaction with AZT in infected human lymphocytes is unclear, and may be related to the inhibition of nucleoside transport and regulation of *de novo* synthesis of nucleotides. Clinical trials with DHEA (EL10) are now ongoing as both an antiviral agent and potential immunomodulator. Studies with other steroids on chronic HIV infections have been reported.[75]

Trichosanthin (GLQ223)

Trichosanthin is a ribosomal inhibitor protein that blocks replication of HIV in lymphocytes and macrophages[76] (Table 1). The structure is similar to the A-chain of the ricin toxin. GLQ223 has been used in China as an abortifacient and for the treatment of trophoblastic tumors. *In vitro* reduction of HIV was observed when GLQ223 was incubated with infected monocyte-derived macrophages. Five days after a single 3 h treatment with the drug, concentration-dependent decreases were observed in all measurable HIV parameters; the effects persisted for at least 28 days.[76] The therapeutic index of this protein varies considerably depending on the source of the drug, its degree of purification, and the cell line used for testing the compound.[77] Rats injected with a high dose of trichosanthin extract (200 µg/kg) for 10 days showed mild to persistent intolerances. Clinical studies have been performed to evaluate the safety and efficacy of the drug. In a recent study, decreases in p24 antigen and increases in CD4+ cell levels were noted when patients were administered the drug in a dose range of 10-30 µg/ml over a 9-21 day period.[78] In a Phase I study where 18 subjects with AIDS or ARC were administered the drug, no sustained changes were observed in the CD4+ lymphocyte population or HIV antigen levels.[79] The associated toxic effects of GLQ223 when used in the clinic have been reported.[80]

MAP-30

This single chain 30 kDa protein, isolated from a plant indigenous to China called *Momordica charantia*, inhibits HIV-1, HSV-1, and poliovirus I replication in culture (Table 1).[81] Extracts from the fruit have antitumor activity and immunopotentiating ability. Cell culture studies indicate that MAP-30 affects HIV-1 infection as well as transmission of viral gene products through cell contact or release of extracellular virus. The compound has considerable homology with the ricin A chain (34%) and GLQ223 (57%) when both identical and conserved residues are considered. In our laboratory, MAP-30 was as active and toxic as GLQ223 in a mitogen proliferation assay in human PBM cells (Table 1). The concern with this class of compounds is that they may not penetrate into the brain where HIV can be detected. A potential problem with MAP-30 and related compounds[81a] is the development of antibodies to these proteins which may neutralize their antiviral activity on prolonged treatment.

TAP-29

This 29 kDa protein, isolated from *Trichosanthes kirilowii*, is related to GLQ223. It has been reported to be as effective in cell culture as GLQ223, but with less cytotoxicity. Its therapeutic index is 100-1000 times greater than its sister compound.[82] Analyses of the N-terminal amino acid sequences of both proteins show that TAP-29 sequence -Lys-Lys-Lys-Val-Tyr- at position 12-16 differs from the -Ser-Ser-Tyr-Gly-Val- at the corresponding position of GLQ-223. TAP-29, like GLQ223, is a single chain ribosome inactivating protein (SCRIP) that catalytically inhibits *in vitro* eukaryotic protein synthesis. TAP-29 and MAP-30, unlike GLQ223, have 2 N-terminal lysine residues that can accommodate tryptic cleavage. No *in vivo* studies with this compound have been reported.

GAP-31, DAPs 30 and 32

These proteins were extracted and purified from Euphorbiaceae Himalaya seeds and carnation leaves and reported to have anti-HIV-1 activity in CEM cells.[83]. GAP-31 had the highest therapeutic index in cell culture. The exact mechanism of action and the reasons for the low toxicity of these compounds to normal cells remain to be elucidated. Interestingly, GAP-31 and DNA topoisomerase II have extensive sequence homology.

Porphyrins

The potential utility of porphyrin analogues as anti-HIV-1 compounds was reported first by our group.[84,85] Porphyrins have been known for over a decade to inhibit virion-associated RT.[86,87] A report by Asanaka *et al.*[88] confirmed that protoporphyrin has selective anti-HIV activity in MT-4 cells. Their exact mechanism of action is still not known, nor is it known why they are selective for virus infected cells. It is well established that (a) iron porphyrins are a component of: hemoglobin, myoglobin, peroxidases, catalases, and cytochromes; (b) porphyrins and metalloporphyrins may either up- or downregulate heme synthesis; (c) porphyrins may potentiate the effect of other drugs; (d) certain porphyrins inhibit microtubule assembly; and (e) cationic porphyrins exhibit high affinity for nucleic acids.[89] For example, TMpyP(3) can bind to "GC" regions of DNA by intercalation and "AT" regions by outside binding.[90] In addition, metalloporphyrins have been shown to have immune stimulatory properties that could be beneficial in AIDS patients.[91] Certain porphyrins have been shown to localize predominantly in the mitochondria and bind strongly to peripheral benzodiazepine receptors located on the outer mitochondrial membrane.[92-95] The relationship between inhibition of HIV-1 and binding to benzodiazepine receptors has yet to be made. Many porphyrins themselves are photoactive and can cause cell damage when irradiated. However, we have found that certain porphyrins are equieffective against HIV-1 in culture in light and dark,[84] and that certain non-photochemical porphyrins are selective anti-HIV agents. It is clear that more work is necessary to determine the value of this class of compounds. Treatment of AIDS-related oral Kaposi's sarcoma with photophrin II phototherapy produced "dramatic early partial and complete responses" in 5 patients.[96]

Sulfated Compounds

Dextran sulfate, cyclodextrin sulfates, glycyrrhizin, and pentosan sulfate are potent inhibitors of HIV in acutely infected cells. These compounds block binding to CD4+ cells and inhibit syncytia formation.[97-100] More recent studies by Callahan et al. have indicated that dextran sulfate blocks antibody binding to the principal neutralizing domain of HIV-1 without interfering with gp120-CD4 interactions.[101]. This suggests that binding of the V3 loop and blocking post-receptor binding uptake events may be a potential mechanism of dextran sulfate and related compounds. Treatment of 34 patients with stage 3 or 4 HIV infection for 8 weeks with dextran sulfate had little effect on HIV levels, although no significant side effects were observed.[102] One problem may be that dextran sulfate is poorly absorbed in the gastrointestinal tract and may be completely degraded.[103,104] Other sulfated compounds have also been proven to be potent inhibitors *in vitro*. Sulfated bacterial glycosaminoglycan (Org 31518) and chemically degraded heparin (Org 31733) were shown to have 50% inhibitory effects at 0.67 and 0.52 μM, respectively. These compounds were also effective against human CMV.[105,106] Four kinds of gangliosides (GM1a, GD1a, GD1b, and GT1b) inhibited infection of human T cells with virus at 10 μg/ml and syncytium formation induced by HIV-1 at 30 μg/ml.[107] A newly synthesized polysaccharide, curdlan sulfate, was found to inhibit binding of HIV-1 to the H9 cell surface.[108] Various naphthalenesulfonic acids have been shown to inhibit HIV-1 by 50% at concentrations greater than or equal to 7 μM, but most of them have narrow therapeutic indices.[109]. Also, sulfated porphyrins have been shown to inhibit HIV-1 in T cells.[85] N-carboxy–methylchitosan-N,O-sulfate (NCMCS) is a sulfated polysaccharide derivative of chitin that inhibits HIV-1 replication in primary human lymphocytes at about 10 μg/ml.[110] NCMCS also inhibits HIV-1 RT (competitively) and the virus binding in Jurkat cells with no significant toxic effects. This compound has a broad spectrum of antiretroviral activity, including activity against murine retroviruses. Since this class of compounds has anticoagulant activity, analogs will have to be designed which do not have this untoward effect as well as good absorption through the gastrointestinal tract.

Other polyanionic compounds, such as Evans blue and aurin tricarboxylic acid, were reported by various groups to inhibit the fusion process and HIV-1 replication in culture.[111] It appears that these compounds bind selectively to the cellular CD4 receptor. Changes in membrane potential or ion fluxes were not noted with these and other membrane active compounds.

A remarkable sulfated polysaccharide-peptidoglycan compound (SP-PG) produced by a specific species of the bacterium *Arthrobacter sp.* (strain AT-25) was recently reported to inhibit the development of Kaposi's sarcoma-related lesions in *in vitro* and *in vivo* model systems.[112] It is interesting that this bacterial cell wall complex was more effective and less toxic than other inhibitors of angiogenesis such as protamine sulfate, heparin and related analogues. SP-PG was also found more effective than suramin, IFN-α, or anticancer agents such as vinblastine or bleomycin in these systems.

Saponins

Soybean saponins isolated from soybean seeds inhibited HIV-1 replication in MT-4 cells at 0.5 μg/ml.[113] These saponins had a narrow therapeutic index and did not inhibit HIV-1 RT. One of the saponins (B1) was found to inhibit HIV-induced cell fusion in MOLT-4 cells.

Illimaquinone

This is a natural marine substance isolated from the Red Sea sponge *Smenospongia sp.* that was found to inhibit specifically RNase H.[114] The X-ray structure of the RNase-H from the RT of HIV-1 was recently reported by Davies et al.[115] It appears that divalent cations are necessary for RNase H activity (e.g., Mn^{2+}). Compounds that chelate these ions may be expected to inhibit this enzyme. Inhibition of HIV-1 RNase-H by sulfated polyanions such as heparin, dextran sulfate, and xylan polysulfate has also been reported.[116]

Avarol and Avarone

These compounds were isolated from the sponge *Dysidea avara*. They have been described as potential antileukemic as well as antiviral agents. Controlled studies failed to confirm their utility in the treatment of HIV-1 infections.[117] Analogues of these compounds have been prepared but none were found to be effective against HIV-1 in C8166 cells.

Soluble CD4

The variable third domain of gp120, known as the V3 loop, has been implicated as the neutralizing domain of HIV-1. Because of its high variability, soluble CD4 may not be able to interact with this glycoprotein, and virus may escape its neutralizing effect. *In vitro* studies have demonstrated inhibition of HIV in lymphoid cells by sCD4.[118-121] In monocytes and macrophage cells, sCD4 was shown to inhibit HIV infection when cell-cell contact was prevented, but not where cell-cell contact predominated.[122] A chimeric related molecule, CD4 immunoadhesin (CD4-IgG), has been reported to protect chimpanzees and monkeys from infection.[7,123] The human half-life of the molecule increases by a factor of ~25 when the IgG portion is added. However, this molecule like sCD4 may also be more effective against laboratory strains of HIV than wild type virus.[5,6,124] In addition, large amounts of the protein are needed for protection in animal models. It also appears that competitive blocking HIV infection with sCD4 depends on the cell density and viral stock age.[125] These findings may explain the lack of efficacy of sCD4-based drug trials demonstrated to date in humans.

Studies have demonstrated that sCD4 causes the virus to release the gp120 spikes, which in turn leads to exposure of an epitope near gp41 resulting in inactivation.[126] Interestingly, mutations in the glycosylation of gp120 proteins produced delays in viral production. It was demonstrated that removal of 3-N linked glycosylation sites was sufficient in reducing infectivity in transfected MT-2 cells.[127]

AL 721

Active lipid 721 extracted from egg yolk was reported to inhibit HIV-1 *in vitro* .[128] The mechanism of action was hypothesized to be one of extraction of cholesterol from the viral envelope or host-cell membrane, thereby interfering with the membrane fluidity and the viral-cell attachment phenomenon. Mildvan et al. recently reported that AL 721 was ineffective in a major study with persons with persistent generalized lymphadenopathy.[129] As a consequence of therapy, marked increases in serum lipids were observed.

Sulfolipids

A number of lipid extracts of cultured cyanobacteria (blue-green algae) were found to inhibit HIV-1 in various cells.[130] The most potent compounds were sulfonic acid-containing glycolipids (sulfoquinovosyl diacylglycerols). These lipids are structural components of chloroplast membranes, and occur widely in higher plants, algae, and photosynthetic microorganisms.

Lipoic Acid

The naturally occurring coenzyme alpha-lipoic acid has been reported to inhibit HIV-1 replication in T-cells and in HeLa-CD4+ cells at non-toxic concentrations of 35-70 µg/ml.[131] An "augmentation of the antiviral activity" was noted when this compound was combined with AZT.

Lectins

α-(1-3)- and α-(1-6)-D-Mannose-specific plant lectins isolated from plants of the *Amaryllidaceae* family were found to inhibit HIV-1, HIV-2, and SIV in MT-4 cells.[132] The plant lectins suppressed syncytium formation between persistently infected HUT-78 cells and uninfected MOLT-4 cells. Unlike dextran sulfate (mol. wt. 5,000), the plant lectins did not appear to inhibit HIV-1 binding to CD4+ cells. Combinations of the plant lectins and dextran sulfate were reported to produce antiviral synergy. The authors propose that these compounds interfere with the virus replication cycle that is subsequent to virus attachment.

Hansen and colleagues demonstrated that several lectins, including concanavalin A, PHA, wheat germ agglutinin, *Lens culinaris* agglutinin, and *Vinca faba* agglutin can bind to HIV-1 gp120.[133] They determined that lectins which bound to the glycans on gp120 prevented virus fusion, and suggested that binding of certain parts of the gp120 molecule could be useful for therapeutic or prophylactic chemotherapeutic approaches. However, it is not clear how these compounds would be delivered. In addition, the activation of cells infected with HIV-1 by these agglutinins causing stimulation and cellular proliferation could be detrimental to the patient.

During studies of sugar recognizing lectins, neoglycoproteins (modified albumins) were shown to inhibit HIV-1 replication by inhibiting the fusion process. This activity correlated with the net negative charge on the proteins.[134] Conjugates of these proteins with AZT have been synthesized, thus allowing the design of drug carrier conjugates with intrinsic anti-HIV activity.[135]

Flavonoids

5,6,7-Trihydroxyflavone (baicalein) was reported to selectively and competitively inhibit HIV-1 RT with respect to template-primer.[136] The compound appears to be specific since it did not inhibit DNA polymerase α and β. Unfortunately, data were not provided on inhibition of this and related flavonoids, which are commonly found in plants in cell culture against HIV-1.

Myristic Acid Analogs

A variety of sulfur- and oxygen-substituted myristic acid analogs have been claimed to inhibit HIV-1 in culture.[137] HIV-1 depends on myristoylation of *gag* polyprotein precursors for assembly. These compounds could act as analogs of the natural substrate for the myristoyl-CoA:protein N-myristoyltransferase which catalyzes the cotranslational linkage of myristate to the N-terminal glycine residues of viral proteins. The antifungal antibiotic cerulenin, an inhibitor of fatty acid biosynthesis, has been shown to inhibit HIV-1 myristoylation and proteolytic cleavage of a *gag*-encoded polyprotein.[138,139] However, this compound may be too hydrophobic, unstable, and toxic to be of clinical use.

Polyether Antibiotics

A number of polyether antibiotics were recently reported to inhibit HIV-1 in H9 cells at low concentrations (0.4 -7.1 µg/ml) with therapeutic indeces in the 6-50 range.[113] The compounds included monensin, salinomycin, lasalocid, nigericin, dianemycin, alborixin, and kijimicin. Some of the compounds were also effective in chronically infected U937 cells. The mechanism of action of these polyethers is not known. Unfortunately, the possibility that these compounds could chelate divalent cations was not explored. Recently, Pal et al.[113a] demonstrated that brefeldin A, an antibiotic containing a 13-member macrocyclic lactone ring, inhibited the processing and secretion of envelope glycoproteins of HIV-1 in MOLT-3 cells, whereas other viral proteins such as *gag* were not affected.

Glutathione and N-acetyl-cystein

The antioxidants such as glutathione (GSH), glutathione-monoester (GSE), and N-acetyl-cystein (NAC), suppress phorbol ester PMA and cytokine-induced HIV expression in U1 cells at 1-20 mM. NAC apparently blocks tumor necrosis factor-α (TNF-α) activation of virus in acutely infected cells. High concentrations of NAC (15 mM) almost completely suppressed HIV RNA accumulation.[140,141] The authors suggest that the antiviral effect of NAC seems only in part common to GSH and GSE, and that multiple steps of HIV expression can be affected by this class of reducing agents. Our own experience with U1 cells indicate that HIV-1 replicates very poorly in these cells, and the variability between replicate results is large, raising some questions about the reproducibility of the results reported (P. Feorino, unpublished results). NAC, a relatively nontoxic modified amino acid, generates glutathione in the cell, and this activates NF-κB, thus preventing activation of the HIV-1 enhancer element. Clinical safety and pharmacokinetic trials with NAC are ongoing at the NIH. Our group has demonstrated that NAC has no effect on the prevention of latent virus in a cell culture model for virus reactivation with TNF-α using OM-10.1 cells.[142,143] It is of interest that the anti-HIV-1 activity of AZT is completely abolished in certain infected cells treated with TNF-α.[144]

Desferrioxamine

This trivalent iron chelator has been recently reported to inhibit HIV-1 replication in H9 cells.[145] This compound can be considered a natural product since it is found in certain bacteria (e.g., *Streptomyces, Actinomyces*). Our laboratory recently confirmed its activity and selectivity in human PBM cell (EC_{50} = 2.6-4.7 µM) in our laboratory (Table 1). However, this compound was toxic to rapidly dividing cells such as Vero cells (IC_{50} = 5.2 µM). Although the mechanism of action is not known, Tabor et al.[145] suggest that DNA synthesis is inhibited through the inhibition of iron-dependent ribonucleotide reductase.

Recent studies indicate that desferrioxamine can only enter cells by pinocytosis, and that once internalized it remains in the lysosomes.[146] This chelator has been used effectively for the treatment of iron overload, and is claimed to be useful for the treatment of Alzheimer's disease, but long-term treatment with this drug is not known.[147] It is intriguing that the similarity in neurological disease between HIV and Alzheimer's disease has prompted evaluations of a number of compounds that have been claimed to be useful against the latter disease.[148]

EPILOGUE

Isolation or synthesis of potent natural compounds with good bioavailability, little or no toxicity, and those that will maintain an effective concentration at the target site sustained long enough to exert their antiviral activities are still needed. There is a pressing need for better and more selective anti-HIV agents since most of the currently available compounds have toxicities of varying degrees, may be immunosuppressive, allow emergence of resistance, have a restricted spectrum of viral inhibition, do not eliminate the latent virus, and do not adequately control an established HIV infection. The hallmark of a selective antiviral agent is the ability of viruses to develop resistance to that drug *in vitro* or *in vivo*.[24] The rate at which drug resistance develops is of concern and is an important reason for initiating therapy with a combination early in the course of the infection or disease.[149] Strategies must be developed to combat the virus, not only in helper T cells, but also in monocyte/macrophages, other hematopoietic cells, and chronically or latently infected cells.[142,150,151] An important mechanism for HIV transmission is the virus-infected cell. Several natural products interfere with viral attachment and fusion of infected cells to uninfected healthy cells. Ideally, these antifusion compounds should also be able to penetrate cells and interfere with a secondary step in the viral replication cycle. Rationally designed "smarter" antiviral agents that target specifically the virus "Achilles' heel" or specific cellular compartment will become available in the future through concerted efforts and interactions between the phytochemists and virologists.[3,4,152-155]

Approaches such as gene therapy are now on the horizon, although substantial technical problems remain.[156-159] It is unlikely that gene therapy will completely prevent virus replication or spread. A combined modality with other compounds, such as nucleoside analogues or antiviral natural products, will be necessary because of potential virus breakthrough. In effect, AIDS has propelled the development and use of combined chemotherapeutic approaches to new heights in order to produce antiviral synergy and reduce toxicity.[24] However, in clinical situations, synergy alone should not be the main criterion for using drug combinations.

Three criteria should be used to select the correct clinically relevant combination. First and most importantly, the drugs should have non-overlapping toxicities (no synergistic toxicity and preferably no additive toxicities); second, the drugs should not be cross-resistant; and third, the drugs should have no antagonistic antiviral interactions. As more drugs become available, the urge will be to combine them indiscriminately. This temptation should be avoided in the absence of a clear rationale for using them together. In the next few years, the hope is that non-toxic drugs that inhibit HIV expression in chronically or latently infected patients will become available. If synergistic toxicity is not a problem, then these should be combined with drugs that affect acute infection. Progress is being made in the development of novel compounds that affect different viral infections commonly found in AIDS patients.[160]

AIDS has had a remarkable impact on the development of novel antiviral agents and chemotherapeutic approaches to treat viral infections. Nucleosides continue to provide a rich source of specific antiviral agents, although other compounds such as the TAT and proteinase inhibitors look promising. Research on material obtained from natural products has yet to yield compounds of clinical value as has been done in the field of cancer chemotherapy. Further modification of lead compounds obtained from diverse natural products should produce the desired selectivity, activity, and pharmacologic and pharmacodynamic properties. The diverse structure and mechanism of action of natural products should reduce the likelihood of developing cross-resistance to TIBO-like compounds, nucleoside analogues, and other molecules such as proteinase inhibitors. Based on our limited success to date in the laboratory, it is likely that research on natural products will lead to effective compounds that inhibit virus replication especially in chronically infected cells.

Acknowledgments

Parts of this review were abstracted from a paper to be published by the author in *AIDS and Human Retroviruses* in 1992. The author would like to thank his collaborators for allowing him to share unpublished data with the reader. This work was supported by Public Health Service grants and contracts AI-05078, AI-25899, AI-27196, and AI-28124 from the National Institutes of Health, and the Department of Veterans Affairs.

REFERENCES

1. P.S. Rosenberg, M.H. Gail, L.K. Schrager, et al., National AIDS incidence trends and the extent of zidovudine therapy in selected demographic and transmission groups., *J. Acquir. Immun. Defic. Syndr.* 4:392 (1991).
2. W.A. Haseltine, Development of antiviral drugs for the treatment of AIDS: Strategies and prospects, *J. Acquir. Immun. Defic. Syndr.* 2:311 (1989).
3. W.A. Haseltine, Molecular biology of the human immunodeficiency virus type 1, *FASEB J.* 5:2349 (1991).

4. H. Mitsuya, R. Yarchoan, S. Kageyama, S. Broder, Targeted therapy of human immunodeficiency virus-related disease, *FASEB*. 5:2369 (1991).

5. J.P. Moore, J.A. McKeating, Weiss RA, Q.J. Sattentau, Dissociation of gp120 from HIV-1 virions induced by soluble CD4, *Science* 250:1139 (1990).

6. J.P. Moore, J.A. McKeating, W.A. Norton, Q.J. Sattentau, Direct measurement of soluble CD4 binding to human immunodeficiency virus type 1 virions: gp120 dissociation and its implications for virus-cell binding and fusion reactions and their neutralization by soluble CD4, *J. Virol*. 65(3):1133 (1991).

7. R.H.R. Ward, D.J. Capon, C.M. Jett, et al., Prevention of HIV-1 IIIB infection in chimpanzees by CD4 immunoadhesin, *Nature* 352:434 (1991).

8. J.W. Mellors, G.E. Dutchman, J-J Im, E. Tramontano, S.R. Winkelr, Y-C Cheng, *In vitro* selection and molecular characterization of human immunodeficiency virus-1 resistant to non-nucleoside inhibitors of reverse transcriptase, *Mol. Pharmac*. 41:446 (1992).

9. D. Richman, C-KShih, I. Lowy, et al., HIV-1 mutants resistant to non-nucleosides inhibitors of reverse transcriptase arise in tissue culture, *Proc. Natl. Acad. Sci. USA*, 88:11241 (1991).

10. S.E. Krown, F.X. Real, S. Cunningham-Rundles, et al., Preliminary observations on the effect of recombinant leukocyte A interferon in homosexual men with Kaposi's sarcoma, *N. Engl. J. Med*. 308:1071 (1983).

11. S. Krown, AIDS-associated Kaposi's sarcoma: Pathogenesis, clinical course and treatment, *AIDS* 2:71 (1988).

12. D.D. Ho, K.L. Hartshorn, T.R. Rota, et al., Recombinant human interferon alpha-A suppresses HTLV-III replication *in vitro*, *Lancet* 1:602 (1985).

13. R. de Wit, J.K. Schattenkerk, C.A. Boucher, P.J. Bakker, K.H. Veenhof, S.A.Danner, Clinical and virological effects of high-dose recombinant interferon–α in disseminated AIDS-related Kaposi's sarcoma, *Lancet* 2:1214 (1988).

14. H.C. Lane, J.A. Kovacs, J. Feinberg, et al., Anti-retroviral effects of interferon–α in AIDS-associated Kaposi's sarcoma, *Lancet* 2:1218 (1988).

15. V. Davey, J.A.Kovacs, B. Herpin, J.A. Metcalf, A.S. Fauci, H.C Lane, A placebo controlled trial of interferon alpha-2b in asymptomatic HIV infection, V International Conference on AIDS, Montreal (1989).

16. G.H. Friedland, R.S. Klein, B.R. Saltzman, et. al., A randomized placebo-controlled trial of recombinant human interferon alpha 2a in patients with AIDS, *J. Acquir. Immun. Defic. Syndr*. 1:111 (1988).

17. K.L. Hartshorn, D. Neumeyer, M.W. Vogt, R.T. Schooley, M.S. Hirsch, Activity of interferons alpha, beta, and gamma against human immunodeficiency virus replication *in vitro*, *AIDS Res. Human Retrov*. 3:125 (1987).

18. P.M. Pitha, J.A. Bilello, C.H. Riggin, Effect of interferon on retrovirus replication, *Texas Rep. Biol . Med*. 41:603 (1981).

19. J.K. Yamamoto, F. Barre-Sinoussi, F. Bolton, N.C. Pedersen, M.B. Gardner, Human alpha and beta interferon but not gamma suppress the *in vitro* replication of LAV, HTLV-III and ARV-2, *J. Interferon Res.* 6:143 (1986).

20. Y. Yasuda, S. Miyake, S. Kato, et al., Interferon–α treatment leads to accululation of virus particles on the surface of cells persistently infected with the human immunodeficiency virus type 1, *J. Acquir. Immun. Defic. Syndr.* 3:1046 (1990).

21. M.S. Smith, R.J. Thresher, J.S. Pagano, Inhibition of human immunodeficiency virus type 1 morphogenesis in T cells by alpha interferon, *Antimicrob. Agents Chemother.* 35:62 (1991).

22. H.E. Geldenman, R. M. Friedman, S. Joe, et al., A selective defect of interferon alpha production in human immunodeficiency virus-infected monocytes, (published erratum appears in *J. Exp. Med.* 173:277 (1991), *J. Exp. Med.* 172:1433 (1990).

23. H. Geldenman, L. Baca, J. Turpin, et al., Regulation of HIV replication in infected monocytes by INF–α, *J. Immunol* .145:2669 (1990).

24. R.F. Schinazi. Combined therapeutic modalities for viruses: Rationale and clinical potential, *in:* "Molecular Mechanisms of Chemotherapeutic Synergism, Potentiation, and Antagonism," T.C. Chou, D.C. Rideout, eds., Academic Press, Orlando, FL (1972).

25. J.A. Kovács, L. Deyton, R. Davey, et al., Combined zidovudine and interferon a therapy in patients with Kaposi's sarcoma and the acquired immunodeficiency syndrome (AIDS), *Ann. Intern. Med.* 111:280 (1989).

26. M.W. Vogt, A.G. Durno, T.C. Chou, et al., Synergistic interaction of 2',3'-dideoxycytidine and recombinant interferon–α-A on replication of human immunodeficiency virus type 1, *J. Infect. Dis.* 158:378 (1988).

27. D.T. Scadden, H.A. Bering, J.D. Levine, et al., Granulocyte-macrophage colony-stimulating factor mitigates the neutropenia of combined interferon alpha and zidovudine treatment of acquired immune deficiency syndrome-associated Kaposi's sarcoma, *J. Clin. Oncol.* 9:802 (1991).

28. D.T. Scadden, H.A. Bering, J.D. Levine, et al., CM-CSF as an alternative to dose modification of the combination zidovudine and interferon-alpha in the treatment of AIDS-associated Kaposi's sarcoma, *Am J. Clin. Oncol.* 14:S40 (1991).

29. R. Datema, S. Olofsson, P.A. Romero, Inhibitors of protein glycosylation and glycoprotein processing in viral systems, *Pharmacol. Ther.* 33:221 (1987).

30. R.M. Ruprecht, L.D. Bernard, R. Bronson, M.A. Gama Sosa, S. Millaney, Castanospermine *vs.* its 6-O-butanoyl analog: A comparison of toxicity and antiviral activity *in vitro* and *in vivo*, *J. Acquir. Immun. Defic. Syndr.* 4:48 (1991).

31. R.M. Ruprecht, Murine models for antiretroviral therapy, *Intervirology* 30:2 (1989).

32. P.S. Sunkara, T.L. Bowlin, P.S. Liu, A. Sjoerdsma., Antiretroviral activity of

castanospermine and deoxynojirimycin, specific inhibitors of glycoprotein processing, *Biochem. Biophys. Res. Commun.* 1:206 (1987).

32a. D. Schols, R. Pauwels, M. Witvrouw, J. Desmyter, E. De Clercq, Differential activity of polyanionic compounds and castanospermine against HIV replication and HIV-induced syncytium formation depending on virus strain and cell type, *Antiviral Chem. Chemother.* 3:23 (1992).

33. R.M. Ruprecht, S. Mullaney, J. Andersen, R. Bronson, *In vivo* analysis of castanospermine, a candidate antiretroviral agent, *J. Acquir. Immun. Defic. Syndr.* 2:149 (1989).

34. M.G. Hollingshead, L. Westbrook, B.J. Toyer, L.B. Allen, Failure of castanospermine to reduce spleen or serum virus titers in Rauscher murine leukaemia virus-infected mice, *Antiviral Chem. Chemother.* 2:119 (1991).

35. P.S. Sunkara, M.S. Kang, T.L. Bowlin, P.S. Liu, A.S. Tyms, A. Sjoerdsma, Inhibition of glycoprotein processing and HIV replication by castanospermine analogues, *Ann. NY Acad. Sci.* 616:90 (1990).

36. A. Karpas, G.W.J. Fleet, R.A. Dwek, et al., Aminosugar derivatives as potential anti-human immunodeficiency virus agents, *Proc. Natl. Acad. Sci. USA* 85:9229 (1988).

37. H. Shimizu, H. Tsuchie, K. Yoshida, et al., Inhibitory effect of novel 1-deoxynojirimycin derivatives on HIV-1 replication, *AIDS* 4:975 (1990).

38. I.M. Jones, G.S. Jacob, Anti-HIV drug mechanism, *Nature* 352:198 (1991)

39. R.F. Arakaki, J.A. Hedo, E. Collier, P. Gorden, Effects of castanospermine and 1-deoxynojirimycin on insulin receptor biogenesi: Evidence for a role of glucose removal from core oligosaccharides, *J. Biol. Chem.* 262:1186 (1987).

40. R. Saul, J.J. Ghidoni, R.J. Molyneux, A.D. Elbein, Castanospermine inhibits alpha-glucosidase activities and alters glycogen distribution in animals, *Proc. Natl. Acad. Sci. USA* 82:93 (1985).

41. V.A. Johnson, B.D. Walker, M.A. Barlow, T.J. Paradis, T-C Chou, M.S. Hirsch, Synergistic inhibition of human immunodeficiency virus type 1 and type 2 replication *in vitro* by castanospermine and 3'-azido-3'-deoxythymidine, *Antimicrob. Agents Chemother.* 33:53 (1989).

42. R.S. Chang, L. Ding, C. Gai-Qing, P. Qi-Choa, Z. Zi-Lin, K.M. Smith, Dehydroandrographolide succinic acid monoester as an inhibitor against the human immunodeficiency virus, *Proc. Soc. Exp. Biol. Med.* 197:59 (1991).

43. S-Z Qian, Z-G Wang, Gossypol: A potential antifertility agent for males, *Ann. Rev. Pharmacol. Toxicol.* 24:329 (1984).

44. E.E. Montamat, C. Burgos, N.M. Gerez de Burgos, L.E. Rovai, A. Blanco, E.L. Segura, Inhibitory action of gossypol on enzymes and growth of *trypanosoma cruzi*, *Science* 218:288 (1982).

45. J.E. Heidrich, L.A. Hunsaker, Gossypol, an antifertility agent, exhibits antimalarial activity *in vitro*, *IRCS Med. Sci.* 11:304 (1983).

46. P.H. Dorset, E.E. Kerstin, L.J. Powers, Antiviral activity of gossypol and apogossypol, *J. Pharm. Sci.* 64:1073 (1975).

47. K. Wichmann, A. Vaheri, T, Luukkainen. Inhibiting herpes simplex virus type 2 infection in human epithelial cells by gossypol, a potent spermicidal and contraceptive agent, *Amer. J. Obstet. Gynecol.* 142:593 (1982).

48. B. Polsky, J.W.M. Gold, P.A. Baron, et al., Inactivation of human immunodeficiency virus (HIV) by gossypol (GP), *Clin .Res.* 35:487A (1987).

49. D.S. Sampath, P. Balaram, Resolution of racemic gossypol and interaction of individual enantiomers with serum albumens and model peptides, *Biochem. Biophys. Acta* 2:183 (1986).

50. T-S Lin, R.F. Schinazi, B.P. Griffith, et al., Selective inhibition of human immunodeficiency virus type 1 replication by the (-) but not the (+) enantiomer of gossypol, *Antimicrob. Agents. Chemother.* 33:2149 (1989).

51. D. Meruelo, G. Lavie, D. Lavie, Therapeutic agents with dramatic antiretroviral activity and little toxicity at effective doses: Aromatic polycyclic diones hypericin and pseudohypericin, *Proc. Natl. Acad. Sci. USA* 85:5230 (1988).

52. D. Meruelo, G. Lavie, Mechanisms of the antiretroviral activity of aromatic polycyclic diones hypericin and pseudohypericin, *Int. Conf. AIDS* 5:662 (1989).

53. I. Takahashi, S. Nakanishi, E. Kobayashi, H. Nakano, K. Suzuki, T. Tamaoki, Hypericin and pseudohypericin specifically inhibit protein kinase C: Possible relation to their antiretroviral activity, *Biochem. Biophys. Res. Commun.* 165:1207 (1989)

54. J. Tang, J.M. Colacino, S.H. Larsen, W. Spitzer, Virucidal activity of hypericin against enveloped and non-enveloped DNA and RNA viruses, *Antiviral Res.* 13:313 (1990).

55. G.A. Kraus, D. Pratt, J. Tossberg, S., Carpenter. Antiretroviral activity of synthetic hypericin and related analogs, *Biochem . Biophys. Res. Commun.* 172:149 (1990).

56. J.B. Hudson, I. Lopez-Bazzocchi, G.H.N. Towers, Antiviral activities of hypericin, *Antiviral Res.* 15:101 (1991).

57. D.O. Andersen, N.D. Weber, S.G. Wood, B.G. Hughes, B.K. Murray, J.A. North, *In vitro* virucidal activity of selected anthraquinones and anthraquinone derivatives, *Antiviral Res.* 16:185 (1991).

58. F.T. Valentine, V. Itri, N. Kudler, R. Georgescu, Synthetic hypericin enters blood lymphocytes and monocytes *in vitro* and decreases culturable HIV in blood obtained from infected individuals, VII International Conference on AIDS, Florence, 97 (1991).

59. Y. Hafuri, E. Takemori, K. Oogose, et al., Mechanism of inhibition of reverse transcriptase by quinone antibiotics II. Dependence on putative quinone pocket on the enzyme molecule, *J. Antibiotics* 41:1471 (1988).

60. K.V.R. Dhananjaya, A. Antony, Inhibition of avian myeloblastosis virus reverse transcriptase and its associated activities by daunomycin and adriamycin, *Ind. J. Biochem. Biophys.* 24:265 (1987).

61. H. Nakashima, N. Yamamoto, Inhibition by doxorubicin of human immunodeficiency virus (HIV) infection and replication *in vitro*, *J. Antibiotics* 40:396 (1987).

62. G. Lavie, F. Valentine, B. Levin, et al., Studies of the mechanisms of action of the antiretroviral agents hypericin and pseudohypericin, *Proc. Natl. Acad. Sci.* 86:5963 (1989).

63. R.F. Schinazi, C.K. Ch, J.R. Babu, et al., Anthraquinones as a new class of antiviral agents against human immunodeficiency virus, *Antiviral Res.* 13:265 (1990).

64. H. Higuchi, K. Mori, A. Kato, et al., Antiretroviral activities of anthraquinones and their inhibitory effects on reverse transcriptase, *Antiviral Res.* 15:205 (1991).

65. D.I. Abrams, Alternative therapies in HIV infection, *AIDS* 4:1179 (1990).

66. K.P. Manfredi, J.W. Blunt, J.H.I. Cardellina, et al., Novel alkaloids from the tropical plant *Ancistrocladus abbreviatus* inhibit cell killing by HIV-1 and HIV-2, *J. Med. Chem.* 34:3402 (1991).

66a. E. Priel, S.D. Showalter, D.G. Blair, Inhibition of human immunodeficiency virus (HIV-1) replication *in vitro* by noncytotoxic doses of camptothecin, a topoisomerases I inhibitor, *AIDS Res. Hum. Retrov.* 7:65 (1991).

66b. C. Schweitzer, F. Keller, M.P. Schmitt, et al., Morphine stimulates HIV replication in primary cultures of human Kupffer cells, *Res. Virol.* 142:189 (1991).

67. P.K. Lai, J. Donovan, H. Takayama, et al., Modification of human immunodeficiency viral replication by pine cone extracts, *AIDS Res. Hum. Retrov.* 6:205 (1990).

68. H. Takayama, G. Bradley, P.K. Lai, et al., Inhibition of human immunodeficiency virus forward and reverse transcription by PC6, a natural product from cones of pine trees, *AIDS Res. Hum. Retrov.* 7:349 (1991).

69. Y. Tamura, P.K. Lai, W.G. Bradley, K. Konno, A. Tanaka, M. Nonoyama, A soluble factor induced by an extract from *Pinus parviflora* Sieb et Zucc can inhibit the replication of human immunodeficiency virus *in vitro*, *Proc. Natl. Acad. Scc.* 88:2249 (1991).

70. D.C. Montefiori, J. Zhou, Selective antiviral activity of synthetic soluble L-tyrosine and L-DOPA melanins against human immunodeficiency virus *in vitro*, *Antiviral Res.* 15:11 (1991).

71. M. Nishizawa, T. Yamagishi, G.E. Dutschman, et al., Anti-AIDS agents, 1. Isolation and characterization of four new tetragalloylquinic acids as a new class of HIV reverse transcriptase inhibitors from tannic acid, *J. Nat. Prod.* 52:762 (1989).

72. H.C. Schröder, H. Merz, R. Steffen, et al., Differential *in vitro* anti-HIV activity of natural lignans, *Z. Naturforsch*. (C)45:1215 (1990).

72a. H. Nakashima, T. Murakami, N. Yamamoto, et al., Inhibition of human immunodeficiency viral replication by tannins and related compounds, *Antiviral Res*. 18:91(1992).

73. W. Kreis W, M.H. Kaplan, J. Freeman, D.K. Sun, P.S. Sarin, Inhibition of HIV replication by *Hyssop officinalis* extracts, *Antiviral Res*. 14:323 (1990).

74. R.F. Schinazi. Antiviral activity of dehydroepiandrosterone, *in:* "The biological role of dehydroepiandrosterone (DHEA)," M. Kalimi, W. Regelson, eds, Walter de Gruyter & Co, New York (1990).

75. J.M. Laurence, B. Sellers, S.K. Sikder, Effect of glucocorticoids on chronic human immunodeficiency virus (HIV) infection and HIV promoter-mediated transcription, *Blood* 14:291 (1989).

76. M.S. McGrath, S. Santulli, I. Gaston, Effects of GLQ223(tm) on HIV replication in human monocyte/macrophages chronically infected *in vitro* with HIV, *AIDS Res. Hum. Retrov*. 6:1039 (1990).

77. P. Ferrari, M. Trabaud, M. Rommain, et al., Toxicity and activity of purified trichosanthin, *AIDS* 5:865 (1991).

78. V.S. Byers, A.S. Levin, L.A. Waites, et al., A phase I/II study of trichosanthin treatment of HIV disease, *AIDS* 4:1189 (1990).

79. J.O. Kahn, L.D. Kaplan, J.G. Gambertogilo, et al., The safety and pharmacokinetics of GLQ223 in subjects with AIDS and AIDS-related complex: A Phase I study, *AIDS* 4:1197 (1990).

80. J. Palaca, Trials and tribulations of AIDS testing, *Science* 247:14406 (1990).

81. S. Lee-Huang, P.L. Huang, P.L. Nara, et al., MAP 30: A new inhibitor of HIV-1 infection and replication, *FEBS Lett*. 272:12 (1990).

81a. M.C. Olson, S. Ramakrishnan, Anand R, Ribosomal inhibitory proteins from plants inhibit HIV-1 replication in acutely infected peripheral blood mononuclear cells, *AIDS Res Hum Retrovir* 7:1025(1991).

82. S. Lee-Huang, P.L. Huang, H-F. Kung, et al., TAP 29: An anti-human immunodeficiency virus protein from *Trichosanthes kirilowii* that is nontoxic to intact cells, *Proc. Natl. Acad. Sci*. 88:6570 (1991).

83. S/ Lee-Huang, H-F Kung, P.L. Huang, et al., A new class of anti-HIV agents: GAP 31, DAPs 30 and 32, *FEBS Lett*. 291(1):139 (1991).

84. D.W. Dixon, M.S. Kim, V. Kumar, G. Obara, L. Marzilli, R.F. Schinazi, Structure-activity relationship of amino and hydroxytetraphenylporphyrin derivatives with anti-HIV activity, *Antiviral Chem. Chemother.*, in press (1992).

85. D.W. Dixon, R.F. Schinazi, L.G. Marzilli, Porphyrins as agents against the human immunodeficiency virus, *Ann NY Acad. Sci*. 616:511 (1990).

86. B.R. Munson, R.J. Fiel, Hematoporphyrin-sensitized photodynamic inactivation of viral RNA-dependent DNA polymerase, *Res. Commun. Chem. Pathol. Pharmacol.* 16:175 (1977).

87. K. Tsutsui, G.C. Mueller, Hemin inhibits virion-associated reverse transcriptase of murine leukemia virus, *Biochem. Biophys. Res. Comm. Pharmacol.* 149:628 (1987).

88. M. Asanaka, T. Kurimura, H. Toya, J. Ogaki, Y. Kato, Anti-HIV activity of protoporphyrin, *AIDS* 3:403 (1989).

89. R. Kuroda, E. Takahashi, C.A. Austin, L.M. Fisher, DNA binding and intercalation by novel porphyrins: Role of charge and substituents probed by DNase I footprinting and topoisomerase I unwinding, *FEBS Lett.* 262:293 (1990).

90. K. Boekelheide, J. Eveleth, A.H. Tatum, J.W. Winkelman, Microtubule assembly inhibition by porphyrins and related compounds, *Photochem. Photobiol.* 46:657 (1987).

91. A. Novogrodsky, M. Suthanthiran, K.H. Stenzel, Immune stimulatory properties of metalloporphyrins, *J. Immunol.* 143:3981 (1989).

92. S.H. Snyder, A. Verma, R.R. Trifilietti, The peripheral-type benzodiazepine receptor: A protein of mitochondrial outer membranes utilizing porphyrins and endogenous ligands, *FASEB* 1:282 (1987).

93. K. Ikezaki, K. Black, Stimulation of cell growth and DNA synthesis by peripheral benzodiazepine, *Cancer Lett* . 49:115 (1990).

94. S.L. Gibson, R. Hilf, Interdependence of fluence, drug dose and oxygen on hematoporphyrin derivative induced photosensitization of tumor mitochondria, *Photochem. Photobiol.* 42:367 (1985).

95. S.L. Gibson, R. Hilf, Photosensitization of mitochondrial cytochrome c oxidase by hematoporphyrin derivative and related porphyrins *in vitro* and *in vivo*, *Cancer Res.* 43:4191 (1983).

96. V.G. Schweitzer, D. Visscher, Photodynamic therapy for treatment of AIDS related oral Kaposi's sarcoma, *Otolaryngol. Head Neck Surg.* 102:639 (1990).

97. M. Baba, H. Tanaka, E. De Clercq, et al., Highly specific inhibition of human immunodeficiency virus type 1 by novel 6-substituted acyclouridine derivative, *Biochem. Biophys. Res. Commun.* 165:1375 (1989).

98. H. Mitsuya, D.J. Looney, S. Kuno, R. Ueno, F. Wong-Staal, S. Broder, Dextran sulfate suppression of viruses in the HIV family: inhibition of virion binding to $CD4^+$ cells, *Science* 240:646 (1988).

99. T. Moriya, H. Kurita, K. Matsumoto, et al., Potent inhibitory effect of a series of modified cyclodextrin sulfates (mCDS) on the replication of HIV-1 *in vitro*, *J. Med. Chem.* 34:2301 (1991).

100. H. Nakashima, T. Matsui, O. Yoshida, et al., A new anti-human immunodeficiency virus substance, glycyrrhizin sulfate; endowment of glycyrrhizin with reverse

transcriptase-inhibitory activity by chemical modification, *Jpn. J. Cancer Res.* (Gann)78:767 (1987).

101. L.N. Callahan, M. Phelan, M. Mallinson, M.A. Norcross, Dextran sulfate blocks antibody binding to the principal neutralizing domain of human immunodeficiency virus type 1 without interfering with gp120-CD4 interactions, *J. Virol.* 65:1543 (1991).

102. D.I. Abrams, S. Kuno, R. Wong, et al., Oral dextran sulfate (UA001) in the treatment of the acquired immunodeficiency syndrome (AIDS) and AIDS-related complex, *Ann. Intern. Med.* 110:183 (1989).

103. K. Lorentsen, C. Hendrix, J. Collins, R. Eckel, B. Petty, P. Lietman, Dextran sulfate is poorly absorbed after oral administration, *Ann. Intern. Med.* 111:561 (1989).

104. N.R. Hartman, D.G. Johns, H. Mitsuya, Pharmacokinetic analysis of dextran sulfate in rats as pertaining to its clinical usefulness for therapy of HIV infection, *AIDS Res. Human Retrovir.* 6:805 (1990).

105. M. Baba, E.De Clercq, S. Iida, et al., Anti-human immunodeficiency virus type 1 activities and pharmacokinetics of novel 6-substituted acyclouridine derivatives, *Antimicrob. Agents Chemother.* 34:2358 (1990).

106. M. Baba, E. De Clercq, D. Schols, et al., Novel sulfated polysaccharides: dissociation of anti-human immunodeficiency virus activity from antithrombin activity, *J. Infect. Dis.* 161:208 (1990).

107. A. Handa, H. Hoshino, K. Nakajima, et al., Inhibition of infection with human immunodeficiency virus type 1 by sulfated gangliosides, *Biochem. Biophys. Res. Commun.* 175:1 (1991).

108. T. Aoki, Y. Kaneko, M.S. Stefanski, T. Nguyen, R.C.Y. Ting, Curdlan sulfate and HIV-1. I. *In vitro* inhibitory effects of curdlan sulfate on HIV-1 infection, *AIDS Res. Hum. Retrov.* 7:409 (1991).

109. P. Mohan, R. Singh, M. Baba, Potential anti-AIDS agents: Synthesis and antiviral activity of naphthalenesulfonic acid derivatives against HIV-1 and HIV-2, *J. Med. Chem.* 34:212 (1991).

110. M.A. Gama Sosa, F. Fazely, J.A. Koch, S.V. Vercellotti, R.M. Ruprecht, N-Carboxymethylchitosan-N,O-sulfate as an anti-HIV-1 agent, *Biochem. Biophys. Res. Commun.* 174:489 (1991).

111. J.L. Weaver, P. Gergely, P.S. Pine, E. Patzer, A. Aszalos, Polyionic compounds selectively alter availability of CD4 receptors for HIV coat protein rgp 120, *AIDS Res. Hum. Retrov.* 6:1125 (1990).

112. S. Nakamura, S. Sakurada, S.Z. Salahuddin, et al., Inhibition of development of Kaposi's sarcoma-related lesions by a bacterial cell wall complex, *Science* 255:1437 (1992).

113. M. Nakamura, S. Kunimoto, Y. Takahashi, et al., Inhibitory effects of polyethers on human immunodeficiency virus replication, *Antimicrob. Agents. Chemother.* 36:492 (1992).

113a. R. Pal, S. Mumbauer, G.M. Hoke, A. Takatsuki, M.G. Sarngadharan, Brefeldin A inhibits the processing and secretion of envelope glycoproteins of human immunodeficiency virus type 1, *AIDS Res. Hum. Retrovir.* 7:707 (1991).

114. S. Loya, R. Tal, Y. Kashman, A. Hizi, Illimaquinone, a selective inhibitor of the RNase H activity of human immunodeficiency virus type 1 reverse transcriptase, *Antimicrob. Agents. Chemother.* 34:2009 (1990).

115. J.F. Davies, Z. Hostomska, Z. Hostomsky, S.R. Jordan, D.A. Matthews, Crystal structure of the ribonuclease H domain of HIV-1 reverse transcriptase, *Science* 252:88 (1991).

116. K. Moelling, T. Schulze, H. Diringer, Inhibition of human immunodeficiency virus type 1 RNase H by sulfated polyanions, *J. Virol.* 63:5489 (1989).

117. A. De Giulio, S. De Rosa, G. Strazzullo, et al., Synthesis and evaluation of cytostatic and antiviral activityies of 3' and 4'-avarone derivatives, *Antiviral Chem. Chemother.* 2:223 (1991).

118. D.H. Smith, R.A. Byrn, S.A. Marsters, T. Gregory, J.E. Groopman, D.J. Capon, Blocking of HIV-1 infectivity by a soluble, secreted form of the CD4 antigen, *Science* 238:1704 (1987).

119. R. Fisher, J. Bertonis, W. Meier, et al., HIV infection is blocked *in vitro* by recombinant soluble CD4, *Nature* 331:76 (1988).

120. R. Hussey, N. Richardson, M. Kowalski, et al., A soluble CDr protein selectively inhibits HIV replication and syncytium formation, *Nature* 331:82 (1988).

121. A. Traunecker, W. Lueke, K. Karjalainen, Soluble CD4 molecules neutralize human immunodeficiency virus type 1,. *Nature* 331:84 (1988).

122. M.A. Harbison, J.M. Gillis, P. Pinkston, R.A. Byrn, R.M. Rose, S.M. Hammer, Effects of recombinant soluble CD4(rCD4) on HIV-1 infection of monocyte/macrophages, *J. Infect. Dis.* 161:1 (1990).

123. J.P. Moore, R.A. Weiss, Passive primate protection, *Nature* 352 (1991), 376 (1 August 1991).

124. W.A. O'Brien, I.S.Y. Chen, D.D. Ho, E.S. Daar, Mapping genetic determinants for human immunodeficiency virus type 1 resistance to soluble CD4, *J. Virol.* 66:3125 (1992).

125. S.P. Layne, M.J. Merges, J.L. Spouge, M. Dembo, P.L. Nara, Blocking of human immunodeficiency virus infection depends on cell density and viral stock age, *J. Virol.* 65:3293 (1991).

126. T.K. Hart, R. Kirsh, H. Ellens, et al., Binding of soluble CD4 proteins to human immunodeficiency virus type 1 and infected cells induces release of envelope glycoprotein GP120, *Proc. Natl. Acad. Sci. USA* 88:2189 (1991).

127. L. Dirckx, D. Lindemann, R. Ette, C. Manzoni, D. Moritz, J. Mous, Mutation of conserved N-glycosylation sites around the CD4-binding site of human immunodeficiency virus type 1 GP120 affects viral infectivity, *Virus Res.* 18:9 (1990).

128. P.S. Sarin, R.C. Gallo, D.I. Scheer, F. Crews, A.S. Lippa, Effects of a novel compound (AL 721) on HTLV-III infectivity *in vitro*, *N. Engl. J. Med.* 313:1289 (1985).

129. D. Mildvan, J. Buzas, D. Armstrong, et al., An open-label, dose ranging trial of AL721 in patients with persistent generalized lymphadenopathy and AIDS-related complex, *J. Acquir. Immun. Defic. Syndr.* 4:945 (1991).

130. K.R. Gustafson, J.H. Cardellina, R.W. Fuller, et al., AIDS-antiviral sulfolipids from cyanobacteria (blue-green algae), *J. Natl. Cancer Inst.* 81:1254 (1989).

131. A. Baur, T. Harrer, M. Peukert, G. Jahn, J.R. Kalden, B. Fleckenstein, Alpha-lipoic acid is an effective inhibitor of human immunodeficiency virus (HIV-1) replication, *Klin. Wochenschr.* 69:722-724 (1991).

132. J. Balzarini D. Schols, J. Neyts, E. Van Damme, W. Peumans, E. De Clercq, a-(1-3)- and a-(1-6)-D-Mannose-specific plant lectins are markedly inhibitory to human immunodeficiency virus and cytomegalovirus infections *in vitro*, *Antimicrob. Agents. Chemother.* 35:410 (1991).

133. J.E.S. Hansen, C.M. Nielsen, C. Nielsen, P. Heegaard, L.R. Mathiesen, J.P. Nielsen, Correlation between carbohydrate structures on the envelope glycoprotein gp120 of HIV-1 and HIV-2 and syncytium inhibition with lectins, *AIDS* 3:635 (1989).

134. R.W. Jansen, G. Molema, R. Pauwels, D. Schols, E. De Clercq, D.K.F. Meijer, Potent *in vitro* anti-HIV activity of modified human serum albumins, *Mol. Pharmacol.* 39:818 (1991).

135. G. Molema, R.W. Jansen, J. Visser, P. Herdewijn, F. Moolenaar, D.K.F. Meijer, Neoglycoproteins as carriers for antiviral drugs: synthesis and analysis of protein-drug conjugates, *J. Med. Chem.* 34:1137 (1991).

136. K. Ono, H. Nakane, M. Fukushima, J.C. Chermann, F. Barre-Sinoussi, Inhibition of reverse transcriptase activity by a flavonoid compound, 5,6,7-trihydroxyflavone, *Biochem. Biophys. Res. Commun.* 160:982 (1989).

137. M.L. Bryant, R.O. Heuckeroth, J.T. Kimata, L. Ratner, J.I. Gordon, Replication of human immunodeficiency virus 1 and Moloney murine leukemia virus is inhibited by different heteroatom-containing analogs of myristic acid, *Proc. Natl. Acad. Sci.* 86:8655 (1989).

138. R. Pal, R.C. Gallo, M.G. Sarngadharan, Processing of the structural proteins of human immunodificiency virus type 1 in the presence of monensin and cerulenin, *Proc. Natl. Acad. Sci. USA* 85:9283 (1989).

139. J.J. Blumenstein, T.D. Copeland, S. Oroszlan, C.J. Michejda, Synthetic non-peptide inhibitors of HIV protease. *Biochem Biophys Res Commun* 163:980 (1989).

140. R. Buhl, H.A. Jaffe, K.J. Holroyd et al., Glutathione deficiency and HIV, *Lancet* 335:546 (1990).

141. A.L. Kinter, T. Kalebic, G. Poli, A.S. Fauci, Suppression of cytokine-induced human immunodeficiency virus (HIV) expression by N-acetyl-cystein (NAC), glutathione (GSH) and glutathione-monoester (GSE), *FASEB* Abstr II:A1265 (1991).

142. S.T. Butera, V.L. Perez, B-Y Wu, G.J. Nabel, T.M. Folks, Oscillation of the human immunodeficiency virus surface receptor is regulated by the state of viral activation in a $CD4^+$ cell model of chronic infection, *J. Virol.* 65:4645 (1991).

143. P. Feorino, S. Butera, T.M. Folks, R.F. Schinazi, Antiviral agents that prevent reactivation of latent HIV-1 in OM 10.1 cells. National Collaborative Drug Discovery Group, Frontiers in HIV Therapy, San Diego, CA (Nov. 3-7,1991).

144. M. Ito, M. Baba, S. Mori, et al., Tumor necrosis factor antagonizes inhibitory effect of azidothymidine on human immunodeficiency virus (HIV) replication *in vitro*, *Biochem. Biophys. Res. Commun.* 166:1095 (1990).

145. E. Tabor, J.S. Epstein, I.K. Hewlett, S.F. Lee, Inhibition by desferrioxamine of *in-vitro* replication of HIV-1, *Lancet,* 337:795 (March 30, 1991).

146. J.V. Lloyd, H. Cable, C. Rice-Evans, Evidence that desferrioxamine cannot enter cells by passive diffusion, *Biochem. Pharmacol.* 41:1361 (1991).

147. D.R.C. McLachlan, A.J. Dalton, T.P.A. Kruck, et al., Intramuscular desferrioxamine in patients with Alzheimer's disease, *Lancet* 337:1304 (1991).

148. G. Fredj, F. Dietlin, D. Fredj, et al., Tetrahydroaminoacridine in HIV infections, *Int. J. Clin. Pharmacol. Ther. Toxicol.* 27:408 (1989).

149. D.D. Richman, Zidovudine resistance of human immunodeficiency virus, *Rev. Infect. Dis.* 12:S507 (1990).

150. S.M. Crowe, M.S. McGrath, T. Elbeik, J. Kirihara, J. Mills, Comparative assessment of antiretrovirals in human monocyte-macrophages and lymphoid cell lines acutely and chronically infected with the human immunodeficiency virus, *J. Med. Virol.* 29:S145 (1989).

151. R. Yarchoan, J.M. Pulda, C.F. Perno, H. Mitsuya, S. Broder, Antiretroviral therapy of human immunodeficiency virus infection: current strategies and challenges for the future, *Blood* 78:859 (1991).

152. S.G. Ayre, New approaches to the delivery of drugs to the brain, *Med. Hypoth.* 29:283 (1989).

153. R.F. Schinazi, Strategies and targets for anti-human immunodeficiency virus type 1 chemotherapy, *in:* "AIDS in Children, Adolescents and Heterosexual Adults: An Interdisciplinary Approach to Prevention," R.F. Schinazi, A.J. Nahmias, eds., Elsevier, New York, 126 (1988).

154. R.F. Schinazi, C.K. Chu, B.F. Eriksson, et al., Antiretroviral activity, biochemistry, and pharmacokinetics of 3'-azido-2',3'-dideoxy-5-methylcytidine, *Ann. NY Acad. Sci* .616:385 (1990).

155. F.C. Szoka, The future of liposomal drug delivery, *Biotechnol. Appl. Biochem.* 12:496 (1990).

156. X. Lin, A. Dashti, R.F. Schinazi, J. Tang, Intracellular diversion of envelope glycoprotein gp 160 of human immunodeficiency virus to lysosomes for degradation, American Society for Biochemistry and Molecular Biology/Biophysical Society Houston, TX (February 9-13, 1992).

157. B.E. Huber, C.A. Richards, T.A. Krenitsky, Retroviral-mediated gene therapy for the treatment of hepatocellular carcinoma: An innovative approach for cancer therapy, *Proc. Natl. Acad. Sci. USA* 88:8039 (1991).

158. W. James, Towards gene-inhibition therapy: a review of progress and prospects in the field of antiviral antisense nucleic acids and ribozymes, *Antiviral Chem. Chemother.* 2:191 (1991).

159. J.A.T. Young, P. Bates, K. Willert, H.E. Varmus, Efficient incorporation of human CD4 protein into avian leukosis virus particles, *Science* 250:1421 (1990).

160. R.F. Schinazi, D.C. Liotta, W.B. Choi, et al., Selective inhibition of human immunodeficiency virus and hepatitis B virus by 2',3'-dideoxy-5-fluoro-3'-thiacytidine (FTC), National Collaborative Drug Discovery Group, Frontiers in HIV Therapy, San Diego, CA (Nov. 3-7, 1991).

STRUCTURE-ACTIVITY CORRELATIONS OF NATURAL

PRODUCTS WITH ANTI-HIV ACTIVITY

Mohamed Nasr, James Cradock and Margaret Johnston

Division of AIDS
National Institute of Allergy
  and Infectious Diseases (NIAID)
National Institutes of Health (NIH)
Rockville, MD 20892

INTRODUCTION

The search for safe and effective therapies to treat infections caused by the human immunodeficiency virus (HIV) and related opportunistic infections (OI's) are among the highest priorities of the National Institutes of Health.[1,2] The Division of AIDS (DAIDS), National Institute of Allergy and Infectious Diseases (NIAID) supports rational drug design and targeted drug discovery for HIV and the OI's through investigator-initiated research grants including the National Cooperative Drug Discovery Group programs.[1]

Computerized data bases containing chemical structures and biological data have been established at DAIDS to track developments in the chemotherapy of HIV and OI's and to serve as an information base available to researchers. Information on compounds evaluated preclinically for HIV and the OIs is acquired by continuous surveillance of primary literature sources. In this review, the structure-activity relationships have been analyzed among more than 300 compounds in the HIV data base that are either of natural origin per se or synthetic analogs.   These analyses of the structure activity correlations may assist investigators to elucidate what has been done with a particular chemical class and identify new semi-synthetic targets with potential anti-HIV activity. Steps in the replication of HIV that have been identified with specific chemotherapeutic natural products and analogs are indicated in Figure 1. The sites of action of some of the natural products described in this report are not presently known.

This review will focus on the following chemical classes: nucleosides, quinones, terpenoids, phenolic compounds, lipids, cholic acid derivatives, flavonoids, peptides, pyrones, polyethers and alkaloids.   Synthetic modifications which have resulted in increased or decreased anti-HIV activity within the chemical classes will be discussed. Classes reviewed by other investigators in this volume will not be discussed in detail. Data are listed in tabular format using the parameters of ID50 (concentration to inhibit viral replication by 50% of control value), TD50 (concentration to inhibit viability of uninfected cells by 50%, and selectivity index or SI (ratio of TD50 to ID50). Since this paper summarizes results from the published literature, it should be noted that a variety of methodologies were used to obtain the cited data. In vitro anti-HIV activity is

Natural Products as Antiviral Agents, Edited by C.K. Chu
and H.G. Cutler, Plenum Press, New York, 1992

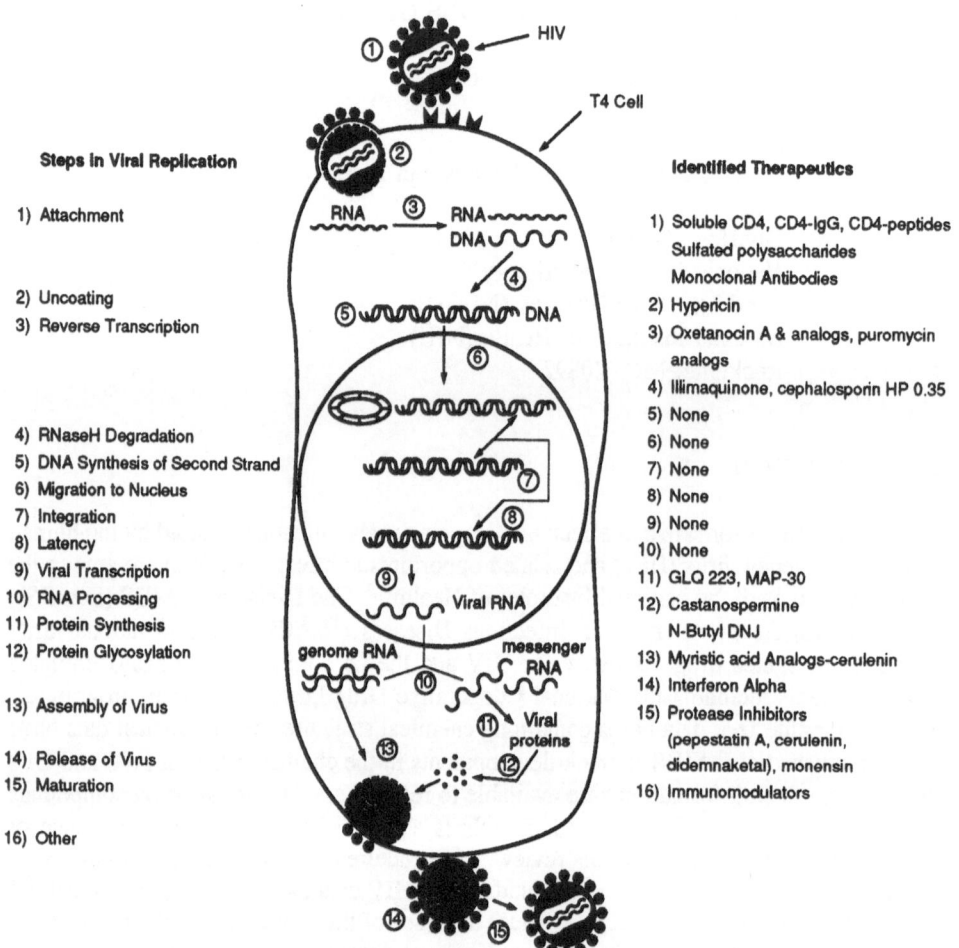

Steps in Viral Replication

1) Attachment

2) Uncoating
3) Reverse Transcription

4) RNaseH Degradation
5) DNA Synthesis of Second Strand
6) Migration to Nucleus
7) Integration
8) Latency
9) Viral Transcription
10) RNA Processing
11) Protein Synthesis
12) Protein Glycosylation

13) Assembly of Virus

14) Release of Virus
15) Maturation

16) Other

Identified Therapeutics

1) Soluble CD4, CD4-IgG, CD4-peptides
   Sulfated polysaccharides
   Monoclonal Antibodies
2) Hypericin
3) Oxetanocin A & analogs, puromycin
   analogs
4) Illimaquinone, cephalosporin HP 0.35
5) None
6) None
7) None
8) None
9) None
10) None
11) GLQ 223, MAP-30
12) Castanospermine
   N-Butyl DNJ
13) Myristic acid Analogs-cerulenin
14) Interferon Alpha
15) Protease Inhibitors
   (pepestatin A, cerulenin,
   didemnaketal), monensin
16) Immunomodulators

Figure 1. Potential Therapeutic Targets to Stop Life Cycle of HIV

32

indicated by (+) if the compound has shown activity in a particular cell line or enzyme assay, (+ -) if the compound has shown equivocal behavior, and (-) if there was no activity. References to the primary information source are listed in the tables and the data are grouped according to cell type or enzyme assay.

NUCLEOSIDES

Since nine of the eleven clinically approved antiviral agents are nucleoside analogs (table 1), this chemical class is very important therapeutically.[3-6] Vidarabine (ara A) is the only natural product among the clinically active antiviral nucleosides. It was first synthesized in 1960[7], and was later isolated from marine and microbial sources.[8,9] The first arabinofuranosyl containing nucleosides, spongothymidine (araT) and spongouridine (araU) were obtained from the sponge Cryptotethya crypta.[10] The identification of these compounds as arabinosyl nucleosides[11] stimulated the synthesis of other ara nucleosides.

Many synthetic pyrimidine and purine 2',3'-dideoxynucleoside (ddN) analogs have been shown to inhibit replication and pathogenic effects of HIV-1 in cell culture.[12,13] Other nucleoside analogs are currently undergoing clinical evaluation (Figure 2). Over 800 purine and pyrimidine nucleosides, predominantly synthetic compounds, are included in the NIAID HIV data base.

Table 1. Antiviral drugs approved by The US-FDA

Generic Name	Chemical Name	Clinical Use
1- Zidovudine	3'-azido-3'-deoxythymidine (AZT)	HIV
2- Didanosine	2',3'-dideoxyinosine (ddI)	HIV
3- Zalcitabine	2',3',-dideoxycytidine (ddC)	HIV
4- Idoxuridine	5-iodo-2'-deoxyuridine (IDU)	herpes keratitis
5- Trifluridine	5-trifluoromethyl-2'-deoxyuridine(TFT)	herpes keratitis
6- Acyclovir	9-(2-hydroxyethoxymethyl)-guanine	herpes keratitis genital herpes
7- Vidarabine	9-B-D-arabinofuranosyladenine(araA)	herpes encephalitis
8- Ribavirin	1-B-D-ribofuranosyl-1,2,4-triazole -3-carboxamide	respiratory syncytial virus (RSV)
9- Amantadine	1-adamantanamine hydrochloride	influenza A virus
10- Ganciclovir	[9-(1,3-dihydroxy-2-propoxymethyl) guanine]	CMV retinitis
11- Foscarnet	phosphonoformate sodium salt	CMV retinitis

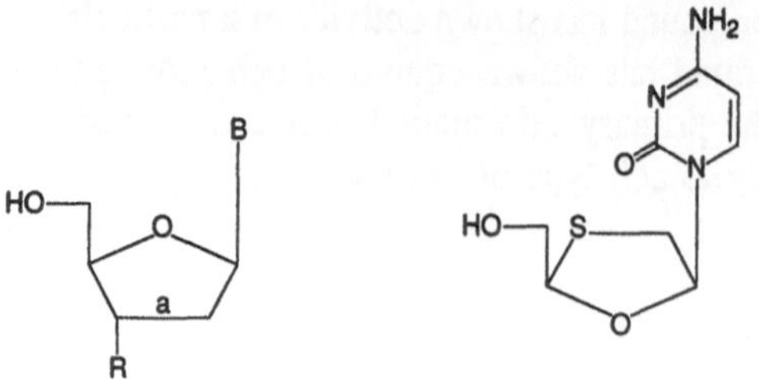

(–) BCH189, (3TC)

Drug	B	R	Clinical Status
AZT	Thymine	N_3	Approved adults and children
ddI	Hypoxanthine	H	Approved
ddC	Cytosine	H	Phase II/III
D4T	Thymine (a double bond)	H	Phase I/II
FLT	Thymine	F	Phase I
AZdU (CS-87)	Uracil	N_3	Phase I/II
(–) BCH189, 3TC (IAF)			Phase I

Figure 2

Oxetanocin A [9-(2-deoxy-2-hydroxymethyl-B-D-erythro-oxetanosyl)adenine]was the first naturally occurring nucleoside exhibiting in vitro anti-HIV activity.[14-17] This compound was isolated from the culture filtrates of Bacillus megaterium. Oxetanocin A, as well as similar oxetanosyl N-glycosides with different base moieties, have also shown activity against other viruses.[18] The order of anti-HIV activity in MT-4 cells for the oxetanocins with different purine bases is: hypoxanthine (OXT-H) > guanine (OXT-G) > 2,6-diaminopurine (2-amino-OXT-A) > adenine (OXT-A) > xanthine (OXT-X). The activity of oxetanocin G was equivalent to that of 9-[(1,3-dihydroxy-2-propoxy)methyl]guanine (DHPG) against human cytomegalovirus (HCMV) in vitro.[18]

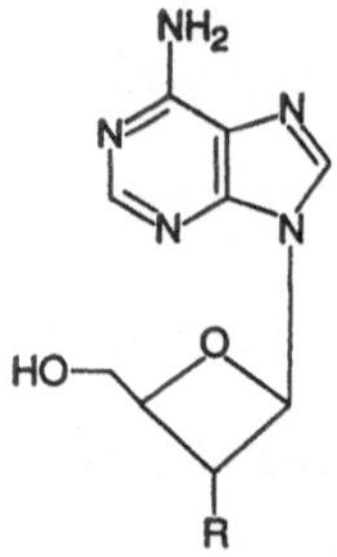

In Vitro Anti-HIV Activity MT-4 cells

Compound	R	ID_{50}	TD_{50}(ug/ml)	SI*
OXT–A	CH_2OH	1.4	16	11
OXT–A analog	OH ↑	0.10	35	355
OXT–A analog	H	0.03	6	210
OXT–A analog	OH ↓, CH_2OH	1.00	130	130
OXT–A analog	CH_2N_3	0.75	75	100
OXT–A analog	$= CH_2$	0.60	11	18
2–NH_2–OXT–A	CH_2OH	4.7	130	28
OXT–G	CH_2OH	7.3	210	29
OXT–H	CH_2OH	2.2	330	150
OXT–X	CH_2OH	>100	>300	3
ddI	–	0.21	260	1,200

*Ref. 17, 19

OXT–A Analogs

34

Deletion or modification of the 2'-substituent on the oxetane ring of OXT-A resulted in improved activity when tested against HIV-1 infected MT-4 cells.[19] With respect to selectivity index: 2'-OH (up) was > 2'-H > 2'-OH > 2'-CH$_2$OH > 2'-CH$_2$N$_3$ > 2'- =CH$_2$ > 2'-2'-CH$_2$OH (OXT-A). None of the oxetanocin derivatives exhibited an in vitro selectivity index greater than ddI.

Replacement of the oxetane ring by a cyclobutyl ring yielded compounds that showed broad spectrum antiviral activities. Both cyclobut-A and cyclobut-G provided significant protection to HIV infected ATH8 cells at 1 uM concentration.[20] Cyclobut-G was comparable to acyclovir against the herpes viruses HSV-1 and HSV-2 in vitro but clearly superior to this drug against the thymidine kinase deficient strains of the same viruses. Cyclobut-G is also active in vivo against duck hepatitis B virus and a lethal challenge of HSV-1 in mice.[20]

Aristeromycin, a naturally occurring analog of adenosine in which the ribose oxygen is replaced by a methylene group, was inactive but served as a lead for the synthesis of other carbocyclic nucleoside analogs. Among a series of carbocyclic anti-HIV active nucleoside analogs, Carbovir, carbocyclic 2',3'-didehydrodideoxyguanosine, showed the best in vitro anti-HIV activity.[21]

Puromycin, a naturally occurring nucleoside, did not show in vitro anti-HIV activity, but puromycin aminonucleoside was found to be active.[22] The anti-HIV activity of nucleocidin, a naturally occurring purine nucleoside containing 4'-fluoro substitution, has not been described but analogs of nucleocidin were reported to possess anti-HIV activity in vitro.[23] The recent observation of the anti-HIV activity of 4'-azidothymidine[24], suggests that 4'-fluorodideoxynucleoside analogs represent a potential area for future investigation.

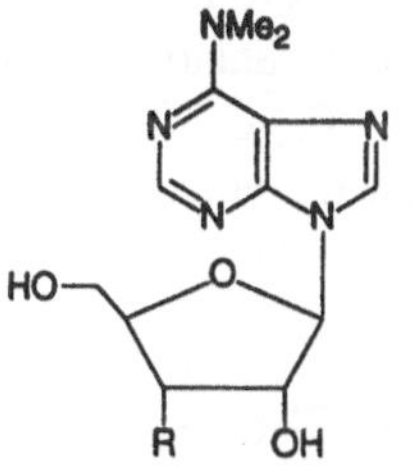

Compound	R	Anti-HIV activity
Puromycin	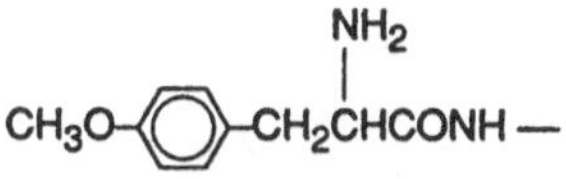 CH$_3$O—⟨⟩—CH$_2$CHCONH— (NH$_2$)	–
Puromycin aminonucleoside	NH$_3^+$	+*
N6,N6,–dimethyladenosine	OH	–

*95% inhibition of HIV-1 replication and complete inhibition of syncytia formation in H9 cells at 2-10 µg/ml

Anti-HIV Activity of Puromycin Analogs

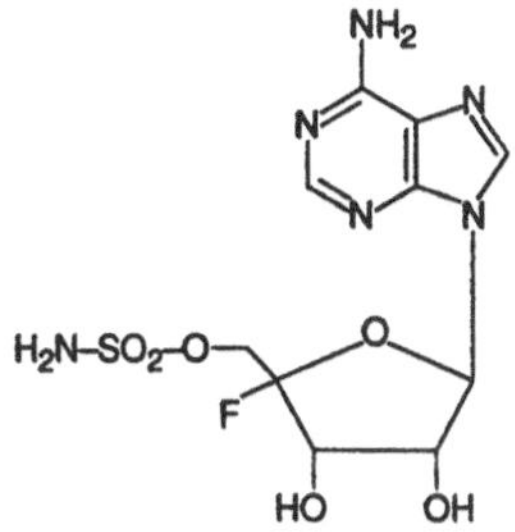

Nucleocidin

35

Most nucleosides are progressively phosphorylated by cytoplasmic enzymes to nucleoside 5'-triphosphates that compete with the natural nucleoside triphosphate substrates for binding to cellular DNA polymerase and the viral RT.[13] Modification of naturally occurring nucleosides in either the base or the sugar can remarkably alter the biological activity. In the area of HIV chemotherapy, certain changes in the sugar moiety have resulted in the most active derivatives. In contrast, modification of the base but not the sugar of natural nucleosides yielded active antitumor agents.

QUINONES (Table 2)

Quinones represent a chemical class with diverse pharmacological properties including antitumor and antimicrobial activities.[26] Quinones have been isolated from fermentation broths and/or plant materials and evaluated for antiviral effects either in cell based assays or in prescreens using HIV-specific enzymes (Table 2).[27,28] Hypericin and pseudo-hypericin, polycyclic quinones isolated from the St. Johnswort plant, exhibited activity versus retroviral infections in mice[30,31] and inhibited both replication of HIV and enveloped DNA viruses in vitro.[46] Hypericin has been reported to confer high stability on the HIV capsid structure and may exhibit antiviral activity by inhibiting the uncoating process.[32] This class of compounds will be discussed in detail elsewhere in this volume. The anthraquinone antitumor agents doxorubicin, daunorubicin, and aclarubicin have been reported to exhibit modest anti-HIV activity in vitro.[36,37] Also pradimicin A, and the structurally related antibiotics benanomicin A and B, have been reported to inhibit HIV replication in cell culture.[34,35]

Doxycycline and minocycline were reported to possess anti-HIV activity in CEM cells when tested with other tetracycline analogs.[49] A substituted naphthoquinone, 19-O-n-pentyldamavaricin Fc (n-pentyl-DvFc) exhibited modest in vitro anti-HIV activity (SI < 5).[38] The authors suggested that the antiviral activity of n-pentyl-DvFc may be due to an effect on host cell metabolism rather than HIV replication per se.

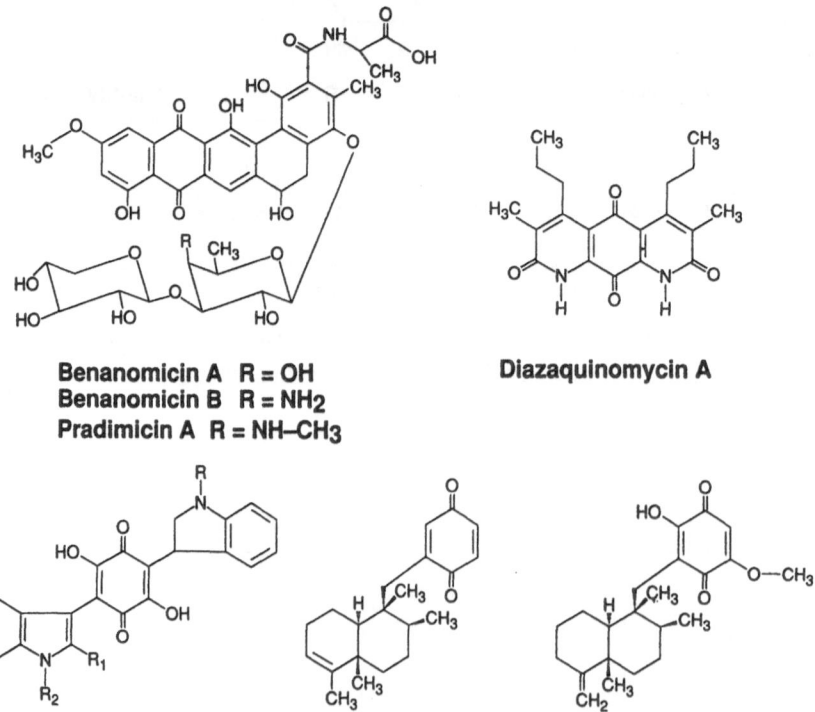

Benanomicin A R = OH
Benanomicin B R = NH₂
Pradimicin A R = NH–CH₃

Diazaquinomycin A

Asterriquinone   **Avarone**   **Illimaquinone**

Table 2. In vitro activity of quinones in cell culture or HIV/RT

Compound Name	Act.	cell line	ID50 ug/ml	TD50	Reference
POLYCYCLIC QUINONES					
Hypericin	+	PBMC	0.22	>50	31,46
Daunorubicin	+	HUT-78	0.01		37
Doxorubicin	+	MT-4	0.01	1.0	36,47
Pradimicin A	+	MT-4	>3.5*	30	48
Benanomicin A	+	MT-4	30-100*	100	35
Benanomicin B	+	MT-4	10-100**		35
Emodin	-	PBMC	9.7	>27	49
Doxycycline	+	CEM	0.31	3.0	45
Minocycline	+	CEM	0.20	2.8	45
NAPHTHOQUINONES					
Damavaricin	+	HUT-78	3	16.0	38
BENZOQUINONES					
Avarone	+	H9	< 0.1	> 1.0	40
Asterriquinone	+	RT	0.5		39
6'-Hydroxy-4'-methoxyavarone	+	RT	1.0		43
Illimaquinone	+	RNaseH	5.37		44

*Concentration producing complete inhibition

Other naturally-occurring quinones/quinols including the marine natural products avarone and avarol have shown anti-HIV activity in cell culture.[40] Several other quinones including chelocardin and diazaquinomycin A have been reported to inhibit HIV reverse transcriptase but effects on HIV replication in a cell based assays were not described.[39,41]

The observation that a semiquinone intermediate may influence inhibition of reverse transcriptase (RT) was demonstrated more than ten years ago.[29] Levodopa, dopamine and their analogs have been shown to be potent inhibitors of avium myeloblastosis virus RT in the presence of tyrosinase (copper containing polyphenol oxidase). This inhibition was attributed to the generation of quinol-quinone oxidation reduction systems. Both the reduction of a quinone and the oxidation of a quinol, which have the semiquinone as a common intermediate, resulted in RT inhibition.

Illimaquinone, a marine natural product and analog of avarone, was reported to selectively inhibit the RNase H activity of HIV-1.[44] Few specific inhibitors of the RNase H portion of HIV-1 RT have been reported to date but this area will likely be more actively investigated since the crystal structure of the RNase H portion of HIV-1 RT was recently published.[33]

In summary, several members of the quinone family have demonstrated anti-HIV activity in cell culture. Unfortunately, the selectivity indices of these compounds have generally been low and their mechanisms of action are unresolved. Thus, enthusiasm for developing quinones for clinical evaluation has remained low. However, one of the most potent and specific representatives of this class, hypericin, is now in phase I clinical trial, the results of which could impact this enthusiasm.

TERPENOIDS (Table 3)

Gossypol, a sesquiterpene obtained from cottonseeds has been reported to exhibit antiviral activity against herpes simplex type 2, and influenza virus.[51,52] Recently (-)gossypol but not the (+) enantiomer exhibited activity against HIV infected PBM cells.[53] Several analogs of racemic gossypol have been evaluated against HIV in vitro.[54] Gossylic iminolactone was found to be less cytotoxic and four-fold more active than

Gossypol   **Gossypol lactones**

X = NH, O

racemic gossypol. The lactone of gossypol was essentially inactive (SI = 4). Dehydroandrographolide is the major diterpenoid lactone found in Andrographis panticulata, a medicinal herb used in Chinese traditional medicine to treat a variety of illnesses including hepatitis, meningitis, dysentery, etc. Although devoid of anti-HIV activity, its succinic acid monoester (DASM) exhibited in vitro anti-HIV activity in H9 and PBM cells with a SI of $\geq$ 60.[55] The in vivo stability of this ester and HIV activity of analogs has apparently not been assessed. Diterpenes isolated from Homalanthus acuminatus and Chrysobalanus icaco, trees indigenous to tropical rain forest have also been reported to possess in vitro anti-HIV activity.[56]

Glycyrrhizin, a triterpenoid isolated as aqueous extract from the plant, Glcyrrhiza radix, and its derivatives exhibited anti-HIV activity in vitro.[57,58] Inhibition of protein kinase C (PKC) activity by glycyrrhizin has been discussed in relation to its anti-HIV activity.[59] It has been reported that the binding of HIV-1 to CD4[+] induces rapid phosphorylation of the CD4 receptors which involve protein kinase C (PKC).[60] However, it is not known if the HIV inhibition by glycyrrhizin is mediated through blocking the CD4 receptor.

The anti-HIV activity of soybean saponin fraction B1[63] and the triterpenoid from Patrinia scabiosaefolia seeds has also been reported. Didemnaketals A and B are linear heptaprenoids, a very rare terpenoid class. These compounds were isolated from the Ascidian didemnum sp. and reported to inhibit HIV-1 protease.[62] Assessment of these compounds in a cell based assay for anti-HIV activity was not reported. Synthetic modifications of naturally occuring terpenoids offer the potential for enhancement of anti-HIV activity. Based on available information, DASM and gossypol lactones would be candidates for synthetic modifications.

Recombinant CD4, Peptides and Reagents to Interfere with Viral Attachment

Human CD4+ T lymphocytes and monocytes are the primary cellular targets for HIV-1 infection. The CD4 membrane antigen present on these cells is the receptor for HIV-1 gp 120 envelope glycoprotein. This role of CD4 has prompted considerable interest in the therapeutic use of soluble, truncated forms of the CD4 protein as "receptor decoys" to block HIV-1 infection.[68] To date the use of CD4 congeners or synthetic polymers that block attachment has not resulted in a therapeutically useful approach in man. Limitations due to short biologic half-life and poor tissue penetration have been tackled by covalent attachment of soluble CD4 to the Fc fragment of IgG.[69] However, recent studies have brought to light another problem, whether pharmacologically relevant concentrations can be achieved and maintained in vivo. The sensitivity of primary viral isolates to inhibition by soluble, truncated CD4 proteins (176-178 amino acids) is much less than for laboratory adapted strains. To date, neither rsCD4 nor CD4-Fc have demonstrated clinical activity.

Table 3. Terpenoids

Compound Name	Act.	cell line	ID50 uM	TD50	SI	Reference
(-)-Gossypol	+	PBMC	5.2	>100	>20	53,54,66
(+)-Gossypol	-	PBMC	50.7	52	1	53,61
Gossypol iminolactone	+	H9	0.25	50	200	54
Gossypol lactone	-	H9	63	250	4	54
Dehydroandrographolide ester	+	H9	3.1	>200	>60	55
Andrographolide	-	H9	>50	50		55
Glycyrrhizin	+	MT-4	0.15	2.6	17	57,58,64
Glycyrrhizin derivative	-	MT-4	<160	>620	>4	58
Diterpene	+	CEM	19	38	2	56
Soybean saponin B1	+	MT-4	0.125	>1	>8	63

Anionic polysaccharides, such as dextran sulfate and pentosan polysulfate, have been found to impair the interaction of HIV-1 with CD4 in cultured cells, but no clinical activity was noted when dextran sulfate was administered intravenously to p24 positive patients.[70]

FLAVONOIDES (Table 4)

Flavonoides, which are widely distributed in plants, were initially found to be active against HIV and other viral transcriptases. Baicalein was evaluated as a potent inhibitor of HIV RT. Baicalein was not inhibitory to cellular DNA polymerases alpha and beta, but DNA polymerase gamma was sensitive.[71-73] The differential inhibitory effects of various flavonoids on activities of reverse transcriptase and cellular DNA and RNA polymerases has been reviewed recently.[71] Catechin like compounds, similar in structure to flavonoids but lacking the pyrone ring, were much less active against the HIV RT. Quercetin and myricetin were found to be active against the viral RT but were also potent inhibitors of the cellular alpha and beta DNA polymerases.[71] In general, the

Table 4. Flavonoides

Compound Name	RT/activity[*]	Reference
5,6,7-Trihydroxyflavone; Baicalein	+	50;74
3,3',4',5,7-Pentahydroxyflavone; Quercetin	+	75;76
5,7-Dihydroxyflavone; Chrysin	-	71
6,7-Dihydroxyflavone	-	71
7,8-Dihydroxyflavone	-	71
4',5,7-Trihydroxyflavone; Apigenin	-	71
3',4',5,7-Tetrahydroxy flavone; Luteolin	+	71
2',3,4',5,7-Pentahydroxy flavone; Morin	+	71
Quercetin-3-L-rhamnoside;Quercitin	+	71
Quercetin-3-rutinoside; Rutin	+	71
3,3',4',5,6,7-Hexahydroxyflavone;Quercetagetin	+	71;76
3,3',4',5,5',7-Hexahydroxyflavone; Myricetin	+	71;76
Flavones of Viscum coloratum	+	50
Isoflavone	-	50

[*] RT refers to HIV/RT and other retroviral RT's.

potency of these compounds against the HIV RT has been less than the activity of nucleoside analogs such as AZTTP and non-nucleoside RT inhibitors. The anti-HIV activity of flavonoids (Table 4) in cell culture assays has not yet been reported. Further investigation of this class of compounds in cell culture assays should be pursued to gain additional information on the predictability of polymerase data for inhibition of HIV replication.

PYRONES

The isolation and structural elucidation of phenoxan from cell cultures of Myxobacteria, has been recently reported.[77] Phenoxan exhibited potent anti-HIV activity in vitro (ID50 = 6.6 nM) and low cytotoxicity (TD50 >6600 nM). Other structurally related natural products that have shown anti-HIV activity include, aureothin, isolated from Streptomyces luteus, and spectinabilin, a co-metabolite of the streptovaricins. These three compounds are closely related not only in their pyrone residues but also in the side chain which consists of a phenyl residue coupled via a C3, C4, or C7 aliphatic chain. Relative to other classes of potential anti-HIV agents, pyrones have been minimally studied. Further evaluation of other derivatives is needed to assess the potential of this class as antiviral agents.

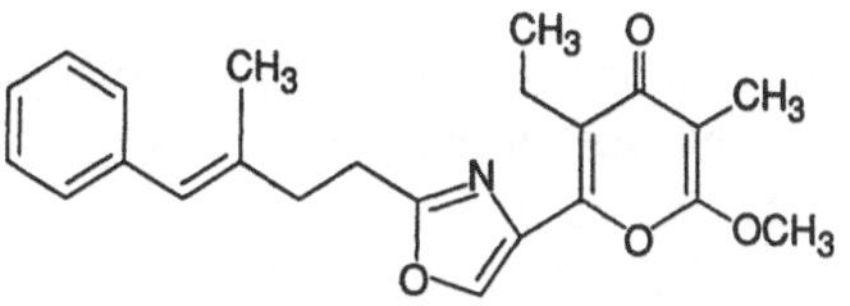

Phenoxan

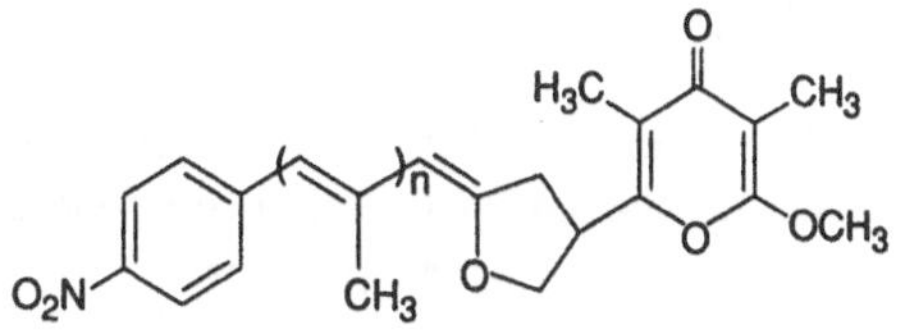

Aureothin   n = 1
Spectinabilin n = 3

POLYPHENOLS (Table 5)

Gallic acid derivatives have been evaluated for anti-HIV activity or RT inhibition. Examples of compounds that have been tested in cell culture are listed in Table 5. Most of these compounds will be discussed in a chapter concerning anti-HIV activity of tannins.[79-83]

Table 5. Polyphenols

Compound Name	Act.	cell line	ID50 uM	TD50	SI	Reference
Punicalin	+	H9	<6	>30	>5	79
(-)-Epigallocatechin gallate	+	MT4	2.1			81
(-)-Epicatechin gallate	+	MT4	13.7			81
Chebulinic acid	+	H9	<1	>20	·20	79
Chebulagic acid	+	H9	<1	20	>20	79
Punicalagin	-	H9	<5	>20	>4	79
Punicacortein C	+	H9	<1	<20	<20	79
Etoposide	-	MT2				47
Teniposide	-	MT2				47
Gallotannin	+	Molt-3				82

Glycosylation Inihibitors

Certain viral proteins undergo glycosylation, which has been pursued as a target for therapeutic intervention.[84-86] Castanospermine and deoxynojirimycin, naturally occurring polyhydroxyalkaloids and inhibitors of glucosidase-I, have served as lead compounds for the development of chemical analogs with more potent and selective inhibition of HIV. N-butyl deoxynojirimycin (butyl-DNJ), 6-acetyl and 6-butyl-castanospermine are more potent analogs.[87] Clinical trials of butyl-DNJ are in progress. Substantial gastrointestinal toxicities have been observed (diarrhea, bloating, cramping) that are due to the non-selective effects of these inhibitors against other glucosidases. These inhibitors will be discussed in more detail elsewhere in this volume.

Myristoylation Inhibitors

Assembly of infectious virions are dependent on the action of: (a) an aspartyl protease encoded by the viral *pol* gene and responsible for cleaving *gag* and *gag-pol* precursors into mature proteins[88,89] and (b) cellular N-protein myristoyl transferase (NMT) which adds myristic acid to the N-terminus of the viral *gag, gag-pol and nef* proteins.[90]

Natural products that have been reported to inhibit HIV protease include the peptide pepstatin A,[91] and the nonpeptides, cerulenin[92] and didemnaketal B.[62] These compounds showed low selectivity relative to synthetic, peptide based inhibitors and are therefore not being pursued further.

Design of agents that interfere with the process of myristoylation of HIV proteins is being pursued. Heteroatom substituted analogs of myristate were reported to inhibit HIV-1 replication in acutely infected $CD4^+$ H9 cells at concentrations that were not toxic to uninfected cultured cells.[93] This inhibition was associated with a dramatic reduction in the rate of proteolytic processing of the polyprotein precursor by viral protease, probably resulting from an inability of the gag and gag-pol precursors containing the myristic acid analog to associate with the plasma membrane. Because of their novel mechanism of action, potency and apparent selectivity, further research to bring a candidate to clinical trial is actively underway.

Lipids

Several phospholipid formulations including unsaturated phosphatidylcholine and phosphatidylethanolamine (purified from soybean) exhibited anti-HIV activity in H9 and PBMC test systems at relatively high but apparently nontoxic concentrations (IC50 =16-20 uM).[94] Phospholipids containing two sites of unsaturation in the fatty acid are active while fully saturated fatty acids did not exhibit anti-HIV activity. The mode of action of these compounds is probably not inhibition of HIV binding since they do not inhibit syncytia formation. The mechanism of action is not known but may relate to the accumulation of toxic metabolic products after the hydrolysis of these compounds by phospholipase A2.

A group of sulfonic acid containing glycolipids occurring in higher plants and algae have shown anti-HIV activity in a variety of cell lines including CEM, MT-2, LDV-7 and C3-44 using p24 measurements or syncytium formation as an index of antiviral effects.[95] It was noted that all of the sulfolipids tested had similar levels of activity in a particular cell line.

In summary the class of lipids as anti-HIV has not been sufficiently investigated to make a conclusion regarding their therapeutic potential.

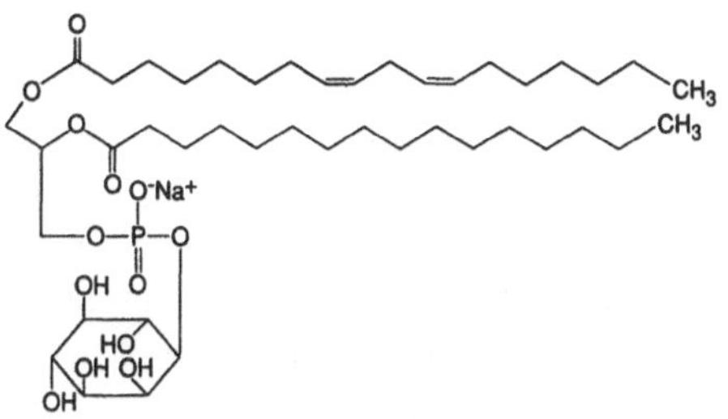

Plant phosphatidylinositol

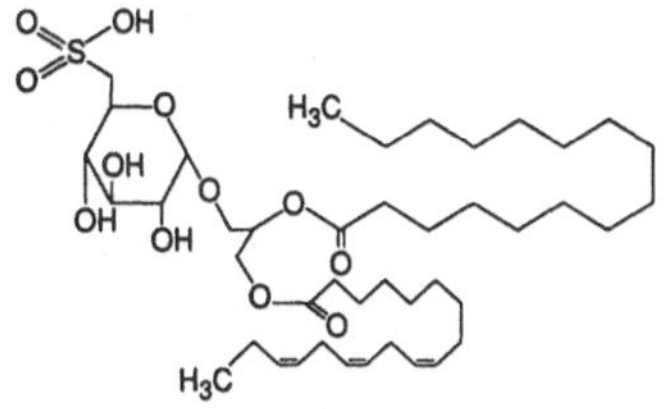

Sulfoglycolipids

Bile Acids and Steroidal Compounds (Table 6)

Lithocholic acid, and taurolithocholic acid occur in human and animal bile. When tested in vitro for anti-HIV activity in the MT-4 cell line, lithocholic acid did not exhibit selectivity toward HIV, while its 3-sulfate derivative showed weak anti-HIV activity. Taurolithocholic acid 3-sulfate showed better selectivity than taurolithocholic acid. Sulfation of these compounds contributed to their overall anti-HIV activity but in general, high concentrations of compound were required for activity and they were not markedly selective.

Weinbersterol disulfate A, an antiviral steroid sulfate from the sponge Petrosia weinbergi, was found active against HIV in vitro. Another steroid structure, fusidic acid, has been reported to possess anti HIV activity in vitro in different cell lines but cytotoxicity was observed in uninfected cells.[96,98] Since sulfate conjugation of certain steroids resulted in improved anti-HIV activity, it may be of interest to evaluate sulfated derivatives of fusidic acid for anti-HIV activity.

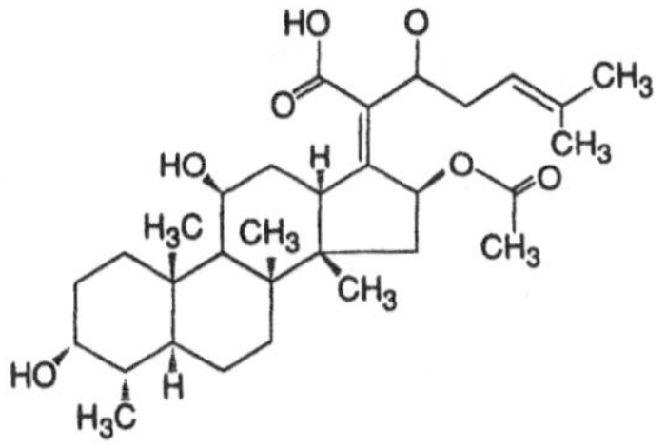

Fusidic acid

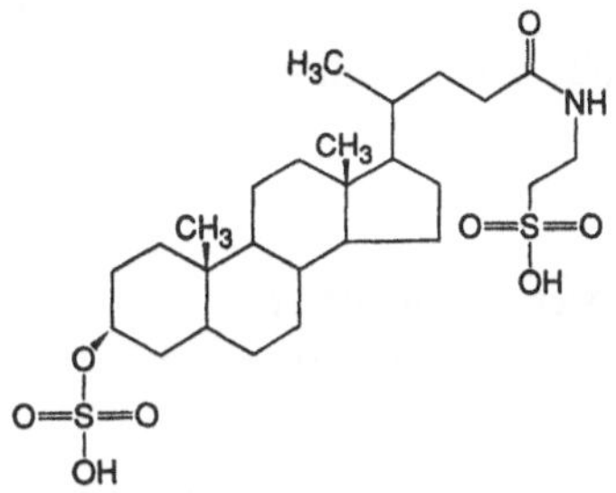

Taurolithocholic acid 3-sulfate

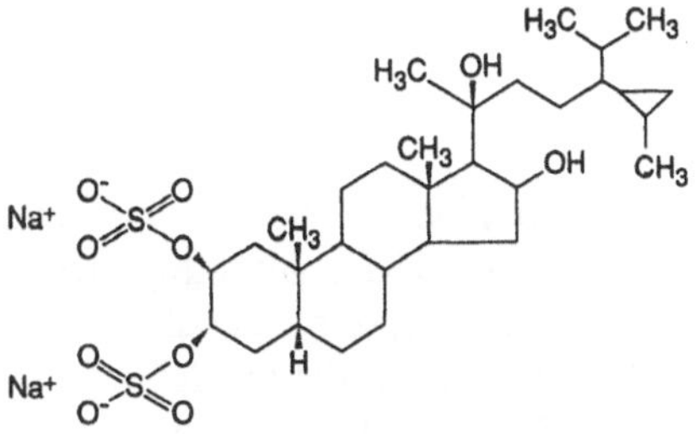

Weinbersterol disulfate A

Polyethers

The anti-HIV activity of monensin,[101,102] a naturally occurring, polyether monovalent carboxylic ionophore, has been reported to suppress HIV replication by blocking proteolytic conversion of the env-coded polyprotein gp160 to gp120. Monensin has previously been reported to inhibit the entry process of picornaviruses by increasing intravesicular pH.[103] It has been suggested that HIV entry via the CD4 receptor mechanisms requires a low pH of the endosome.[104]

A new polyether antibiotic, kijimicin, has recently been identified in the culture filtrate of Actinomadura sp. M1215-NF3. This agent exhibits potent anticoccidial activity. Its anti-HIV activity in both acute (SI=30, H9 cells) and chronic infection (SI=10, U937 cells) has also been reported.[105] Although these compounds showed moderate selectivity indices in vitro in cells infected with HIV, further in vitro studies in animal cell lines and animal models of retroviral disease will be required to adequately assess their potential therapeutic relevance.

Table 6. Bile acids and Steroids

Compound Name	Act.	cell line	ID50ug/ml	TD50	SI	Reference
Dehydro-epi-androsterone (DHEA)	+	PBMC	4.8	138	28.0	97
Taurolithocholic acid 3-sulfate	+	MT4	68	500	7.3	99
Taurolithocholic acid	+	MT4	77	370	4.8	99
Lithocholic acid	-	MT4	47	47	1.0	99
Lithocholic acid 3-sulfate	+	MT4	71	276	3.9	98
Fusidic acid	+	HUT-7850				96
Weinbersterol disulfate A	+	HIV	1			100

Miscellaneous (Table 7)

Several classes of heterocyclic compounds including isoquinoline alkaloids, coumarins, camptothecin, caffeic acid and a thiazole derivative (HP 0.35) are grouped under Table 7. HP 0.35, a cephalosporin degradation product was found to be a specific inhibitor of RNase H activity of HIV-1 and feline immunodeficiency virus (FIV) reverse transcriptases.[115] HP 0.35 was active in vitro against FIV in a cell culture assay but no anti-HIV testing was reported.[115] A list of medicinal herbs used by AIDS patients has been recently compiled.[113]

Table 7. Miscellaneous

Compound Name	Act.	cell line	ID50 ug/ml	TD50	SI	Reference
Pine cone extract	+	CEM	3	> 100	> 33	106;107
Caffeic acid	+	MT4	5	88	17.6	82;108
Camptothecin	+	H9	0.002	0.0008		109
COUMARINS						
Coumermycin A1	-	MT4	1.5	3.3	2.2	110
Novobiocin	-	MT4	109	109	1.0	110
ISOQUINOLINE ALKALOIDS						
Michellamine A	+	CEM	15	150	10	116
Papaverine RS 47	+	H9	6.0	30	5.0	111;112
Ellipticine	-	MT2				47

Table 8. Anti-HIV Activity of Natural and Synthetic Modified Products

Class	Natural products	Act.	thetic modifications	Act.
Nucleosides	Oxetanocin A	+	Cyclobut-G	+++
	Puromycin	-	Puromycin aminonucleoside	+
	Nucleocidin	-	5'-Iodonucleocidin	+
	Aristeromycin	-	Carbovir	+
Quinones	Hypericin	+	None	
	Daunorubicin/Doxorubicin		None	
	Avarone	+	None	
	Damavaricin Fc (DvFc)	-	n-Pentyl-DvFc	+
	Pradimicin A	+	None	
Lipids	Phospholipids	+	Saturated analog	-
	Sulfoglycolipids	+	None	
Terpenoids	(-) Gossypol	+	None	
	Gossypol	+	Gossylic iminolactone	++
	Dehydro-andrographolide	-	DASM	+
	Glycyrrhizin	+	11-Deoxyglycyrrhizin	+
Bile acids	Lithocholic acid	-	3-Sulfate derivative	+
Alkaloids	Castanospermine	+	6-Alkyl derivatives	++
	1-Deoxynojirimycin	+	N-butyl derivative	++

CONCLUSION

In the field of chemotherapy, natural products have been used as leads for the synthesis of useful therapeutic agents. Unfortunately, some of the most interesting compounds are either difficult to obtain in sufficient quantity or their activity is associated with toxicity. Although several classes of natural products have shown anti-HIV activity in different assays, their selectivity indices are lower when compared with synthetic analogs in clinical or late stage preclinical evaluation, for example, dideoxynucleosides or protease inhibitors. However, natural products offer potential as lead compounds. Synthetic modification has resulted in improved anti-HIV activity within several classes of natural products, i.e., carbocyclic analogs of oxetanocin A, analogs of castanospermine or deoxynojirmycin and bile acids, etc (Table 8). In one instance a simple succinate esterification of the inactive diterpenoid, andrographolide yielded an active compound. As noted in this review, several investigators have utilized HIV-RT as a prescreen to evaluate compounds from natural sources. It is anticipated that this approach will be expanded using other viral enzymes: RNaseH, integrase, Tat, protease, etc., to assay for potential lead compounds from natural sources. Once activity is established in these enzymatic assays, evaluation in cell culture assays and in animals leads to optimization of the series and identification of a candidate suitable for entry into clinical trials.

REFERENCES

1. J. McGowan, and D. Hoth, AIDS drug discovery and development, *J. AIDS* 2:355 (1989).

2. O.S. Weislow, R. Kiser, D.L. Fine, J. Bader, R.H. Shoemaker, and M.J. Boyd, New soluble-formazan assay for HIV-1 cytopathic effects: Application to high-flux screening of synthetic and natural products for AIDS-antiviral activity, *J. Nat. Cancer Inst.* 81:577 (1989).

3. H. Mitsuya, K.J. Weinhold, P.A. Furman, M.H. St. Clair, S.N. Lehrman, R.C. Gallo, D. Bolognesi, D.W. Barry, and S. Broder, 3'-Azido-3'-deoxythymidine (BW A509U): An antiviral agent that inhibits the infectivity and cytopathic effect of Human T-lymphotropic Virus Type III/lymphadenopathy-Associated Virus In Vitro, *Proc. Natl. Acad. Sci.* USA 82:7096 (1985).

4. M.A. Fischl, D.D. Richman, M.H. Grieco, M.S. Gottlieb, P.A Volberding, O.L. Laskin, J.M. Leodom, J.E. Groopman, D. Mildvan, R.T. Schooley, G.G. Jackson, D.T. Durack, and D. King, The efficacy of azidothymidine (AZT) in the treatment of patients with AIDS and AIDS-related complex. A double-blind, placebo-controlled trial, *N. Engl. J. Med.* 317:185 (1987).

5. W. Ostertag, G. Roesler, C.J. Krieg, J. Kind, T. Cole, T. Crozier, G. Gaedicke, G. Steinheider, N. Kluge, and S. Dube, Induction of endogenous virus and of thymidine kinase by bromodeoxyuridine in cell cultures transformed by Friend virus, *Proc. Natl. Acad. Sci.*, USA, 71:4980 (1974).

6. S.K. Dube, I.B. Pragnell, N. Kluge, G. Graedicke, G. Steinheider, and W. Ostertag, Induction of endogenous and of spleen focus-forming viruses during dimethylsulfoxide-induced differentiation of mouse erythroleukemia cells transformed by spleen focus-forming virus, *Proc. Natl. Acad. Sci.*, USA, 72:1863 (1975).

7. W.W. Lee, A. Benitez, L. Goodman, and B.R. Baker, Potential anticancer agents. XL. Synthesis of the .beta.-anomer of 9-(D-arabinofuranosyl)adenine, *J. Am. Chem. Soc.* 82:2648 (1960).

8. F. Cimino, S. De Rosa and S. De Stefano, Antiviral agents from a gorgonian, Eunicella cavolini, *Experientia* 40:339 (1984).

9. J.M. Betz and A.H. Der Marderosian, Antiviral/antineoplastic agents from a jellyfish (Stomolophus meleagris L. Agassiz 1860). *Abstract No. 113, Twenty-*

10. W. Bergmann and R. Feeney, Contributions to the study of marine products XXXII. The nucleosides of sponges, *J. Org. Chem.* 16:981 (1951).

11. W. Bergmann and D.C. Burke, Contributions to the study of marine products XXIX. Spongothymidine and spongouridine, *J. Org. Chem.* 20:1501 (1955).

12. M. Nasr, C. Litterst and J. McGowan, Computer-assisted structure-activity correlations of dideoxynucleoside analogs as potential anti-HIV drugs, *Antiviral Res.*, 14:125 (1990).

13. G. Ahluwalia, D.A. Cooney, H. Mitsuya, A. Fridland, K.P. Flora, Z. Hoa, M. Dalal, S. Broeder and D.J. Johns, Initial studies on the cellular pharmacology of 2',3'-dideoxyinosine, an inhibitor of HIV infectivity, *Biochem. Pharmacol.*, 36:3797 (1987).

14. N. Shimada, S. Hasegawa, T. Harada, T. Tomisawa, A. Fujii and T.J. Takita, Oxetanocin, a novel nucleoside from bacteria, *J. Antibiot.* 39:1623 (1986).

15. T. Maruyama, Y. Sato, T. Horii, H. Shiota, K. Nitta, T. Shirasaka, H. Mitsuya and M. Honjo, Synthesis and antiviral activities of carbocyclic oxetanocin analogues, *Chem. Pharm. Bull* (Tokyo) 38(10):2719 (1990).

16. H. Hoshino, N. Shimizu, N. Shimada, T. Takita and T.J. Takeuuchi, Inhibition of infectivity of human immunodeficiency virus by oxetanocin, *J. Antibiot.* 40:1077 (1987).

17. J. Seki, N. Shimada, K. Takahashi, T. Takita, T. Takeuchi and H. Hoshino, Inhibition of infectivity of human immunodeficiency virus by a novel nucleoside, oxetanocin, and related compounds, *Antimicrob. Agents Chemother.* 33(5):773 (1989).

18. Y. Nishiyama, N. Yamamoto, K. Takahasi and N. Shimada, Selective inhibition of human cytomegalovirus replication by a novel nucleoside, oxetanocin G, *Antimicrob. Agents Chemother.* 32:1053 (1988).

19. J. Seki, T. Takeuchi, N. Shimada, K. Takahasi, T. Takita, H. Hoshino, Anti-HIV activities of a novel nucleoside oxetanocin and its derivatives, *5th International Conference on AIDS*, Montreal, June 4-9, 1989; M.C.P. 121:562 (1989).

20. D.W. Norbeck, E. Kern, S. Hayashi, W. Rosenbrook, H. Sham, T. Herrin, J.J. Plattner, J. Erickson, J. Clement, R. Swanson, N. Shipkowitz, D. Hardy, K. Marsh, G. Arnett, W. Shannon, S. Broder and H. Mitsuya, Cyclobut-A and cyclobut-G: Broad-spectrum antiviral agents with potential utility for the therapy of AIDS, *J. Med. Chem.* 33:1281 (1990).

21. R. Vince, M. Hua, J. Brownell, S. Daluge, F. Lee, W.M. Shannon, G.C. Lavelle, J. Qualls, O.S. Weislow, R. Kiser, P.G. Canonico, H. Schultz, V. Narayanan, J.G. Mayo, R.H. Shoemaker and M.R. Boyd, Potent and selective activity of a new carbocyclic nucleoside analog (carbovir: NSC 614846) against human immunodeficiency virua in vitro, *Biochem. Biophys. Res. Commun.* 156(2):1046 (1988).

22. N.V. Heyden, C. Rodi and L. Ratner, Analogs of 3'-deoxyadenosine inhibit HIV-1 replication, *AIDS Res. Hum. Retroviruses* 5(6):647 (1989).

23. A.R. Maguire, W.-D. Meng, S.M. Roberts and A.J. Willetts, The preparation of 4'-fluoronucleoside Analogues with potential anti-HIV activity, *Abstracts: MRC AIDS Directed Programme Workshop*, University of Warwick (1991).

24. H. Maag, N. Chu, D. Crawford-Ruth, E. Eugui, M.J. McRoberts, A. Mirkovich, M. Pettibone, E.J. Prisbe, R.M. Rydzewski and J.P.H. Verheydin, 4'-Azidothymidine: Synthesis and in vitro anti-HIV activity, *Antiviral Res. Suppl.* 1:43 (1991).

25. M. Baba, E. De Clercq, H. Tanaka, et al., Highly potent and selective inhibition of HIV-1 by a novel series of 6-substituted acyclouridine derivatives, *Mol. Pharmacol.* 39:805 (1991).

26. U. Weiss, L. Merlini and G. Nasini, Naturally occurring perylenequinones, *Progress in the Chemistry of Organic Natural Products*, 52:1 (1987).

27. D.E. Pisani, A.J. Elliott, D.R. Hinman, L.M. Aaronson and R.S. Pardini, Relationship between inhibition of mitochondrial respiration by naphthoquinones, their antitumor activity, and their redox potential, *Biochem. Pharmacol.* 35(21):3791 (1986).

28. H.W. Moore, R. Czerniak and A. Hamdan, Natural quinones as quinonemethide precursors - ideas in rational drug design, *Drugs Exptl. Clin. Res.* XII(6/7):475 (1986).

29. M.M. Wick and G. Fitzgerald, Inhibition of reverse transcriptase by tyrosinase generated quinones related to levodopa and dopamine, *Chem.-Biol. Interactions* 38:99 (1981).

30. G. Lavie, F. Valentine, B. Levin, Y. Mazur, G. Gallo, D. Lavie, D. Weiner and D. Mervolo, Studies of the mechanisms of action of the antiretroviral agents hypericin and pseudohypericin, *Proc. Natl. Acad. Sci.* USA, 86:5963 (1989).

31. R. Schinazi, C.K. Chu, J.R. Babu, B.J. Oswald, V. Sallaman, D.L. Cannon, B. Eriksson and M. Nasr, Anthraquinones as a new class of antiviral agents against human immunodeficiency virus, *Antiviral Res.* 13:265 (1990).

32. G. Lavie, Y. Mazur, D. Lavie, B. Levin, Y. Ittah and D. Meruelo, Hypericin as an antiretroviral agent: mode of action and related analogues, *Proc. Natl. Acad. Sci.* USA, 616:556 (1990).

33. J.F. Davies, Z. Hostomska, S.R. Jordan and D.A. Matthews, Crystal structure of the ribonuclease H domain of HIV-1 reverse transcriptase, *Science*, 252:88 (1991).

34. A. Tanaka, H. Nakashima, O. Yoshida, N. Yamamoto, O. Tenmyo and T. Oki, Inhibitory effect of new antibiotic, pradimicin A on infectivity, cytopathic effect and replication of human immunodeficiency virus in vitro, *J. Antibiot.* XLI(11):1708 (1988).

35. T. Takeuchi, T. Hara, M. Hamada, H. Yamamoto, S. Gomi, Y. Orikasa, M. Sezaki, S. Kondo and H. Yamaguchi, Benanomicins A and B, novel antifungal antibiotics, *Abstracts of the 28th ICAAC (1988)*.

36. H. Nakashima, N. Yamamoto, Y. Inouye and S. Nakamura, Inhibition by doxorubicin of human immuno-deficiency virus (HIV) infection and replication in vitro, *J. Antibiot.* XL(3):396 (1987).

37. I. G. Filion and R. Gaudreault, Effect of daunorubicin on HIV-1 Infected U937 and HUT 78 cells, *Abstracts of the 1990 ICAAC*, 175 (1990).

38. S.-I. Ito, G. Gilljams, B. Wahren, H. Wigzell, N. Yamamoto, K. Sasaki and K. Onodera, Inhibition of HIV replication by 19-O-n-pentyldamavaricin Fc in vitro, *J. Antibiot.* 43(8):1045 (1990).

39. K. Ono, H. Nakane, S. Shimizu and S. Koshimura, Inhibition of HIV-reverse transcriptase activity by asterriquinone and its analogues, *Biochem. Biophys. Res. Commun.* 174(1):56 (1991).

40. P.S. Sarin, D. Sun, A. Thornton and W.E.G. Muller, Inhibition of replication of the etiologic agent of acquired immune deficiency syndrome (human T-T-Lymphotropic retrovirus/lymphadenopathy-associated virus) by Avarol and Avarone. *J. Nat. Cancer Inst.* 78(4), 663-666. (1987).

41. F. Reusser, W. Tarpley and I.W. Althaus, Inhibitors of HIV-reverse transcriptase to treat AIDS. *U.S. Patent - Upjohn Co. PCT/US88/03114* (1989).

42. M. Asanaka, T. Kurimura, H. Toya and K. Kato, Screening of anti-HIV activities in existing drugs which are capable of long-term oral administration. *Chemotherapy*, 38(3):249 (1990).

43. S. Loya and A. Hizi, The inhibition of human immunodeficiency virus type 1 reverse transcriptase by avarol and avarone derivatives, *FEBS Letters* 269(1):131 (1990).

44. S. Loya, R. Tal, Y. Kashman and A. Hizi, Illimaquinone, a selective inhibitor of the RNase H activity of human immunodeficiency virus type 1 reverse transcriptase, *Antimicrob. Agents Chemother.*, 34(10):2009 (1990).

45. M. Lemaitre, D. Guetard, Y. Henin, L. Montagnier and A. Zerial, Protective activity of tetracycline analogs against the cytopathic effect of the human immunodeficiency viruses in CEM cells, *Res. Virol.* 141:5 (1990).

46. J. Tang, J.M. Colacino, S.H. Larsen and W. Spitzer, Virucidal activity of hypericin against enveloped and non-enveloped DNA and RNA viruses, *Antiviral Res.* 13:313 (1990).

47. G. Tachedjian, D. Tyssen, S. Locarnini, I. Gust and C. Birch, Investigation of topoisomerase inhibitors for activity against human immunodeficiency virus; inhibition by coumermycin A1, *Antiviral Chem. Chemother.* 1(2):131 (1990).

48. A. Tanabe-Tochikura, T.S. Tochikura, O. Yoshida, T. Oki and N. Yamamoto, Pradimicin A inhibition of human immunodeficiency virus: attenuation by mannan, *Virology* 176:467 (1990).

49. G.A. Kraus, D. Pratt, J. Tossberg and S. Carpenter, Antiretroviral activity of synthetic hypericin and related analogs, *Biochem. Biophys. Res. Commun.* 172(1):149 (1990).

50. T. Xiaoshan, C. Hongshan and Z. Xingquan, Inhibition of human immunodeficiency virus reverse transcriptase by Chinese medicines in vitro, *Proc. Chin. Acad. Med. Sci.* 5(3):140 (1990).

51. P.H. Dorsett, E.E. Kerstine and L.J. Powers, Antiviral activity of gossypol and apogossypol, *J. Pharm Sci.* 64:1073 (1975).

52. K. Wichmann, A. Vaheri and T. Luukkainen, Inhibiting herpes simplex virus type 2 infection in human epithelial cells by gossypol, a potent spermicidal and contraceptive agent, *Am. J. Obestet. Gynecol.* 142:593 (1982).

53. T.-S. Lin, B.P. Schinazi, B.P. Griffith, E.M. August, B.F. Eriksson, D.-K. Zheng, L. Huang, W.H. Prusoff, Selective inhibition of human immunodeficiency virus type 1 replication by the (-) but not the (+) enantiomer of gossypol, *Antimicrob. Agents Chemother.* 33(12):2149 (1989).

54. R.E. Royer, R.G. Mills, L.M. Deck, G.J. Mertz and D.L. Vanderjagt, Inhibition of human immunodeficiency virus type I replication by derivatives of gossypol, *Pharmacol. Res.* 24(4):407 (1991).

55. R.S. Chang, L. Ding, C. Gai-Qing, P. Qi-Choa, Z. Ze-Lin and K.M. Smith, Dehydroandrographolide succinic acid monoester as an inhibitor against the human immunodeficiency virus, *Proc. Soc. Exp. Biol. Med.* 197(1):59 (1991).

56. K.R. Gustafson, M.H.G. Munro, J.W. Blunt, J.H. Cardellina II, J.B. McMahon, R.J. Gulakowski, C.M. Cragg, P.A. Cox, L.S. Brinen, J. Clardy and M.R. Boyd, HIV inhibitory natural products. 3. Diterpenes from Homolanthus acuminatus and Chrysobalanus icaco, *Tetrahedron* 47(26):4547 (1991).

57. M. Ito, H. Nakashima, M. Baba, R. Pauwels, E. De Clercq, S. Shigeta and N. Yamamoto, Inhibitory effect of glycyrrhizin on the in vitro infectivity and cytopathic activity of the human immunodeficiency virus [HIV (HTLV-III/LAV)], *Antiviral Res.* 7:127 (1987).

58. K. Hirabayashi, S. Iwata, H. Matsumoto, T. Mori, S. Shibata, M. Baba, M. Ito, S. Shigeta, H. Nakashima and N. Yamamoto, Antiviral activities of glycyrrhizin and its modified compounds against human immunodeficiency virus type I (HIV-1) and herpes simplex virus type 1 (HSV-1) in vitro, *Chem. Pharm. Bull.* 39(1):112 (1991).

59. M. Ito, A. Sato, K. Hirabayashi, F. Tanabe, T.S. Shigeta, M. Baba, E. DeClercq, H. Nakashima and N. Yamamoto, Inhibitory effect of glycyrrhizin on replication of HIV, *Antiviral Res.* 10:289 (1988).

60. H. Kazuhiro, I. Susamu, M. Hiroatsu, M. Takeo, S. Shoji, B. Masanori, I. Masahiko, S. Shiro, N. Hideki and N. Yamamot, Antiviral activities of glycyrrhizin and its modified compounds against human immunodeficiency virus type 1 (HIV-1) and herpes simplex virus type 1 (HSV-1) in vitro, *Chem. Pharm. Bull.* 39(1):112 (1991).

61. B. Polsky, S.J. Segal, P.A. Baron, J.W. Gold, H. Ueno and D. Armstrong, Inactivation of human immunodeficiency virus in vitro by gossypol, *Contraception* 39(6):579 (1989).

62. B.C.M. Potts, D.J. Faulkner, J.A. Chen, G.C. Simolike, P. Offen, M.E. Hemling and T.A. Francis, T.A., Didemnaketals A and B, HIV-1 protease inhibitors from the Ascidian Didemnum sp., *J. Am. Chem. Soc.* 113:6312 (1991).

63. H. Nakashima, K. Okubo, Y. Honda, T. Tamura, S. Matsuda, N. Yamamoto, Inhibitory effect of glycosides like saponin from soybean on the infectivity of HIV in vitro, *AIDS* 3(10):655 (1989).

64. E. DeClercq, New selective antiviral agents active against the AIDS virus, *TIPS Revs.* 8:339 (1987).

65. Z.G. Wang, J. Ren and R. Zhang, Immunoregulatory studies of ammonium glycyrrhizinate in mice, *Asia Pacific J. Pharmacol.* 5:157 (1990).

66. W.E.G. Muller, B.E. Weiler, R. Charubala, W. Pfleiderer, L. Leserman, R.W. Sobol, R.J. Suhadolnik and H.C. Schroder, Cordycepin analogues of 2',5'-oligoadenylate inhibit human immunodeficiency virus infection via inhibition of reverse transcriptase, *Biochemistry* 30:2027 (1991).

67. T. Nakanishi, A. Inada and T. Kato, Antiviral cyclic triterpenoids from Patrinia, *Patent to Sawai Pharmaceutical Co., Ltd., Jpn. Kokai Tokkyo Koho JP 01207262* (1989).

68. T.J. Smith, M.J. Kremer, M. Luo, et al., The site of attachment in human retrovirus for antiviral agents that inhibit uncoating, *Science* 233:1286 (1989).

69. D.J. Capon and R.H.R. Ward, Immune antiviral effects of CD4 derivatives, *Current Opin.* 2:433 (1990).

70. C. Flexner, P. Barditch-Crovo, D. Kornhauser, H. Farzadegan, L. Nerhood, R.E. Chaisson, K.M. Bell, K.J. Lorentsen, C.W. Hendrix, B. Petty and P. Leitman, Pharmacokinetics, toxicity, and activity of intravenous dextran sulfate in human immunodeficiency virus infection, *Antimicrob. Agents Chemother.* 35(12):2544 (1991).

71. K. Ono, H. Nakane, M. Fukushima, J.-C. Chermann and F. Barre-Sinoussi, Differential inhibitory effects of various flavonoids on the activities of reverse transcriptase and cellular DNA and RNA polymerases, *Eur. J. Biochem.* 190:469 (1990).

72. H. Nakane and K. Ono, Differential inhibitory effects of some catechin derivatives on the activities of human immunodeficiency virus reverse transcriptase and cellular deoxyribonucleic and ribonucleic acid polymerases, *Biochemistry* 29(11):2841 (1990).

73. K. Ono, H. Nakane, M. Fukushima, J.-C. Chermann and F. Barre-Sinoussi, Inhibition of reverse transcriptase activity by a flavonoid compound, 5,6,7-trihydroxyflavone, *Biochem. Biophys. Res. Commun.* 160(3):982 (1989).

74. K. Ono, H. Nakane, M. Fukushima, J.C. Chermann and F. Barre-Sinoussi, 1989, Inhibition of HIV-reverse transcriptase by a Kampo medicine, Sho-saiko-to (TJ-9), *5th Int. AIDS Conference* June 4-9 (1989).

75. G. Spedding, A. Ratty, and E. Middleton, Jr., Inhibition of reverse transcriptases by flavonoids, *Antiviral Res.* 12:99 (1989).

76. K. Ono, Herbal extracts and their components as a novel class of inhibitors for HIV-reverse transcriptase, *Antiviral Res.*, 0(Suppl. 1):53 (1990).

77. R. Jansen, B. Kunze, V. Wray, H. Reichenbach, E. Jurkiewicz, G. Hunsmann, and G. Hofle, Phenoxan: A novel inhibitor of HIV-1 infection in cell cultures from Polyangium sp., strain P1 VO19 (myxobacteria), *Liebigs Ann. Chem.* 707 (1991).

78. W.B. Parker, M. Nishizawa, M.H. Fisher, N. Ye, K.-H. Lee and Y.-C. Cheng, Characterization of a novel inhibitor of human DNA polymerases: 3,4,5-tri-O-galloylquinic acid, *Biochem. Pharmacol.* 38(21):3759 (1989).

79. G.-I. Nonaka, I. Nishioka, M. Nishizawa, T. Yamagishi, Y. Kashiwada, G.E. Dutschman, A.J. Bodner, R.E. Kilkuskie, Y.-C. Cheng, and K.-H. Lee, Anti-

AIDS agents, 2: Inhibitory effects of tannins on HIV reverse transcriptase and HIV replication in H9 lymphocyte cells, *J. Nat. Prods.* 53(3):587 (1990).

80. T. Tanaka, G.-I. Nonaka and I. Nishioka, Tannins and related compounds, XL. Revision of the structures of punicalin and punicalagin, and isolation and characterization of 2-O-galloylpunicalin from the bark of Punica granatum, L. *Chem. Pharm. Bull.* 34:650 (1986).

81. H. Nakane and K. Ono, Differential inhibitory effects of some catechin derivatives on the activities of human immunodeficiency virus reverse transcriptase and cellular deoxyribonucleic and ribonucleic acid polymerases, *Biochemistry* 29:2841 (1990).

82. W. Kreis, M.H. Kaplan, J. Freeman, D.K. Sun and P.S. Sarin, Inhibition of HIV replication by Hyssop officinalis extracts, *Antiviral Res.* 14:323 (1990).

83. H. Nakane, M. Fukushima and K. Ono, Differential inhibition of reverse transcriptase and various DNA polymerases by digallic acid and its derivatives, *J. Nat. Prods.* 53(5):1234 (1990).

84. T. Jacks, M.D. Power, F.R. Masiarz, P.A. Luciw, P.J. Barr, H.E. Varmus, Characterization of ribosomal frameshifting in HIV-1 *gag-pol* expression. *Nature (London)* 331:280 (1988).

85. V.A. Johnson, B.D. Walker, M.A. Barlow, T.J. Paradis, T.C. Chou and M.S. Hirsch, Synergistic inhibition of human immunodeficiency virus type 1 and type 2 replication in vitro by castanospermine and 3'-azido-3'-deoxythymidine, *Antimicrob. Agents Chemother.* 33:53 (1989).

86. R.M. Ruprecht, S. Mullaney, J. Andersen and R. Bronson, In vivo analysis of castanospermine, a candidate antiretroviral agent, *J AIDS* 2:149 (1989).

87. P.S. Sunkara, D. Taylor, M. Kang, T. Bowlin, P. Liu, A. Tyms and A. Sjoerdsma, Anti-HIV activity of castanospermine analogs, *Lancet* 1:1206 (1989).

88. N.R. Kohl, E.A. Emini, W.A. Schleif, L.J. Davis, J.C. Heimbach, R.A. Dixon, E.M. Scolnick I.S. and Sigal, Active human immunodeficiency virus protease is required for viral infectivity, *Proc. Natl. Acad. Sci.*, USA 85:4686 (1988).

89. C. Peng, B.K. Ho, T.W. Chang and N.T. Chang, Role of human immunodeficiency virus type 1-specific protease in core protein maturation and viral infectivity, *J. Virol.* 63:2550 (1989).

90. H.G. Gottlinger, J.G. Sodroski and W.A. Haseltine, Role of the capsid precursor processing and myristoylation in morphogenesis and infectivity of human immunodeficiency virus type 1, *Proc. Natl. Acad. Sci.* USA 86:5781 (1989).

91. K. Von der Helm, L. Gurtler, J. Eberle and F. Deinhardt, Inhibition of HIV replication in cell culture by the specific aspartic protease inhibitor pepstatin A, *FEBS Letters* 247(2):349 (1989).

92. J.J. Blumenstein, T.D. Copeland, S. Oroszlan and C.J. Michejda, Synthetic non-peptide inhibitors of HIV protease, *Biochem. Biophys. Res. Commun.* 163(2):980 (1989).

93. M.L. Bryant, R.O. Heuckeroth, J.T. Kimata, L. Ratner and J.I. Gordon, Replication of human immunodeficiency virus 1 and moloney murine leukemia virus is inhibited by different heteroatom-containing analogs of myristic acid, *Proc. Natl. Acad. Sci.* USA 86:8655 (1989).

94. D. Lewis, A.J. Baran, V. Knight and M. Jett, Inhibitory effects of unsaturated phospholipids on HIV production in vitro, *Fifth International AIDS Conference*, Montreal, Canada (1989).

95. K.R. Gustafson, J.H. Cardellina II, R.W. Fuller, O.S. Weislow, R.F. Kiser, K.M. Smader, G.M.L. Patterson and M.R. Boyd, AIDS-Antiviral sulfolipids from cyanobacteria (blue-green algae), *J. Nat. Cancer Inst.* 81(16):1254 (1989).

96. J. Czajkowski, V. Tozzi and O. Strannegard, Suppression of HIV replication in vitro by fusidic parallels toxicity, *J. Antimicrob. Agents Chemother.* 23:155 (1989).

97. R.F. Schinazi, B.F.H. Eriksson, B.H. Arnold, P. Lekas and M.S. McGrath, Effect of dehydroepiandrosterone in lymphocytes and macrophages infected with human immunodeficiency viruses in "The Biologic Role of Dehydroepiandrosterone (DHEA)" (Kalimi, M. and Regelson, W. eds.), Walter de Gruyter, Berlin, New York, p. 157 (1990).

98. D.D. Richman, H. Mitsuya, S. Broder and K.Y. Hostetler, Fusidic acid, HIV, and host cell toxicity, *Lancet* 1:1051 (1988).

99. M. Baba, D. Schols, H. Nakashima, R. Pauwels, G. Parmentier, D.K.F. Meijer and E. De Clercq, Selective activity of several cholic acid derivatives against human immunodeficiency virus replication in vitro, *J. Acquir. Immun. Def. Synd.* 2(3):264 (1989).

100. H.H. Sun, S.S. Cross, M. Gunasekara and F.E. Koehn, Weinbersterol disulfates A and B, antiviral steroid sulfates from the sponge Petrosia weinbergi, *Tetrahedron* 47(7):1185 (1991).

101. R. Pal, R.C. Gallo and M.G. Sarngadharan, Processing of the structural proteins of human immunodeficiency virus type 1 in the presence of monensin and cerulenin, *Proc. Natl. Acad. Sci.* USA 85:9283 (1988).

102. R.L. Dewar, M.B. Vasudevachari, V. Natarajan and N.P. Salzman, Biosynthesis and processing of human immunodeficiency virus type 1 envelope glycoproteins: effects of monensin on glycosylation and transport, *J. Virol.* 63(6):2452 (1989).

103. S. Olsnes, I.H. Madshus and K. Sandvig, K., Entry mechanisms of picornaviruses In "Virus Attachment and Entry into Cells" Proceedings of an ASM Conference Held in Philadelphia, Pennsylvania, 10-13 April 1985, 171-181 (Crowell, R. L. and Lonberg-Holm, K. eds.) (1986).

104. P.J. Maddon, A.G. Dalgleish, J.S. McDougal, P.R. Clapham, R.A. Weiss and R. Axel, The T4 gene encodes the AIDS virus receptor and is expressed in the immune system and the brain, *Cell* 47:333 (1986).

105. M. Nakamura, T. Ohno, S. Kunimoto, H. Naganawa and T. Takeuchi, Kijimicin: An inhibitor of human immunodeficiency virus in acutely and chronically infected cells, *J. Antibiotics* 44(5):569 (1991).

106. P.K. Lai, J. Donovan, H. Takayama, H. Sakagami, A. Tanaka, K. Konno and M. Nonoyama, Modification of human immunodeficiency viral replication by pine cone extracts, *AIDS Res. Hum. Retrovir.* 6(6):205 (1990).

107. H. Takayama, G. Bradley, P.K. Lai, Y. Tamura, H. Sakagami, A. Tanaka, A., M. Nonoyama, Inhibition of human immunodeficiency virus forward and reverse transcription by PC6, a natural product from cones of pine trees, *AIDS Res. Hum. Retrovir.* 7(3):349 (1991).

108. D. Schols, P. Wutzler, R. Klocking, B. Helbig and E. De Clercq, Selective inhibitory activity of polyhydroxycarboxylates derived from phenolic compounds against HIV replication, *J. Acq. Immun. Defic. Synd.* 4:677 (1991).

109. E. Priel, S.D. Showalter and D.G. Blair, Inhibition of human immunodeficiency virus (HIV-1) replication in vitro by noncytotoxic doses of camptothecin, a topoisomerase I inhibitor, *AIDS Res. Hum. Retrovir.* 7(1):65 (1991).

110. M. Baba, R. Pauwels, J. Balzarini, D. Schols and E. De Clercq, Coumermycin A_1 is a potent inhibitor of human immunodeficiency virus (HIV) replication in vitro, *Intr. J. Exp. Clin. Chemother.* 2(1):15 (1989).

111. A. Turano, G. Scura, A. Caruso, C. Bonfanti, R. Luzzati, D. Bassetti, N. Manca, Inhibitory effect of papaverine on HIV replication in vitro, *AIDS Res. Hum. Retrovir.* 5(2):183 (1989).

112. G.T. Tan, J.M. Pezzuto, A.D. Kinghorn and S.A. Hughes, Evaluation of natural products as inhibitors of human immunodeficiency virus type 1 (HIV-1) reverse transcriptase, *J. Nat. Prods.* 54(1):143 (1991).

113. W.J. Kassler, P. Blanc and R. Greenblatt, The use of medicinal herbs by human immunodeficiency virus-infected patients, *Arch. Intern. Med.* 151:2281 (1991).

114. G.T. Tan, J.M. Pezzuto and A.D. Kinghorn, Evaluation of natural products as inhibitors of human immunodeficiency virus (HIV) reverse transcriptase, *Planta Med.* (short lectures) 56:504 (1990).

115. P. Hafkemeyer, K. Neftel, R. Hobbi, A. Pfaltz, H. Lutz, K. Luthi, F. Focher, S. Spadari and U. Hubscher, HO 0.35, a cephalosporin degradation product is a specific inhibitor of lentiviral RNAses H, *Nucleic Acid Res.* 19:4059 (1991).

116. K.P. Manfredi, J.W. Blunt, J.H. Cardellina, J.B. McMahon, L.L. Pannell, G.M. Cragg and M.R. Boyd, Novel alkaloids from the tropical plant Ancistrocladus abbreviatus inhibit cell killing by HIV-1 and HIV-2, *J. Med. Chem.* 34:3402 (1991).

AIDS-ANTIVIRAL NATURAL PRODUCTS RESEARCH

AT THE U.S. NATIONAL CANCER INSTITUTE

Kirk R. Gustafson, John H. Cardellina II, Kirk P. Manfredi, John A. Beutler, James B. McMahon and Michael R. Boyd

Laboratory of Drug Discovery Research and Development
Developmental Therapeutics Program
National Cancer Institute
Building 1052, Room 121
Frederick, MD   21702-1201

INTRODUCTION

The Laboratory of Drug Discovery Research and Development (LDDRD) is a new intramural NCI laboratory within the Division of Cancer Treatment's Developmental Therapeutics Program. The primary research focus in LDDRD includes the development and refinement of the NCI anti-HIV and antitumor screening models, the bioassay-guided isolation of new active compounds and the detailed biological and pharmacological characterization of new leads. The NCI is currently testing crude extracts and pure compounds in an in vitro primary antiviral screen[1] which detects the inhibition of HIV-induced cell killing. Extracts are additionally assayed in a disease-oriented, human 60 cell line in vitro antitumor screen[2] representing 7 major categories of human cancer. Selected natural products extracts and pure compounds which demonstrate activity in these primary in vitro screens are subjected to further preclinical investigation within LDDRD. The goal of these efforts is to expeditiously identify and perform preclinical development studies of promising new clinical candidates.

The intramural natural products chemistry effort within LDDRD relies heavily on the Developmental Therapeutics Program's Natural Products Branch for management and scientific oversight of a series of extramural contracts for the collection of diverse plant, marine and microbial samples. While all of the collections emphasize biological diversity, available ethnobotanical information is also used to focus the specific collection of species which have histori-cally been utilized in traditional/folk medicines. Collections of terrestrial plant and marine organisms have been made primarily in tropical rainforests or tropical reef areas, respectively. Selected cyanobacterial and fungal isolates have been emphasized in the microbial collections. A contractor-operated extraction laboratory processes and extracts the samples to provide both organic and aqueous extracts of each organism. The extracts, maintained at -20° C in a central repository for long term storage, serve as the primary natural products input source of the NCI anti-HIV and antitumor screens.

The primary NCI anti-HIV screen, utilizing a single target cell

Natural Products as Antiviral Agents, Edited by C.K. Chu
and H.G. Cutler, Plenum Press, New York, 1992

line, currently has a higher throughput capacity (40,000 tests/yr) than the primary 60 cell line human tumor drug screen (5,000 tests/yr). Therefore, a considerably larger number of extracts have thus far been screened for HIV-inhibitory properties rather than antitumor properties. As a result, LDDRD's bioassay-guided natural products chemistry studies of anti-HIV active extracts are the focus of this report. All of the compounds described should be viewed as the combined results of screening assay development, natural products chemistry and biological characterization efforts.

The anti-HIV screen detects virus-induced cell killing and its inhibition by anti-HIV active compounds. The assay employs the CEM-SS human lymphoblastoid cell line as the target of viral infection. This cell line is infectable by HIV and is very sensitive to the cytopathic effects of the virus. The RF variant of the HIV-1 viral strain is used and the assay employs 96 well microculture technology for screening large numbers of samples. After a 6-day incubation of the target cells, the assay end-point is determined by the metabolic reduction of the tetrazolium reagent XTT[3] by viable cells to a colored formazan product which is measured colorimetrically. Uninfected cells or cells which have been protected from the lytic effects of HIV infection produce the colored formazan product and thus yield high optical densities. Cells which have not been protected from the HIV virus are killed and these cultures produce lower optical densities. Each crude extract, chromatographic fraction and pure compound is serially diluted and tested at 8 different concentrations in the presence and absence of the virus. This provides a dose/response profile of HIV inhibition in drug-treated, HIV-infected cells and allows for the evaluation of cytotoxic effects of a compound in drug-treated control cells.

Data and feedback from the anti-HIV screen are crucial to the LDDRD anti-HIV natural products isolation and structural chemistry effort. Crude extracts which show the most promising activity in the screen are prioritized and committed to detailed bioassay-guided fractionation. All of the fractions generated at each separation step are tested in the screen for HIV inhibitory properties. In this fashion, the anti-HIV active constituents of an extract can be tracked through the isolation steps required to produce pure, active constituents. The selected research results reviewed in the remainder of this chapter illustrate LDDRD's approach to the elucidation of new active anti-HIV chemotypes and other potential drug development candidates.

NATURAL PRODUCTS FROM CYANOBACTERIA

Crude cellular extracts from several cultured cyanobacteria (blue-green algae) were found to inhibit the cytopathic effects of in vitro HIV-1 infection in the primary screen. Extracts from Lyngbya lagerheimii and Phormidium tenue were fractionated using the microculture XTT assay to guide the isolation and purification process. Following a combination of gel permeation and reversed-phase chromatographies, a series of HIV-inhibitory sulfonic acid-containing glycolipids were discovered[4]. The structures of compound 1 ($C_{43}H_{76}O_{12}S$) and three other diacylglycerol glycosides, which differed only in their fatty acid acyl constituents, were established by analysis of their spectral characteristics. Sulfonic acid-containing lipids were first isolated by Benson et al.[5,6] and are commonly referred to as sulfolipids. These lipids occur in higher plants, algae and photosynthetic microorganisms[7,8] and are believed to serve a structural role in chloroplast membranes. Sulfolipids are abundant constituents of L. lagerheimii and P. tenue, as the combined sulfolipid fraction constitutes approximately 10% of the cellular organic extract for both species. Sulfolipids have previously been isolated and characterized as mixtures with varying fatty acid ester components. Our studies with cultured

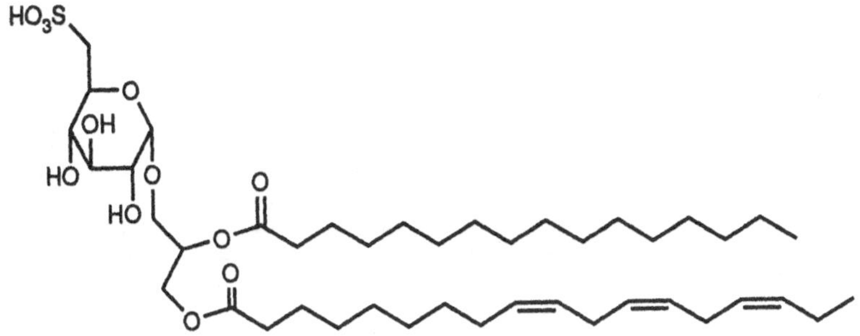

1

cyanobacteria were the first to provide individual purified sulfolipids with fully defined structures. Various cellular extracts
from cyanobacteria species in the genera Phormidium, Oscillatoria,
Scytonema, Calothrix and Anabaena were found to inhibit HIV-1 in the
screen due to the presence of sulfolipids. Purified sulfolipids
tested using the current assay protocol inhibited HIV-1 induced cell
killing with an EC_{50} (50% protection) of 20 μM and an IC_{50} (50%
cytotoxic dose) of 250 μM. A dose-dependent increase in XTT metabolism was observed in sulfolipid-treated, HIV-1 infected target
cells. Essentially 100% protection from HIV-1 infection was observed over a concentration range of 35-125 μM. The increased
cellular viability measured in the XTT assay correlated well with
concomitant reductions in the production of viral p24 antigen and
HIV-1 induced syncytium formation. Sulfolipids were the first class
of HIV-inhibitory natural products identified by LDDRD from a
bioassay-directed fractionation of a crude extract.

NATURAL PRODUCTS FROM TERRESTRIAL PLANTS

Homalanthus nutans and Chrysobalanus icaco

As previously noted, the NCI natural products collection and
screening effort is particularly focused upon the acquisition and
testing of species which have a history of use in traditional
medicinal treatments. An effort to study plants of ethnobotanical
importance to the indigenous population of the islands of Samoa
resulted in the collection and screening of crude extracts of the
tropical tree Homalanthus nutans. H. nutans is an important component of Samoan herbal medicine and various preparations of the plant
are used to treat yellow fever and suppress back pain, abdominal
swelling and diarrhea.[9] An organic extract of the stem wood of H.
nutans showed potent HIV-inhibitory properties in the NCI screen.
By tracking protection from HIV infection through the chromatographic separations of the extract and all resultant fractions, the
principal active agent was isolated and identified as prostratin
(2).[10] Prostratin ($C_{22}H_{30}O_6$), also known as 12-deoxyphorbol-13-
acetate, is a member of the phorbol ester class of diterpenes. The
structure of prostratin was assigned by analysis of its spectral
data and comparison of this data with previously published
spectrochemical values.[11]

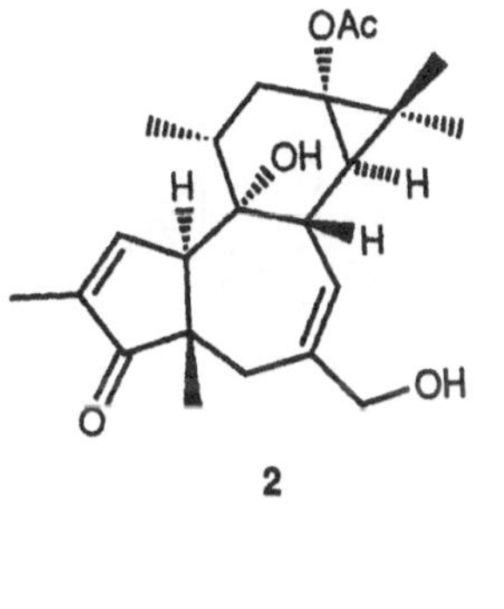

2

Phorbol esters are known to bind to and activate protein kinase C, and many of these compounds show potent tumor-promoting activities.[12] Prostratin protects target CEM-SS cells from the lytic effect of HIV-1, but does not itself appear to have tumor-promoting activity.[13] At concentrations that strongly inhibited the HIV virus, prostratin had a cytostatic effect on both virus infected and noninfected CEM-SS T-lymphoblastoid cells. Over a broad concentration range of prostratin treatments (10 nM - 25 μM), the number of surviving cells was virtually identical in uninfected cell cultures and cultures infected with HIV-1. However, the overall growth rates in these cultures were markedly reduced relative to control cells not treated with prostratin. Cell survival and growth were assessed both in the XTT-formazan assay and in a fluorescence assay utilizing the probe BCECF, (2',7'-biscarboxyethyl-5(6)-carboxyfluorescein acetoxymethyl ester). In the latter assay, cellular viability is established by the hydrolysis of BCECF by cellular esterases to a fluorescent marker that can be readily quantitated.[14] Both the anti-HIV and the anti-proliferative effects of prostratin were vividly apparent by microscopic observation of the cell cultures. Syncytia were not observed in prostratin-treated HIV-1-infected cultures. The corresponding virus-infected control cells, which were not treated with prostratin, showed numerous syncytia and cellular lysis. Correlative assays to determine the effect of prostratin on viral p24 antigen production, reverse transcriptase activity and syncytium formation revealed that over the concentration range where target cells were protected from HIV-1 infection, virus reproduction was apparently eliminated. However, at subprotective concentrations of prostratin, these indices of viral reproduction were enhanced. While the significance of this observation is not clear, it suggests that prostratin's cytoprotective effects in HIV-1 infected cells are not solely related to its effect upon HIV-1 reproduction.

During the bioassay-guided fractionation of the H. nutans extract, it was apparent that several fractions which did not contain phorbol diterpene derivatives also possessed HIV-inhibitory properties. Detailed chemical analysis of these fractions provided a series of diterpenoids with atisane and kaurane carbon skeletons.[15] The structure of compound 3 ($C_{20}H_{28}O_3$), a novel oxygenated atisane derivative, was ultimately proven by x-ray crystallographic analysis. A second atisane diterpene was identified as compound 4 ($C_{20}H_{32}O_3$) by comparison of its spectral data with previously documented literature values[16]. Compounds 5 ($C_{20}H_{32}O_4$) and 6 ($C_{20}H_{32}O_3$)

7 $R^1 = H, R^2 = H$
8 $R^1 = Me, R^2 = OH$

were shown, through extensive spectral analysis, to be novel kaurane type diterpenes. Structural elucidation studies relied heavily on ^{1}H-^{1}H COSY and proton detected heteronuclear correlation NMR experiments (HMQC and HMBC) to establish the carbon skeletons and placements of the functional groups in these compounds. The relative stereochemical assignments in 5 and 6 were defined by difference nOe measurements. Of the four non-phorbol diterpenes isolated from H. nutans, only the atisane derivative 3 inhibited the HIV virus. Compound 3 provided a maximum of 50% protection to HIV-1 infected cells at a concentration of 19 μM; however, cytotoxic effects were apparent in both control and HIV-infected cultures at 38 μM.

Concurrent with the studies on H. nutans, an organic extract from the roots of the tropical tree Chrysobalanus icaco also showed activity in the anti-HIV screen. This activity was tracked to a fraction containing the kaurane carboxylic acid diterpene 7 and its hydroxy methyl ester derivative 8.[15] The structures of 7 and 8 were deduced from spectroscopic analysis and for 7, confirmation of the structure was also made by comparison of our data with published values.[17,18] Compound 8 had not been described before as a natural product; however, it had purportedly been produced as a methylation product of the corresponding hydroxy acid.[19] While carboxylic acid 7 inhibited HIV infection in vitro, compound 8 and the methyl-ester derivative of 7 were completely inactive. Compound 7 was non-cytotoxic with a maximum of 50% protection at 1.6 μM but cytotoxic at 6.3 μM. Thus, kaurane derivative 7 had a similar activity profile, but greater potency than that obtained with the atisane derivative 3.

Ancistrocladus abbreviatus

Organic extracts from the aerial parts of the tropical vine Ancistrocladus abbreviatus were active in the primary anti-HIV screen. Bioassay-guided fractionation of these extracts provided the novel HIV-inhibitory alkaloids, michellamines A (9) and B (10).[20] The pure active compounds were obtained using an acid/base partition scheme, centrifugal partition chromatography and HPLC on an amino-bonded phase column. The genus Ancistrocladus is known to be a rich source of isoquinoline-naphthalene alkaloids.[21,22] However, the molecular formula of the michellamines, $C_{46}H_{48}N_2O_8$ as established by FAB mass spectrometry, suggested that they were previously unknown dimeric derivatives of these alkaloids.

Michellamine A (9) is a symmetrical dimer, as its ^{13}C NMR spectrum contained only 23 resonances. Assignment of the structure, substitution pattern and relative stereochemistry of the tetrahydro-isoquinoline and naphthalene subunits in 9 was aided by ^{1}H-^{1}H coupling constant analysis, heteronuclear correlations and difference

9   10

nOe experiments. Ultimately, the naphthalene-tetrahydroisoquinoline connection and the linkage between the 2 naphthalenes was established by long-range heteronuclear correlations and nOe measurements. Michellamine A represents the first dimeric alkaloid of the naphthalene-isoquinoline class to be discovered.

Michellamine B (10) is an asymmetrical molecule, as its ^{13}C NMR spectrum contained 46 signals. Data from a series of NMR experiments clearly showed that metabolite 10 had the same overall gross structure and relative configuration about the tetrahydroisoquinoline rings as michellamine A (9). The two compounds differed only in the relative configuration of the naphthalene-tetrahydroisoquinoline linkages. NOe interactions revealed that one of these linkages was the same as those observed in 9 while the other linkage was inverted. Michellamines A (9) and B (10) exhibited atropisomerism; that is isomerism caused by restricted rotation about a carbon-carbon single bond. Variable-temperature NMR experiments failed to show evidence of the interconversion between these two compounds; however, interconversion could be induced by prolonged treatment with base.

The HIV-inhibitory properties of the michellamines were evaluated in a sequence of interrelated microculture assays performed concurrently on individual wells of a microtiter plate.[14] Cellular viability in HIV-infected and uninfected cells, with or without michellamine treatment, was estimated by XTT metabolism and by cellular hydrolysis of BCECF to a fluorescent product. Correlative assays also measured total cellular DNA content and the production of viral reverse transcriptase (RT), p24 antigen and syncytium forming units (SFU) [figure 1]. The anti-HIV-1 activity profiles of michellamines A and B were virtually identical. These assays showed an in vitro EC_{50} (50% protection) of approximately 20 μM, providing essentially 100% protection over the concentration range 30-100 μM. Cytotoxicity was apparent only at or above approximately 150 μM. The dose-dependent survival of HIV-1 infected target cells, as determined by both the XTT and BCECF assays, correlated well with measurements of cellular DNA and reductions in indices of viral replication. Both alkaloids inhibited the production of RT, p24 and SFU in HIV-1 infected cells over the same concentration range where target CEM-SS cells were protected from the lytic effects of HIV-1 infection. The michellamines were equally effective against HIV-1 when administered as the free base or the HBr salt. Michellamine B exhibited the same anti-HIV activity profile in fully inhibiting the cytopathic effects of HIV-2 infection in MT-2 target cells. This is notable since very few of the known anti-HIV-1 agents demonstrate any activity against the HIV-2 viral strain. Michellamine A also inhibited HIV-2 but provided only partial protection at comparable concentrations.

Buchenavia capitata

The organic extract of leaves from the tropical tree Buchenavia capitata was cytoprotective in the anti-HIV primary screen. An alkaloid partition and bioassay-guided fractionation of the basic fraction led to three active flavonoid alkaloids (11-13).[23] The

11 $R^1 = CH_3$, $R^2 = H$
12 $R^1 = H$, $R^2 = H$
13 $R^1 = CH_3$, $R^2 = CH_3$

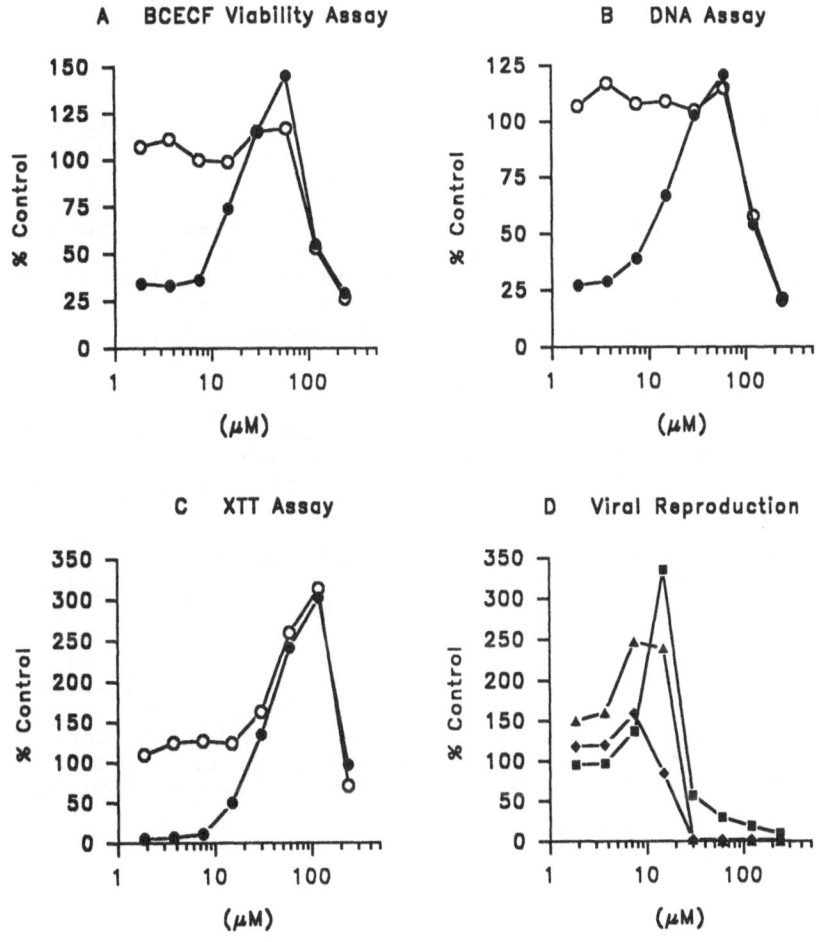

Michellamine B

Fig. 1. Effect of michellamine B on uninfected (O) and HIV-1
infected (●) CEM-SS cells assessed using (A) cellular
viability, (B) total DNA content, (C) metabolic activity,
(D) supernatant reverse transcriptase activity (▲), p24
antigen production (♦), and synthesis of infectious virions (■). All points are graphically represented as the
percentage of the uninfected, non-drug treated control
values in (A), (B) and (C) and virus-infected, non-drug
treated contols in (D).

principal active compound was purified by countercurrent chromatography and identified as O-demethylbuchenavianine (11, $C_{21}H_{21}NO_4$)) by comparison of its spectral data with previously published values.[24] It provided a maximum of approximately 50% protection against HIV-1 at a concentration of 0.26 μM and was cytotoxic at 0.66 μM. Compounds 12 ($C_{20}H_{19}NO_4$) and 13 ($C_{22}H_{23}NO_4$) were identified as N,O-di-demethylbuchenavianine and buchenavianine, respectively. These previously-reported alkaloids occurred as minor constituents and both the potency and maximum level of protection from HIV-1 infection were less than those observed for compound 11.

Pothomorphe peltata

A root extract from the tropical shrub Pothomorphe peltata showed activity in the anti-HIV screen and was additionally active in a phorbol dibutyrate (PDBu) receptor binding assay. The PDBu assay is known to detect compounds which interact with the phorbol receptor [25,26]. Phorbol diterpenes as well as a select number of other compound classes elicit a positive response in the PDBu assay. We have observed that compounds active in the PDBu assay often exhibit HIV-inhibitory properties. Examples of compounds which have tested positively in both the PDBu and anti-HIV assays are phorbol diterpenes such as prostratin,[10] indolactams, aplysiatoxins and bryostatin-1. Both the anti-HIV and PDBu activities were tracked throughout the fractionation procedure, which resulted in an HIV-inhibitory fraction containing peltatols A-C (14-16) and a PDBu-active fraction consisting of 4-nerolidylcatechol (17).[27] Compounds 14-16 represent novel structures while catechol derivative 17 had previously been isolated from P. umbellata.[28]

Peltatol A (14) was the most abundant anti-HIV active compound in the extract with a calculated yield of 0.2% of the dry plant weight. Its molecular formula was established as $C_{42}H_{58}O_4$ by high resolution mass spectrometry, however the ^{13}C NMR spectrum revealed only 21 carbon signals. This indicated that peltatol A was a symmetrical molecule. The ^{1}H and ^{13}C NMR spectra of 14 shared many similar features with the values reported for 17. Both a substituted catechol group and a $C_{15}H_{25}$ nerolidyl substituent were clearly evident. Heteronuclear correlation experiments allowed for complete assignments of the ^{1}H and ^{13}C NMR spectra of 14, revealing that the two halves of the molecule were joined by a carbon-carbon bond between both catechol moieties. Thus, peltatol A consists of two symmetrically substituted catechols joined by a biphenyl linkage. The point of attachment between the catechol subunits was confirmed by long range heteronuclear correlation (HMBC) data.

Peltatols B (15) and C (16) occurred as trace constituents in the extract with a calculated yield of 0.003% of the dry plant weight. The molecular formulae, $C_{42}H_{58}O_4$, and NMR spectral characteristics of peltatols B and C revealed that they were nonsymmetrical isomers of peltatol A (14). They each contained a three proton aromatic spin system similar to that observed in the ^{1}H NMR spectrum of compound 17 and two meta coupled protons similar to those in 14. They also showed only three exchangeable phenolic proton signals. These data indicated that in 15 and 16, the two substituted catechol groups were joined by a diphenyl ether bridge which incorporated one of the catechol oxygens. Heteronuclear correlation experiments confirmed that the position of the ether oxygen attachment to the disubstituted catechol moieties in both 15 and 16 was meta to the nerolidyl substituents. Multiple bond HMBC data demonstrated that in the monosubstituted catechol of 15, the oxygen meta to the nerolidyl group was the ether link of the molecule. In a similar fashion, it was shown that the oxygen para to the nerolidyl group on the monosubstituted catechol of 16 was incorporated into the diphenyl ether bridge.

Peltatol A inhibited the cytopathic effect of in vitro HIV-1 infection over a concentration range of 1.6-16 μM (maximum protection was 75%) and it was cytotoxic at $\geq$ 32 μM. The dose-dependent increase in cellular viability observed in the XTT assay correlated well with reductions in viral p24 and reverse transcriptase measured over a similar concentration range. A 1:1 mixture of peltatols B and C showed a similar activity profile against HIV-1. The monomeric catechol derivative 4-nerolidylcatechol (17) was an abundant metabolite obtained at a yield equal to 2.4% of the dry plant mass. It was inactive against HIV-1, however it did interact with the phorbol receptor binding site with an IC_{50} of 110 μM in the PDBu assay.

NATURAL PRODUCTS AS POTENTIAL ANTI-AIDS AGENTS

There is an urgent requirement for new drug candidates with antiviral activity against HIV. LDDRD's current antiviral research efforts are revealing a wide structural array of HIV-inhibitory compounds from extracts showing activity in the NCI primary anti-HIV screen. These provide intriguing new structural classes of potential AIDS-antiviral compounds which have not previously been associated with HIV-1 inhibitory properties. These natural products may not all show sufficient potency or in vitro therapeutic indices in our primary assays to support their candidacy for further drug development; however, they represent new chemotypes for more detailed biological characterization, structure activity and mechanism of action studies. As prototype anti-HIV compounds, they may provide starting points for synthetic or semisynthetic modifications aimed at enhancing their potency and therapeutic potential.

Current research efforts by LDDRD staff have resulted in the discovery of diverse new nonnucleoside anti-HIV leads of natural product origin. Selected members of this group are being subjected to concerted follow-up studies by the LDDRD team based upon data from detailed in vitro anti-HIV assays. The current protocol for confirmatory in vitro evaluation of new leads provides two different kinds of cellular viability assays in addition to specific assays for total cellular DNA content, supernatant reverse transcriptase activity, p24 core antigen production and synthesis of infectious HIV virions.[14] These assays are all performed concurrently from a single well of a microtiter plate containing human host cells infected with HIV. Follow-up, or stage II testing of new anti-HIV active compounds will aid in their selection and prioritization for more detailed biological and pharmacological investigations. Selections of active natural products for preclinical development are based ultimately upon the collective considerations of in vitro activity, mechanistic profiles and pharmacological and toxicological characteristics of the individual compounds.

REFERENCES

1. O. S. Weislow, R. Kiser, D. L. Fine, J. Bader, R. H. Shoemaker and M. R. Boyd, New Soluble-Formazan Assay for HIV-1 Cytopathic Effects: Application to High-Flux Screening of Synthetic and Natural Products for AIDS-Antiviral Activity, J. Natl. Cancer Inst. 81:577 (1989).

2. A. Monks, D. Scudiero, P. Skehan, R. Shoemaker, K. Paull, D. Vistica, C. Hose, J. Langley, P. Cronise, A. Vaigro-Wolff, M. Gray-Goodrich, H. Campbell, J. Mayo and M. Boyd, Feasibility of a High-Flux Anticancer Drug Screen Using a Diverse Panel of Cultured Human Tumor Cell Lines, J. Natl. Cancer Inst. 83:757 (1991).

3. K. D. Paull, R. H. Shoemaker, M. R. Boyd, J. L. Parsons, P. A. Risbood, W. A. Barbera, M. N. Sharma, D. C. Baker, E. Hand, D. A. Scudiero, A. Monks, M. C. Alley and M. Grote, The Synthesis of XTT - a New Tetrazolium Reagent that is Bioreducible to a Water-Soluble Formazan, J. Heterocyclic Chem. 25:911 (1988).

4. K. R. Gustafson, J. H. Cardellina II, R. W. Fuller, O. S. Weislow, R. F. Kiser, K. M. Snader, G. M. L. Patterson and M. R. Boyd, AIDS-Antiviral Sulfolipids from Cyanobacteria (Blue-Green Algae), J. Natl. Cancer Inst. 81:1254 (1989).

5. A. A. Benson, H. Daniel and R. Wiser, A Sulfolipid in Plants, Proc. Natl. Acad. Sci. USA 45:1582 (1959).

6. M. Lepage, H. Daniel, A. A. Benson, The Plant Sulfolipid. II. Isolation and Properties of Sulfoglycosyl Glycerol, J. Am. Chem. Soc. 83:157 (1961).

7. J. L. Harwood, Sulfolipids, in: "The Biochemistry of Plants. Vol. 4," P. K. Stumpf, ed., Academic Press, New York (1980).

8. J. B. Mudd and K. F. Kleppinger-Sparace, Sulfolipids, in: "The Biochemistry of Plants. Vol. 9," P. K. Stumpf, ed., Academic Press, New York (1987).

9. P. A. Cox, Samoan Ethnopharmacology, in: "Economic and Medicinal Plant Research, Vol. 4: Plants and Traditional Medicine," H. Wagner and N. Farnsworth, eds., Academic Press, London (in press).

10. K. R. Gustafson, J. H. Cardellina II, J. B. McMahon, R. J. Gulakowski, J. Ishitoya, Z. Szallasi, N. E. Lewin, P. M. Blumberg, O. S. Weislow, J. A. Beutler, R. W. Buckheit, Jr., G. M. Cragg, P. A. Cox, J. P. Bader and M. R. Boyd, A Non-Promoting Phorbol from the Samoan Medicinal Plant, Homalanthus acuminatus, Inhibits Cell Killing by HIV-1, J. Med. Chem. (in press).

11. A. R. Cashmore, R. N. Seelye, B. F. Cain, H. Mack, R. Schmidt and E. Hecker, The Structure of Prostratin: A Toxic Tetracyclic Diterpene Ester from Pimelea prostrata, Tetrahedron Lett. 20:1737 (1976).

12. F. J. Evans and S. E. Taylor, Pro-Inflammatory Tumor-Promoting and Antitumor Diterpenes of the Plant Families Euphorbiaceae and Thymelaeceae, in: "Progress in the Organic Chemistry of Naural Products. Vol. 44," W. Herz, H. Grisebach and G. W. Kirby, eds., Springer-Verlag, New York (1983).

13. Z. Szallasi and P. M. Blumberg, Prostratin, a Nonpromoting Phorbol Ester, Inhibits Induction by Phorbol 12-Myristate 13-Acetate of Ornithine Decarboxylase, Edema, and Hyperplasia in CD-1 Mouse Skin, <u>Cancer Res</u>. 51:5355 (1991).

14. R. J. Gulakowski, J. B. McMahon, P. G. Staley, R. A. Moran and M. R. Boyd, A Semiautomated Multiparameter Approach for Anti-HIV Drug Screening, <u>J. Virol. Methods</u> 33:87 (1991).

15. K. R. Gustafson, M. H. G. Munro, J. W. Blunt, J. H. Cardellina II, J. B. McMahon, R. J. Gulakowski, G. M. Cragg, P. A. Cox, L. S. Brinen, J. Clardy and M. R. Boyd, HIV Inhibitory Natural Products. 3. Diterpenes from <u>Homalanthus</u> <u>acuminatus</u> and <u>Chrysobalanus</u> <u>icaco</u>, <u>Tetrahedron</u> 47:4547 (1991).

16. A. R. Lal, R. C. Cambie, P. S. Rutledge and P. D. Woodgate, <u>Ent</u>-Atisane Diterpenes from <u>Euphorbia</u> <u>fidjiana</u>, <u>Phytochemistry</u> 29:1925 (1990).

17. M. Hutchinson, P. Lewer and J. MacMillan, Carbon-13 Nuclear Magnetic Resonance Spectra of Eighteen Derivatives of <u>ent</u>-Kaur-16-en-19-oic Acid, <u>J. Chem. Soc. Perkin Trans. I</u> 2363 (1984).

18. C. M. Hasan, T. M. Healey and P. G. Waterman, Kaurane and Kaurene Diterpenes from the Stem Bark of <u>Xylopia</u> <u>acutiflora</u>, <u>Phytochemistry</u> 21:2134 (1982).

19. P. C. Cheng, C. D. Hufford and N. J. Doorenbos, Isolation of 11-Hydroxylated Kauranic Acids from <u>Adenostemma</u> <u>lavenia</u>, <u>J. Nat. Prod</u>. 42:183 (1979).

20. K. P. Manfredi, J. W. Blunt, J. H. Cardellina II, J. B. McMahon, L. K. Pannell, G. M. Cragg and M. R. Boyd, Novel Alkaloids from the Tropical Plant, <u>Ancistrocladus</u> <u>abbreviatus</u>, Inhibit Cell Killing by HIV-1 and HIV-2, <u>J. Med. Chem</u>. 34:3402 (1991).

21. G. Bringmann, Naphthylisoquinoline Alkaloids, <u>in</u>: "The Alkaloids," A. Brossi, ed., Academic Press, New York (1986).

22. T. R. Govindachari and P. C. Parthasarathy, Alkaloids of Ancistrocladaceae, <u>Heterocycles</u> 7:661 (1977).

23. J. A. Beutler, J. H. Cardellina II, J. B. McMahon, G. M. Cragg and M. R. Boyd, Anti-HIV and Cytotoxic Alkaloids from <u>Buchenavia</u> <u>capitata</u>, <u>J. Nat. Prod</u>. 55:207 (1992).

24. A. Ahond, A. Fournet, C. Moretti, E. Philogène, C. Poupat, O. Thoison and P. Potier, Premiers Alcaloïdes Vrais Isolés de Combrétacées, <u>Buchenavia</u> <u>macrophylla</u> Eichl. et <u>Buchenavia</u> <u>capitata</u> Eichl, <u>Bull. Soc. Chim. Fr.</u> 41 (1984).

25. D. G. DeVries, C. L. Herald, G. R. Pettit and P. M. Blumberg, Demonstration of Sub-Nanomolar Affinity of Bryostatin 1 for the Phorbol Ester Receptor in Rat Brain, <u>Biochem. Pharmacol</u>. 37:4069 (1988).

26. P. M. Blumberg, Protein Kinase C as the Receptor for the Phorbol Ester Tumor Promoters: Sixth Rhoads Memorial Award Lecture, <u>Cancer Res</u>. 48:1 (1988).

27. K. R. Gustafson, J. H. Cardellina II, J. B. McMahon, L. K. Pannell, G. M. Cragg and M. R. Boyd, The Peltatols, Novel HIV Inhibitory Catechol Derivatives from <u>Pothomorphe</u> <u>peltata</u>, <u>J. Org. Chem</u>. 57:2809 (1992).

28. A. Kijjoa, A. M. Giesbrecht, M. K. Akisue, O. R. Gottieb and H. E. Gottlieb, 4-Nerolidylcatechol from <u>Pothomorphe</u> <u>umbellata</u>, <u>Planta Med</u>. 39:85 (1980).

TANNINS AND RELATED COMPOUNDS AS ANTI-HIV AGENTS[1]

Kuo-Hsiung Lee,[2,*] Yoshiki Kashiwada,[2] Gen-ichiro Nonaka,[3]
Itsuo Nishioka,[3] Makoto Nishizawa,[2] Takashi Yamagishi,[2]
Anne J. Bodner,[4] Robert E. Kilkuskie[4] and Yung-Chi Cheng[5]

[2]Natural Products Laboratory
Division of Medicinal Chemistry and Natural Products
School of Pharmacy
University of North Carolina
Chapel Hill, North Carolina 27599
[3]Faculty of Pharmaceutical Sciences
Kyushu University
Fukuoka 812, Japan
[4]Cambridge Biotech Corporation
1600 East Gude Drive
Rockville, Maryland 20850
[5]Department of Pharmacology
Yale University School of Medicine
New Haven, Connecticut 06510

ABSTRACT

Four new tetragalloylquinic acids, 3,5-di-O-galloyl-4-O-digalloylquinic acid (2), 3,4-di-O-galloyl-5-O-digalloylquinic acid (3), 3-O-digalloyl-4,5-di-O-galloylquinic acid (4), and 1,3,4,5-tetra-O-galloylquinic acid (5), were isolated and characterized from a commercial tannic acid as a new class of human immunodeficiency virus (HIV) reverse transcriptase (RT) inhibitor. Compounds 2, 3 and 4 inhibit HIV RT activity 90, 89 and 84% at 100 μM and 73, 70 and 63% at 30 μM, respectively. Compounds 2-5 also inhibit the HIV growth in cells in the range of 61-70% with low cytotoxicity at 25 μM. The HIV cell growth inhibitory effects of these compounds at 25 μM and 6.25 μM (44-57%) are comparable to their effects against the HIV RT at 30 μM and 10 μM, respectively. The inhibitory effect of 3 against DNA polymerases indicates that the selective antiviral action of 3 is determined by more than its action with HIV RT.

Following the aforementioned discoveries, 51 tannins, including gallo-, ellagi- and condensed tannins, were evaluated for their inhibitory effect against HIV RT. The most potent inhibitors were found to be punicalin (32) and punicacortein-C (40) with ID_{50} of 8 and 5 μM, respectively. Several of the tannins including these were further examined for their anti-HIV activity. 1,3,4-Tri-O-galloylquinic acid (13), 3,5-di-O-galloylshikimic acid (16), 3,4,5-tri-O-galloylshikimic acid (17), punicalin (32) and punicalagin (33) inhibited HIV replication in infected H9 lymphocytes with little cytotoxicity. Preincubation of virus with tannins indicated that chebulagic acid (31) and punicalin (32) did not inactivate virus directly. However, 1,3,4-

---

[1]Anti-AIDS Agents 7.
*To whom correspondence should be addressed.

Natural Products as Antiviral Agents, Edited by C.K. Chu
and H.G. Cutler, Plenum Press, New York, 1992

tri-*O*-galloylquinic acid (**13**) and 3,5-di-*O*-galloylshikimic acid (**16**) were more effective inhibitors under these conditions. Thus, in spite of their anti-RT activity, the mechanism by which tannins inhibit HIV may not be associated with this enzyme.

Structure-activity relationship studies among **32** related tannins have led to the synthesis and evaluation of a series of hexahydroxydiphenyl derivatives of ellagic acid as simple analogs of ellagitannins and inhibitors of HIV replication. Among these analogs, 2,2'-dimethoxy-3,3',4,4'-tetrabenzyloxy-1,1'-diphenyl-6,6'-dimethanol (**61**) was found to be a potent inhibitor of HIV replication in infected H9 lymphocytes with little cytotoxicity. Preincubation of **61** with virions before infection also showed that there was no inhibition of HIV. This mechanism of HIV inhibition is different from those ellagitannins, such as **32**, **33** and **40**.

	R$_1$	R$_2$	R$_3$	R$_4$
1	H	G	G	G
2	H	G	G-G	G
3	H	G	G	G-G
4	H	G-G	G	G
5	G	G	G	G

A. INTRODUCTION

Since the discovery of human immunodeficiency virus [HIV (HTLV-III/LAV)], much progress has been made in elucidating the genomic structure of HIV as well as the mechanism of HIV infection.[1-5] The fact that the reverse transcriptase (RT) plays a very important role in controlling the replication of the HIV makes RT one of the most attractive targets in the development of anti-AIDS (acquired immunodeficiency syndrome) drugs. The use of HIV RT as a prime target, coupled with an *in vitro* inhibition assay against HIV replication in H9 lymphocyte cells, has been an approach employed in our laboratories for the detection of potential anti-HIV natural products and their synthetic analogs. The results of these studies on anti-HIV tannins and related compounds are summarized below.

B. ANTI-HIV TETRAGALLOYLQUINIC ACIDS FROM TANNIC ACID

In searching for natural products as potential anti-AIDS agents, the tannic acids were found to show potent inhibitory activity against HIV RT. Among the tannic acids investigated, which were obtained either from a commercial source or extracts prepared from both Turkish and Chinese galls (Table 1), the one from the commercial source showed the strongest inhibitory effect against HIV RT (74% at 100 µg/ml) (Table 1). Subsequent bioassay-directed fractionation of this tannic acid mixture resulted in the isolation of four new tetragalloylquinic acids.

Table 1. Inhibitory Effect of Tannic Acids
Against HIV Reverse Transcriptase

Tannic Acid	Inhibition at 100 µg/ml (%)[a]
Commercial source	74 ± 7
Turkish galls extract	7 ± 8
Chinese galls extract	10 ± 14

[a]Positive control: AZT triphosphate at 100 nM

64 ± 6%, phosphonoformate 0.6 µM 63 ± 9%.

The EtOAc-soluble portion of tannic acid was fractionated using a column of Sephadex LH-20 to yield eight fractions. These fractions were separated according to the degree of galloylation[6] and were composed of mono- to octagalloylated compounds. Among these, fraction G-4, which contained tetragalloylated compounds, showed the strongest inhibitory effect against HIV RT (Table 2). Subsequent repeated preparative HPLC on reversed-phase columns yielded active compounds 2, 3, 4 and 5. Compound 1 was isolated from the less active fraction G-3. The structural determination of 1-5 was based on extensive NMR spectral evidence, especially the ^{13}C-NMR study as described in ref. 7.

Table 2. Inhibitory Effect of the Fractions
from Tannic Acids Against
HIV Reverse Transcriptase

Fraction	Inhibition at 100 µg/ml (%)[a]
G-1	0
G-2	10 ± 15
G-3	49 ± 4
G-4	83 ± 2
G-5	68 ± 5
G-6	56 ± 11
G-7	49 ± 17
G-8	48 ± 14

[a]Positive control: AZT triphosphate 100 nM

64 ± 6%, phosphonoformate 0.6 µM 63 ± 9%.

The effects of 1 [3,4,5-GQA (galloylquinic acid)], 2 (3,5-G-4-diGQA), 3 (3,4-G-5-diGQA), 4 (3-diG-4,5-GQA) and 5 (1,3,4,5-GQA) against HIV RT isolated from infected cells and HIV-infected H9 lymphocytes are summarized in Table 3. Compounds 2, 3, 4 and 5 showed 90, 89, 84 and 94% inhibitory activity, respectively, against HIV RT at 100 µM. Compounds 2, 3 and 4 exhibited more than 50% inhibitory activity at 10 µM, and there was no significant difference in inhibitory activity among these three tetragalloylquinic acids. The inhibitory activity of 1 was shown to be less than that of 2, 3 or 4. This indicated that the depside galloyl group in the molecule of 2, 3 and 4 plays an important role in their inhibitory effects. In view of the difference of the dose response between 4 and 5, the mode of inhibition of these two compounds could be different. This requires further investigation.

These compounds (**1-5**) also exhibited inhibitory activity against the growth of HIV in infected H9 lymphocytes (61-70% inhibition at 25 μM). At this level, these galloylquinic acids showed little cytotoxicity against the uninfected H9 cells (0-25%), and no cytotoxicity was observed at 6.25 and 1.25 μM. The growth of HIV was inhibited even at 6.25 μM (44-59%) and 1.25 μM (13-34%). It is interesting to note that the inhibition of HIV RT at 30 μM and HIV growth at 25 μM by **2, 3** and **4** was found to show good correlation. Good correlation was also found between HIV RT inhibition at 10 μM and HIV growth at 6.25 μM. Whether the anti-RT activity and anti-HIV activity are two independent events, or this class of compounds inhibits HIV replication by acting on sites other than RT, is not clear.

Table 3. Effect of Galloylquinic Acids (GQA) on HIV Cell Growth and HIV Reverse Transcriptase (RT) Activity

Compound	HIV RT Activity			HIV growth		
	(% Inhibition)			(% Inhibition)		
	100 μM	30 μM	10 μM	25 μM	6.25 μM	1.25 μM
3,4,5-TriGQA (**1**)	58 ± 6	48 ± 11	39 ± 10	70 ± 1^a $(14)^b$	59 ± 8	26 ± 1
3,5-G-4-diGQA (**2**)	90 ± 3	73 ± 5	53 ± 3	71 ± 1 (14)	57 ± 2	34 ± 5
3,4-G-5-diGQA (**3**)	89 ± 5	70 ± 9	55 ± 11	63 ± 1 (0)	50 ± 3	28 ± 2
3-DiG-4,5-GQA (**4**)	84 ± 7	63 ± 6	50 ± 11	61 ± 1 (25)	44 ± 4	13 ± 2
1,3,4,5-TetraGQA (**5**)	94 ± 2	84 ± 7	36 ± 6	66 ± 2 (25)	52 ± 1	15 ± 2

[a]The μM concentration of **1** was 30.8, 7.7, and 1.5 μM, respectively.

[b]The figures in parentheses indicate percent inhibition of uninfected H9 cell growth in the presence of drug.

The inhibitory effects of **3** against DNA polymerase α, β and γ, as well as HIV RT, are shown in Figure 1. The sensitivity of these DNA polymerases against **3** was found to be quite different. DNA polymerase α is most sensitive to **3**. The ID_{50} of **3** for DNA polymerase α ($ID_{50} = 0.065 \pm 0.019$ μM) is about 1/450 of that for HIV RT ($ID_{50} = 29 \pm 18$ μM) and the sensitivities are in this order: DNA polymerase α > γ ($ID_{50} = 2.5 \pm 0.9$ μM) > β (21 ± 11 μM) = HIV RT. However, compound **3** was much more potent against HIV growth than against cell growth. These results imply that there are other factors in uninfected cells which render them resistant to this compound. This is under current investigation.

Compounds **2**, **3** and **4** appear to be the first examples of plant products that demonstrate potent inhibition of both HIV RT activity and HIV growth in culture. The structures of these compounds are unique compared to those of the other known HIV RT inhibitors. These compounds deserve further development as potential anti-AIDS drugs.

C. ANTI-HIV TANNINS

With the discovery of tetragalloylquinic acids as inhibitors of HIV RT and HIV replication, investigation on other classes of tannins as inhibitors for the same type of activity was initiated. Fifty-one tannins, including gallo-, ellagi-, condensed and complex tannins as well

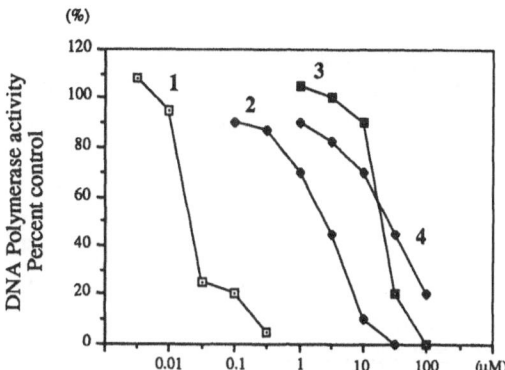

Figure 1. Inhibitory activity of compound **3** on DNA polymerase α, β and γ and HIV reverse transcriptase: 1= DNA polymerase α, 2 = DNA polymerase γ, 3 =DNA polymerase β, 4 = HIV RT

as other related compounds, were examined for their inhibitory effect against HIV RT and HIV replication. The gallotannins were classified into four groups: gallotannins with a glucose core (**6-10**), a quinic acid core (**11-13**), a shikimic acid core (**14-17**) and gallotannins having a hamamelose core (**18-19**). The ellagitannins were classified as ellagitannins with a glucose core (**20-26**), a modified hexa-hydroxydiphenoyl moiety and/or related acyl group (**27-34**), ellagitannin dimers (**35** and **36**), ellagitannin tetramer (**37**), ellagitannins with an open-chain glucose core (**38-41**) and ellagitannins with a triterpenoid moiety (**42** and **43**). The condensed tannins included monomers (**44** and **45**), dimers (**46-48**), trimers (**49** and **50**), tetramers (**51** and **52**), and a pentamer (**53**). Compounds **54** and **55** represent the complex tannins consisting of the component unit [(+)-catechin] of condensed tannins and an ellagitannin moiety linked through a carbon-carbon linkage, while compound **56** is a caffeic acid tetramer.

Tannins described above have been screened as inhibitors of HIV replication in H9 lymphocytes. As demonstrated in Figures 2 and 3, eight tannins (**13**, **16**, **17**, **30-33**, and **40**) showed significant anti-HIV activity (greater than 50% inhibition of virus replication) at concentrations that do not inhibit H9 cell growth. Punicalin (**32**) showed less than 15% growth inhibition even at 30 μM (see Figure 3). One compound, sanguiin H-11 (**37**), did not significantly inhibit HIV replication in H9 cells (data not shown).

Gallotannins (6 - 19)

6 (1-O-galloyl-β-D-glucose)[17]
7 (1,6-di-O-galloyl-β-D-glucose)[18]
8 (1,2,6-tri-O-galloyl-β-D-glucose)[19]
9 (1,2,3,6-tetra-O-galloyl-β-D-glucose)[20]
10 (1,2,3,4,6-penta-O-galloyl-β-D-glucose)[21]

R_1	R_2	R_3	R_4
H	H	H	H
H	H	H	G
G	H	H	G
G	G	H	G
G	G	G	G

11 (4-O-galloylquinic acid)[22]
12 (1,4-di-O-galloylquinic acid)[23]
13 (1,3,4-tri-O-galloylquinic acid)[24]

R_1	R_2	R_3	R_4
H	H	G	H
G	H	G	H
G	G	G	H

14 (3-O-galloylshikimic acid)[24]
15 (3-O-digalloylshikimic acid)[24]
16 (3,5-di-O-galloylshikimic acid)[24]
17 (3,4,5-tri-O-galloylshikimic acid)[25]

R_1	R_2	R_3
G	H	H
G-G	H	H
G	H	G
G	G	G

18 (hamamelitannin)[26]
19 (2',3,5-tri-O-galloylhamamelose)[26]

R
H
G

Ellagitannins (20 - 43)

20 (2,3-HHDP-D-glucose)[27]
21 (strictinin)[28]
22 (pedunculagin)[28]
23 (1(β)-O-galloylpedunculagin)[30]
24 (eugeniin)[31]
32 (punicalin)[34]
33 (punicalagin)[34]

R_1	R_2	R_3
H	S-HHDP	H, H
G(β)	H, H	S-HHDP
H	S-HHDP	S-HHDP
G(β)	S-HHDP	S-HHDP
G(β)	G, G	S-HHDP
H	H, H	Gal
H	S-HHDP	Gal

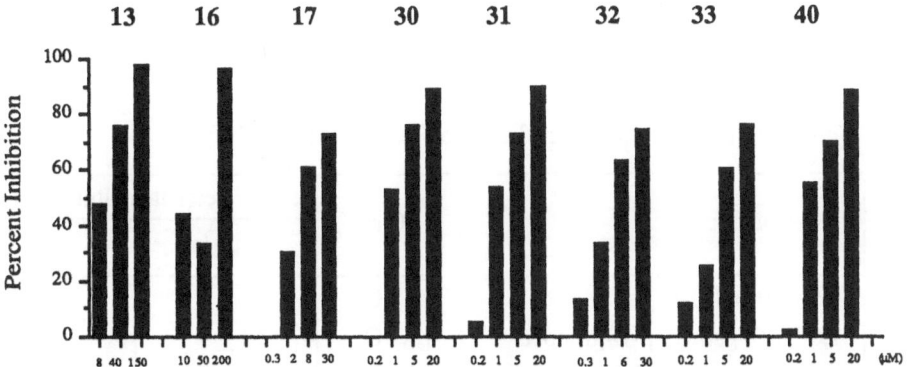

Figure 2. Anti-HIV effect of tannins **13, 16, 17, 30-33** and **40.** This represents a normalized percent HIV replication inhibition as measured by p24 antigen capture. This graph corresponds to data taken from two experiments which showed similar results.

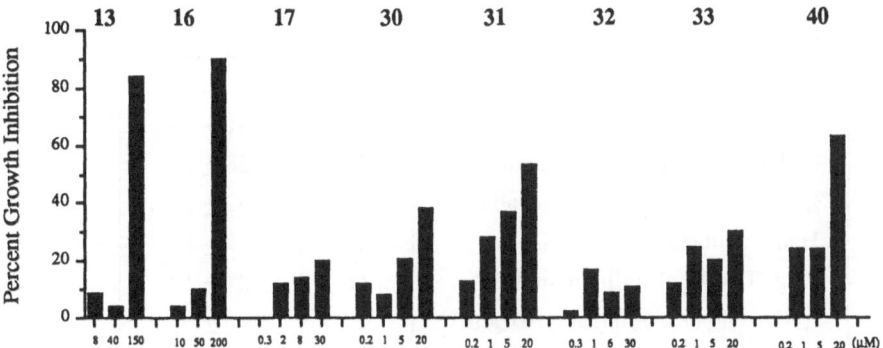

Figure 3. Effect of tannins **13, 16, 17, 30 - 33** and **40** on H9 cell growth. This represents the percent growth inhibition of uninfected H9 cells treated with different tannins.

Table 4. Effect of Tannins on HIV Polymerase[a]

Varied Components	Assay Mixtures		
	A	B	C
Activated DNA (μg/ml)	75	75	375
dTTP (μM)	5	25	5
	Percent Activity (mean ± SD)		
Punicalin [**32**] (64 μM)	11.4 ± 2	34.4 ± 7	45.3 ± 8
Punicacortein-C [**40**] (46 μM)	21.5 ± 3	37.6 ± 7	46.2 ± 4

[a]Punicalin and punicacortein-C were incubated in three reaction conditions as indicated. The rest of the condition is the same as the regular DNA reaction mixture described below: each condition in a 50 μl volume contained 50 mM Tris HCl pH 8.0, 8 mM MgCl$_2$, 100 μg/ml BSA, 100 μM each of dATP, dGTP, and dCTP, and 1 mM DTT. The reactions were incubated 30 min. at 37°C. The rest of the conditions are as in the RT assay.

Table 5. Inhibitory Effect of Tannins Against HIV RT

No.	Compound	Class[+]	Percent Inhibition	
			20 μg/ml	50 μg/ml
6	1-*O*-Galloyl-β-D-glucose	G	18 ± 4	
7	1,6-Di-*O*-galloyl-β-D-glucose	G	14 ± 5	
8	1,2,6-Tri-*O*-galloyl-β-D-glucose	G	16 ± 14	
9	1,2,3,6-Tetra-*O*-galloyl-β-D-glucose	G	2 ± 3	
10	1,2,3,4,6-Penta-*O*-galloyl-β-D-glucose	G	0	
11	4-*O*-Galloylquinic acid	G	0	
12	1,4-Di-*O*-galloylquinic acid	G	0	
13	1,3,4-Tri-*O*-galloylquinic acid	G	53 ± 15	
14	3-*O*-Galloylshikimic acid	G	0	
15	3-*O*-Digalloylshikimic acid	G	31 ± 14	
16	3,5-Di-*O*-galloylshikimic acid	G	58 ± 6	
17	3,4,5-Tri-*O*-galloylshikimic acid	G	42 ± 1	46 ± 1
18	Hamamelitannin	G	16 ± 13	
19	2′,3,5-Tri-*O*-galloylhamamelose	G	25 ± 10	
20	2,3-Hexahydroxydiphenoyl-D-glucose	E	8 ± 9	
21	Strictinin	E	66 ± 10	
22	Pedunculagin	E	33 ± 28	
23	1(β)-*O*-Galloylpedunculagin	E	3 ± 5	
24	Eugeniin	E	27 ± 8	
25	Corilagin	E	57 ± 5	
26	Punicafolin	E	0	
27	Geraniin	E	15 ± 4	
28	Elaeocarpusin	E	23 ± 10	
29	Furosin	E	38 ± 8	
30	Chebulinic acid	E	73 ± 9	94 ± 3
31	Chebulagic acid	E	78 ± 6	94 ± 1
32	Punicalin	E	89 ± 1	98 ± 1
33	Punicalagin	E	56 ± 3	81 ± 8
34	Terchebin	E	13 ± 12	
35	Phillyraeoidin A	E	0	
36	Sanguiin H-6	E	34 ± 16	
37	Sanguiin H-11	E	37 ± 10	
38	Castalagin	E	40 ± 18	
39	Vescalagin	E	45 ± 13	

Table 5. Inhibitory Effect of Tannins Against HIV RT - continued

No.	Compound	Class[+]	Percent Inhibition	
			20 µg/ml	50 µg/ml
40	Punicacortein C	E	83 ± 4	94 ± 2
41	Grandinin	E	27 ± 10	
42	Castanopsinin A	E	17 ± 11	
43	Castanopsiniin A	E	18 ± 21	
44	(-)-Epicatechin	C	22 ± 21	
45	(-)-Epicatechin 3-O-gallate	C	5 ± 5	
46	Procyanidin B-2	C	21 ± 5	
47	Procyanidin B-2 3,3'-di-O-gallate	C	0	
48	Proanthocyanidin A-2	C	18 ± 11	
49	Procyanidin C-1	C	18 ± 12	
50	Procyanidin C-1 3,3',3"-tri-O-gallate	C	13 ± 14	
51	Cinnamtannin A_2	C	18 ± 10	
52	Cinnamtannin B_1	C	13 ± 11	
53	Cinnamtannin B_2	C	19 ± 16	
54	Acutissimin A	EC	24 ± 6	
55	Mongolicain A	EC	23 ± 6	
56	Lithospermic acid B Mg salt	#	18 ± 17	

[+]G: gallotannin, E: ellagitannin, C: condensed tannin, EC: complex tannin,

#: Caffeic acid tetramer

Positive control: AZT triphosphate 0.1 µM 70 ± 6%

Phosphonoformate 0.6 µM 71 ± 6%

Data represent mean of percent inhibition plus a standard deviation rounded to whole numbers.

In order to investigate the mechanism of HIV inhibition, tannins were also evaluated as inhibitors of purified HIV RT (Table 5). Three compounds, sanguiin H-11 (37), punicalin (32) and punicacortein-C (40) were potent inhibitors of HIV RT with ID_{50} of 20, 8 and 5 µM, respectively. The most potent RT inhibitors, 32 and 40, contain a gallagyl moiety, suggesting that this structure may be important for RT inhibition. The RT inhibition by 32 is concentration dependent, and the reaction catalyzed by RT in the presence of drug is linear with time (Figures 4 and 5). The degree of inhibition by 32 and 40 varied with the concentration of substrates in the reaction mixture (Table 4). This could be due to the interaction of the tannins with DNA. DNA binding was measured by the ethidium bromide displacement assay (Figure 6). Compounds 32 and 40 were about sixty-five fold less active in this assay than actinomycin D.

Further investigations of the mechanism of HIV inhibition were conducted in cell culture. The tannins were added to virions before infection, during the infection step, or immediately after infection. Figure 7 shows that the degree of HIV inhibition depends on when the tannin is present in the inhibition assay. Note that although the concentrations of the tannins tested in this experiment varied somewhat (9.2-21 µM), the extent of HIV inhibition by tannins added after infection (our normal protocol) was similar (35-65%). Preincubation of virus with two tannins, chebulagic acid (31) and punicalin (32), did not inhibit HIV replication significantly. However, when five other tannins, 1,3,4-tri-O-galloylquinic acid (13), 3,5-di-O-galloylshikimic acid (16), chebulinic acid (30), punicalagin (33) and punicacortein-C (40), were preincubated with virions before infection, all inhibited HIV. Two of these (13 and 16) were more effective inhibitors when virions were preincubated with tannins than when the tannins were added to the culture post-infection. All tannins virtually eliminated HIV replication when present during the infection step. These results suggest that the tannins might inhibit virus-cell interactions. The tannins that inhibit after preincubation with virions (e.g., 13 and 16) may bind tightly to virions, inactivating them and preventing infection. The two compounds that do not inhibit after preincubation (31 and 32) may interact only weakly with

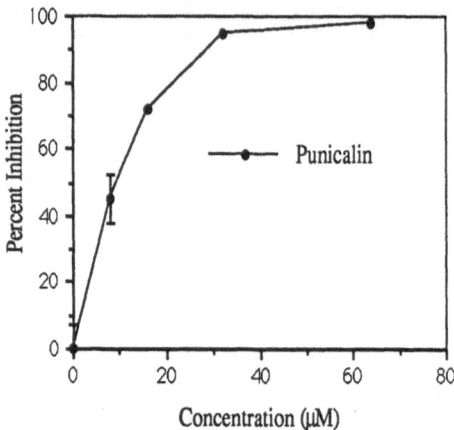

Figure 4.   Effect of punicalin (**32**) on HIV-RT.
Different amounts of punicalin were incubated in the
standard RT mixture for 60 min. After washing and
counting, the drug-treated samples were compared
to drug-free controls

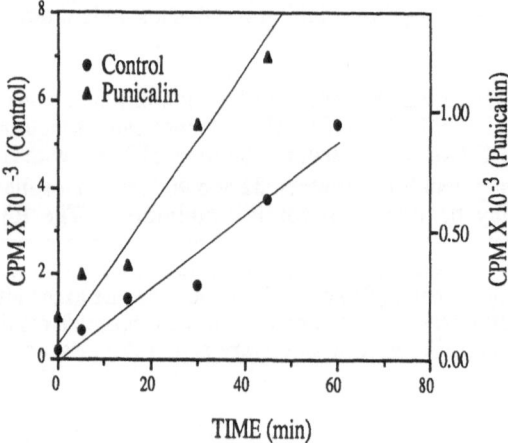

Figure 5.   Time course in the presence of
punicalin (**32**) at 12.5 µM. Drug was incubated
in the standard assay mixture for various times.
After washing and counting, samples were compared to drug-free controls.

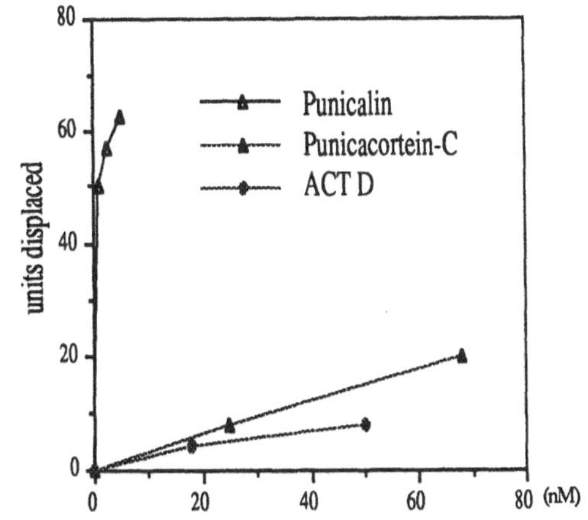

Figure 6. Ethidium displacement by punicalin (**32**). The procedure is described in the Experimental section. Actinomycin D was used as a positive control.

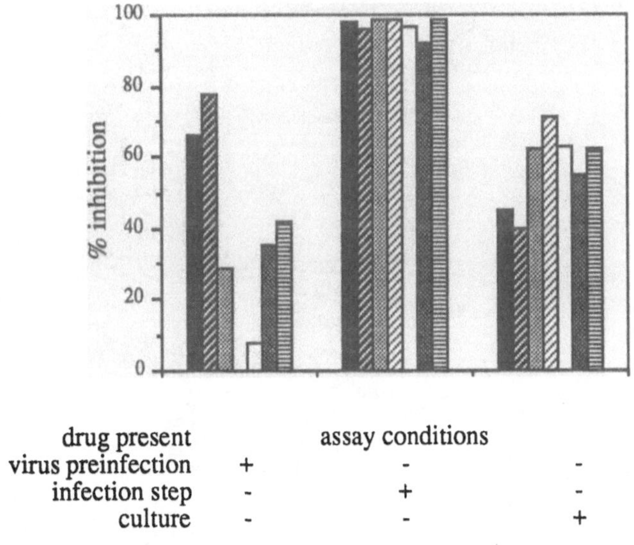

Figure 7. Effect of addition of tannins to cultures at different times. Tannins were assayed for anti-HIV activity under different assay conditions (see Experimental). Tannins were all assayed at 10 μg/ml. Compounds tested and their molar concentrations are: **13** (18 μM) ■ ; **16** (21 μM) ▨ ; **30** (10 μM) ▩ ; **31** (10 μM) ▨ ; **32** (13 μM) □ ; **33** (9.2 μM) ■ ; and **40** (9.2 μM) ▤ .

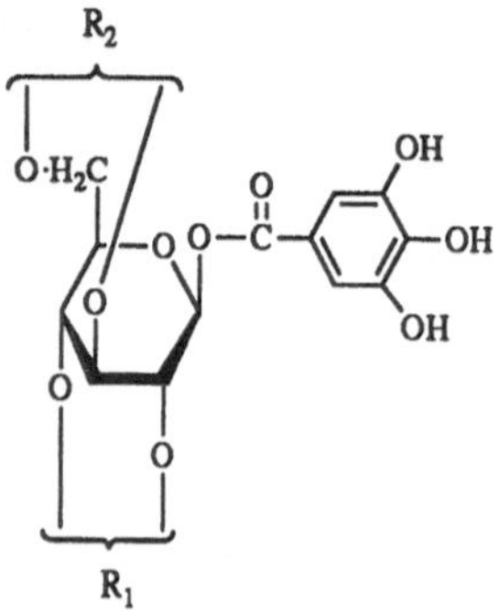

	R_1	R_2
25 (corilagin)[30]	H, H	R-HHDP
26 (punicafolin)[30]	G, G	R-HHDP
27 (geraniin)[32]	DHHDP	R-HHDP
28 (elaeocarpusin)[33]	Ela	R-HHDP
29 (furosin)[35]	DHHDP	H, H
30 (chebulinic acid)[36]	Che	R-HHDP
31 (chebulagic acid)[8]	Che	G, G
34 (terchebin)[35]	DHHDP	G, G

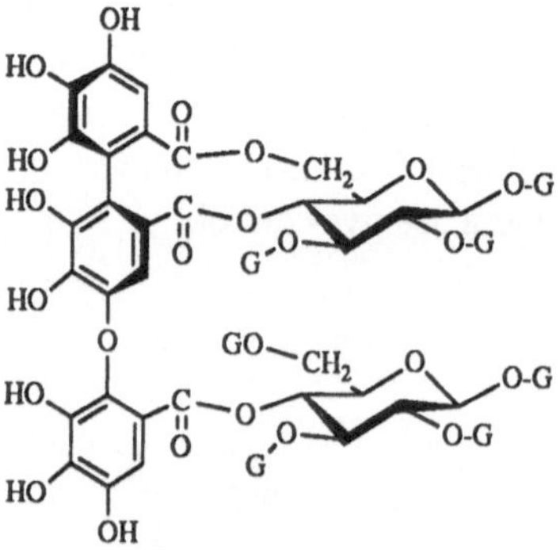

35 (phillyraeoidin A)[37]

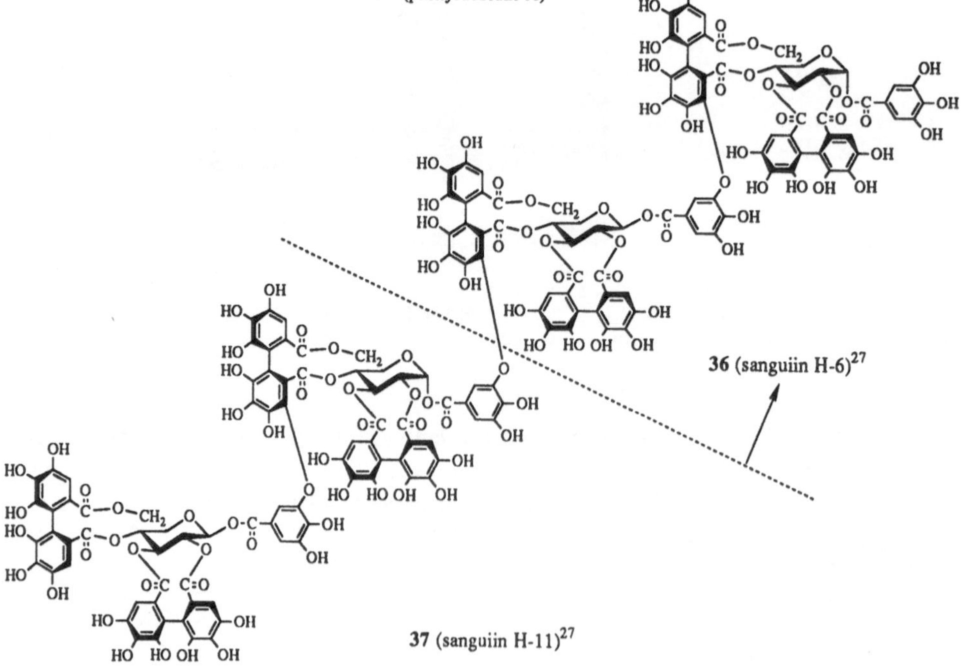

36 (sanguiin H-6)[27]

37 (sanguiin H-11)[27]

80

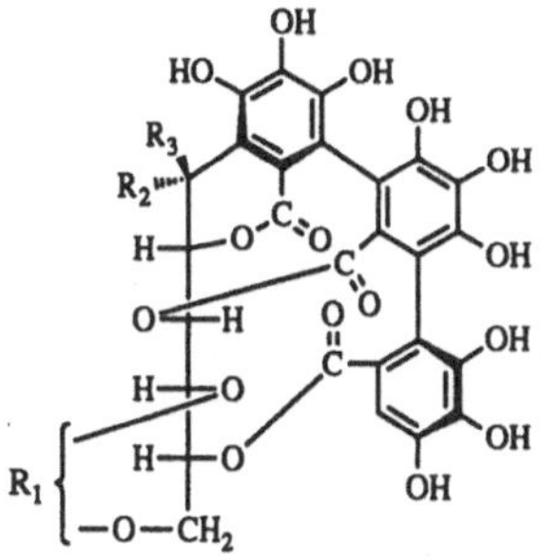

38 (castalagin)[38]

39 (vescalagin)[38]

40 (punicacortein C)[38, 39]

41 (grandinin)[38, 40]

	R_1	R_2	R_3
38 (castalagin)[38]	S-HHDP	OH	H
39 (vescalagin)[38]	S-HHDP	H	OH
40 (punicacortein C)[38, 39]	Gal	OH	H
41 (grandinin)[38, 40]	S-HHDP	H	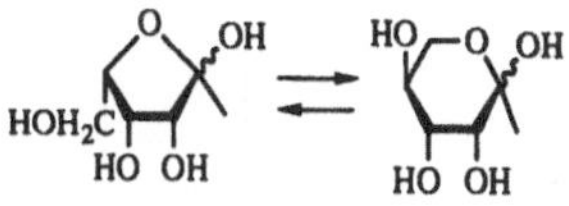

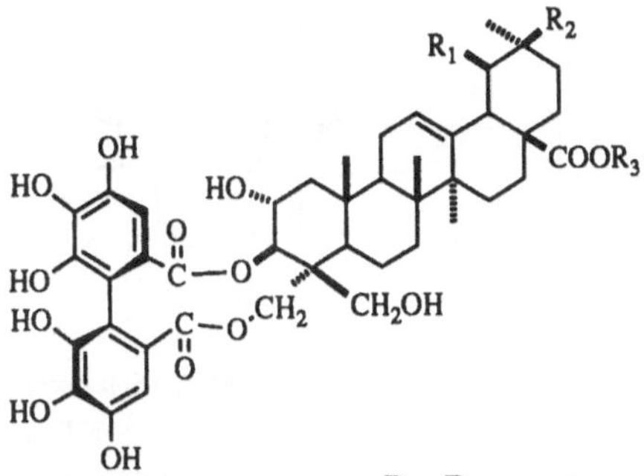

42 (castanopsinin A)[41]

43 (castanopsiniin A)[41]

	R_1, R_2	R_3
42 (castanopsinin A)[41]	H, CH_3	β-D-Glc
43 (castanopsiniin A)[41]	H, CH_3	H

G :

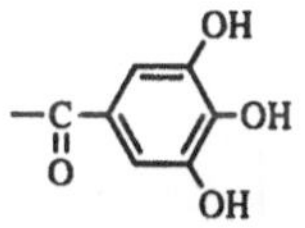

G-G :

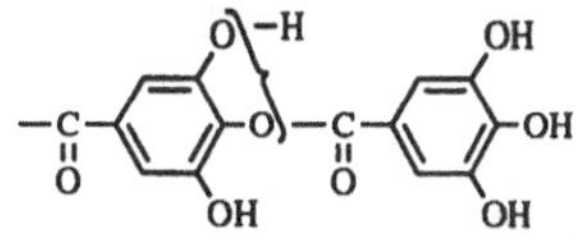

S-HHDP :

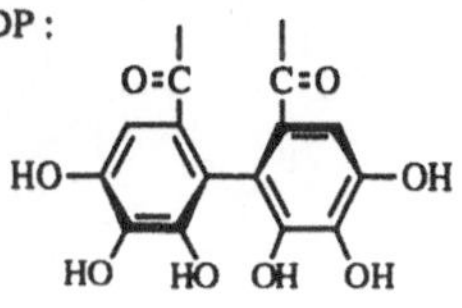

R-HHDP :

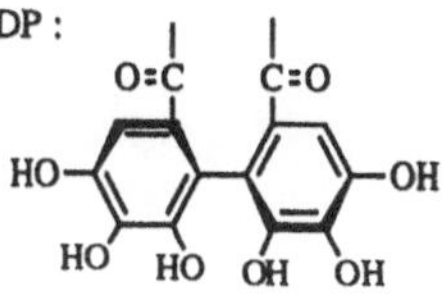

DHHDP :

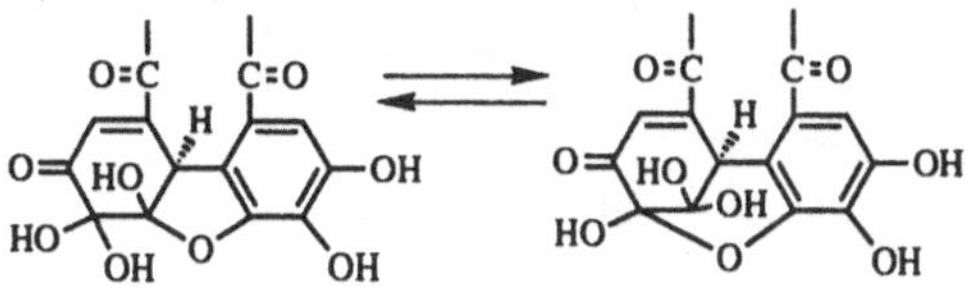

Che :

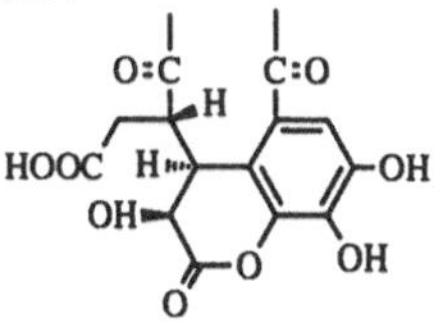

Ela :

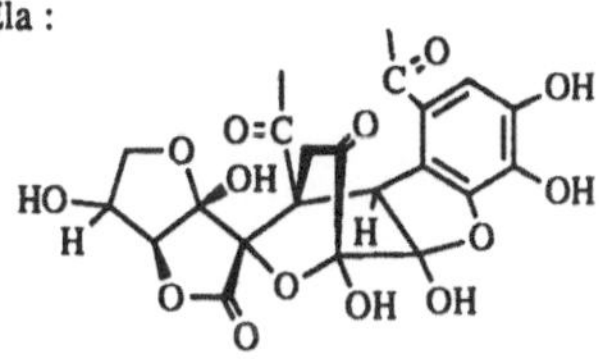

Gal :

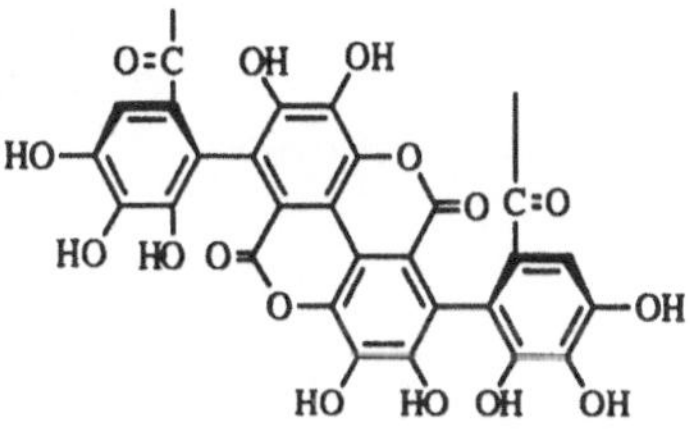

Condensed Tannins (44 - 53)

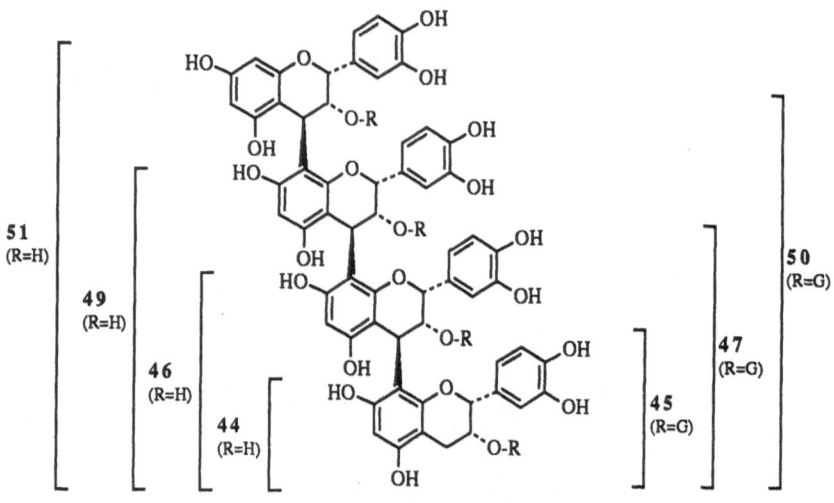

44 ((-)-epicatechin)[42]
46 (procyanidin B-2)[42]
49 (procyanidin C-1)[42]
51 (cinnamutannin A$_2$)[42]

45 ((-)-epicatechin 3-O-gallate)[43]
47 (procyanidin B-2 3,3'-di-O-gallate)[43]
50 (procyanidin C-1 3,3',3"-tri-O-gallate)[43]

48 (proanthocyanidin A-2)[44]
52 (cinnamtannin B$_1$)[44]
53 (cinnamtannin B$_2$)[44]

Complex Tannins (54 and 55)

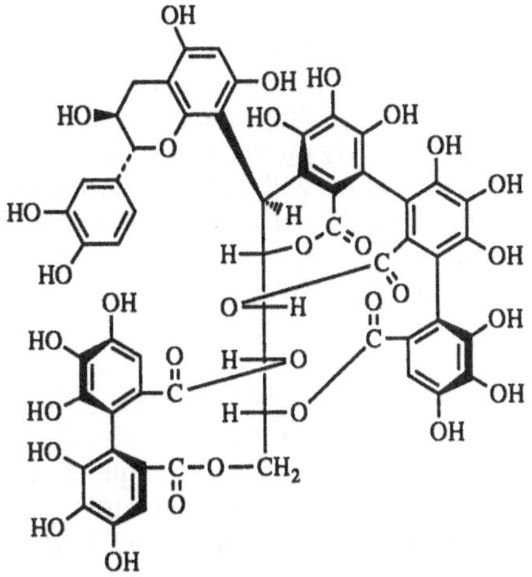

54 (acutissimin A)[38, 45]

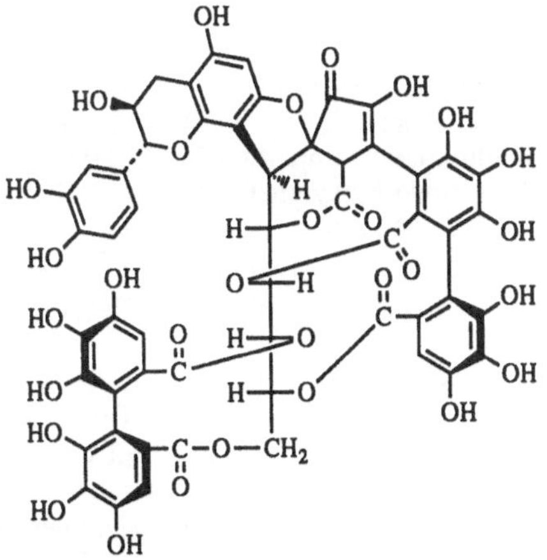

55 (mongolicain A)[38, 46]

virions, so that the tenfold dilution of treated virions prior to the infection step may decrease the virus-tannin interaction. They may still interfere with virus binding to H9 cells when present in the virus-cell mixture at 10 µg/ml.

In summary, our results suggest that a variety of tannins are capable of inhibiting purified HIV RT. They may also inhibit HIV replication in cell culture by additional or other mechanisms, probably by interfering with HIV-cell interactions. We are presently characterizing the nature of the interactions between tannins, HIV and H9 cells.

Caffeic acid Tetramer (56)

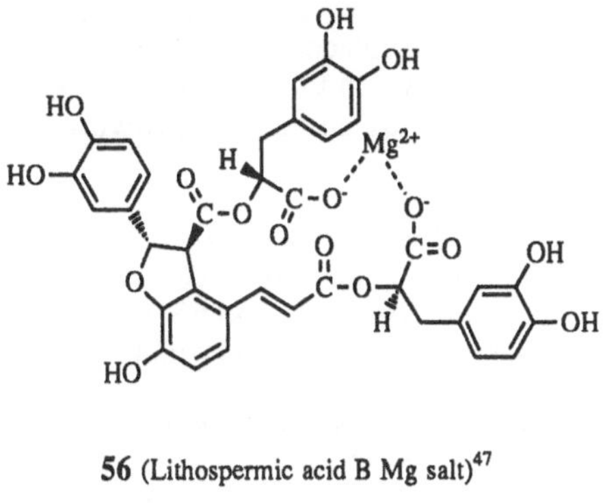

56 (Lithospermic acid B Mg salt)[47]

D. ANTI-HIV HEXAHYDROXYDIPHENYL DERIVATIVES OF ELLAGIC ACID

The discovery of galloylquinic acids, especially tetragalloylquinic acids, and punicalin (**32**), punicalagin (**33**) and punicacortein-C (**40**), as potent HIV inhibitors discussed above has prompted our synthesis of simpler analogs related to these compounds as new anti-HIV agents. Since **32**, **33** and **40** contain a gallagyl (tetraphenoyl) group, it suggests that this structure may be important for the HIV inhibition. In addition, the gallagyl group is considered to be formed biosynthetically from a hexahydroxydiphenoyl (HHDP) group and two galloyl groups.[8] Therefore, the HHDP group is regarded as a basic skeleton for gallagyl group. Based on these observations, we have prepared derivatives of HHDP group from ellagic acid, and investigated their inhibitory effect on HIV replication.[10]

The biphenyl derivatives (**57-67**) were synthesized from ellagic acid (**68**) as shown in Scheme 1. As illustrated in Table 6, only one hexahydroxydiphenyl derivative (**61**) demonstrated potent inhibitory activity against HIV in acutely infected H9 lymphocyte cells (EC_{50} = 10 µM). Toxicity was low, as uninfected H9 cell growth was inhibited by 50% (IC_{50}) at >140 µM. All other derivatives did not inhibit HIV or did so only at toxic concentrations. The two CH_2OH groups at C-6 and C-6' appear to be essential for the specific HIV inhibition. Replacement of these two groups with CHO in **62** greatly increased the toxicity and only slightly increased the anti-HIV activity. When the two groups were replaced by CH_2Br in **63**, toxicity increased with no change in the anti-HIV activity.

Table 6. Biological Evaluation of the Hexahydroxydiphenyl Derivatives (**57-67**)

Compound No.	IC_{50} (µM)[a]	EC_{50} (µM)[b]
57	13	6
58	>120	>120
59	70	45
60	>100	>100
61	>140	10
62	4	2
63	30	13
64	>200	>200
65	>250	65
66	10	5
67	150	90

[a]Concentration which inhibits uninfected cell growth by 50%.
[b]Concentration which inhibits virus replication by 50%.

The OMe groups at C-2 and C-2' were also necessary to retain the selective antiviral activity. When they were replaced by OH in **57**, toxicity was greatly increased. The compound became inactive when OBz replaced the OMe groups in **58**. In **59** and **60**, OBz replaced the OMe groups of **62** and **63**, respectively. Both **59** and **60** are much less active than **62** and **63**.

The replacement of OBz groups at C-3, C-4, C-3' and C-4' with OMe groups in **65** reduced the HIV inhibition. Other derivatives with OBz replaced (**66** and **67**) were also less active than the OMe counterparts (**62** and **63**, respectively). An additional (**64**) containing OMe at C-2, C-3, C-4, C-2', C-3' and C-4' with COOMe at C-6 and C-6' was also inactive.

Previously, it was determined that tannins can inhibit HIV, at least in part, by interfering with virus-cell interactions.[9] Compound **61** was incubated with virions before infection, during the one hour infection step, or added to the infected cells immediately after infection. Compound **61** only inhibited when present after infection (data not shown). This indicates that the mode of inhibition by **61** is different from that of the tannins. The mechanism of inhibition is currently under investigation.

Scheme 1

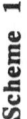

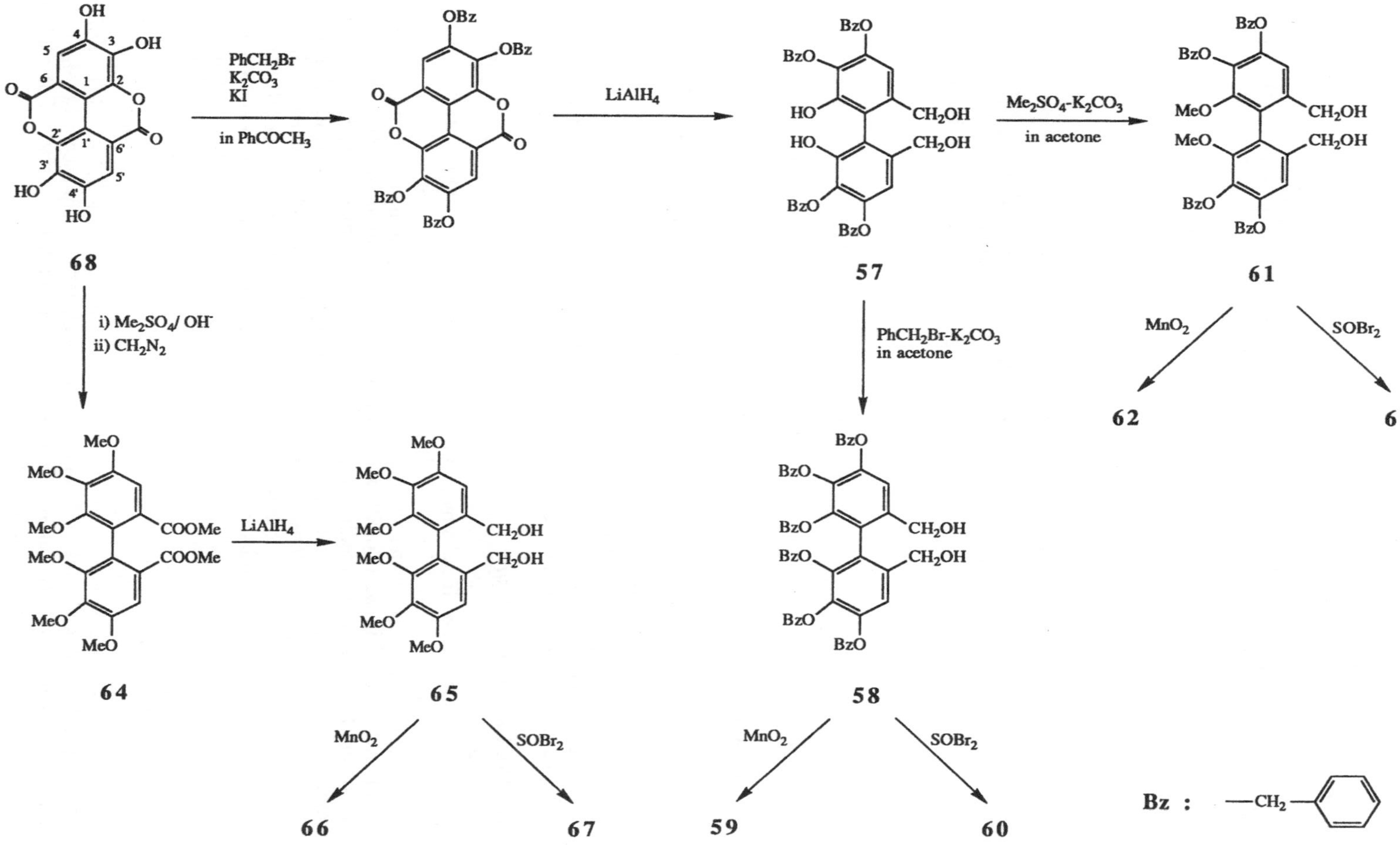

68
PhCH2Br
K2CO3
KI
in PhCOCH3
LiAlH4
57
Me2SO4-K2CO3
in acetone
61
MnO2
SOBr2
62
63
i) Me2SO4/ OH-
ii) CH2N2
PhCH2Br-K2CO3
in acetone
64
LiAlH4
65
58
MnO2
SOBr2
66
67
MnO2
SOBr2
59
60
Bz : —CH2—

Hexahydroxydiphenyl Derivatives (57-67) and Ellagic Acid (68)

57		R		R	R	68

58	CH₂OH	61	CH₂OH	64	COOMe
59	CHO	62	CHO	65	CH₂OH
60	CH₂Br	63	CH₂Br	66	CHO
				67	CH₂Br

58 CH$_2$OH 61 CH$_2$OH 64 COOMe
59 CHO 62 CHO 65 CH$_2$OH
60 CH$_2$Br 63 CH$_2$Br 66 CHO
67 CH$_2$Br   Bz = CH$_2$—⟨phenyl⟩

E. BIOASSAYS

HIV RT Assay

The HIV RT assay was performed according to the method described by Cheng et al.[4] The immuno-affinity purified enzyme from virions used was isolated from extracts of human T cells infected with HIV lymphocytic virus. Poly rA oligo dT$_{10}$ (Pharmacia, Piscataway, New Jersey) was used as the template to measure the incorporation of [^{3}H]dTMP (20 μM). The percentage of inhibition was determined by comparing the RT activity of drug-containing assay to that of the drug-free control.

HIV Growth Inhibition Assay

This assay was performed by incubation of H9 lymphocytes (3.5 x 10^6 cells/ml) in the presence or absence of HIV-1 (HTLV-IIIB) for 1 h at 37°. Cells were washed thoroughly to remove unadsorbed virions and resuspended at 4 x 10^5 cells/ml in culture medium. Aliquots (1 ml) were placed in wells of 24-well culture plates containing an equal volume of test compound (diluted in test medium). After incubation for 3 days at 37°C, cell density of uninfected cultures was determined to assess toxicity of the test compound. A p24 antigen capture assay was used to determine the level of HIV infection in HIV-treated cultures. The ability of test compounds to inhibit HIV replication was measured at four different concentrations of test compound relative to infected, uninfected cultures. Test compounds were considered to be active if p24 levels were less than 70% of infected, untreated cultures (>30% inhibition). The effect of adding compounds to the inhibition assay at different times was also measured. This inhibition assay was conducted as the standard method described above, with the following modifications. Virions were preincubated with 9.1-21 μM of tannins for 1 h at 37° and then diluted tenfold into a suspension of H9 cells for infection. The infection step, using untreated virions, was conducted with tannin included (at 9.2-21 μM). Following infection and washing, the infected cultures were incubated in the presence of tannin (9.2-21 μM) for 3 days at 37° (see above-mentioned normal inhibiton assay). For each assay, compound was added to the assay only at the indicated time.

DNA Polymerase Assay

Human DNA polymerases α,[11] β[12] and γ[12] were purified as described previously.[11] DNA polymerase α activity was assayed in 50 μl of reaction mixture containing 50 mM Tris buffer (pH 8.0), 1 mg/ml of bovine serum albumin, 6 mM $MgCl_2$, 1 mM dithiothreitol, 100 μg/ml of gapped duplex DNA,[13] 10 μM [^{3}H]TTP (1Ci/mmol), 50 μM each of dATP, dCTP and dGTP and approximately 0.01 activity units of DNA polymerase. After incubation at 37° for 1 h, the DNA was precipitated onto glass fiber filters with a 5% trichloroacetic acid/10 mM pyrophosphate solution and counted for radioactivity as previously described.[11] DNA polymerases β, γ and HIV RT were assayed in the same manner except that 100 mM KCl was included in the reaction mixture.

Ethidium Displacement Assay

This assay was performed using an Aminco fluorocolorimeter. Three ml of a solution containing 2.5 μM ethidium bromide, 2 mM Hepes, 9.4 mM NaCl, and 20 mM EDTA was mixed with 20 μl of 300 μM calf thymus DNA (Sigma D1501). This solution was placed into the fluorocolorimeter and it was adjusted to 100% relative intensity. Then additions of the drug in DMSO or DMSO alone were added to the tube. A relative intensity reading was taken after every addition. A change of one on this scale was considered one unit of ethidium displacement.[14-16]

F. CONCLUSIONS

Tannins are polyphenolic substances widely distributed in plants, and have been used as tanning agents, dyes and drugs. They are also ingested by humans and are commonly encountered in beverages and foodstuffs, such as tea, red wine and fruits in milligram to gram quantities.[48] While tannins exhibit various biological activities,[49,50] the discovery of anti-HIV tannins and related compounds discussed above has provided interesting new leads for further research in the development of potential anti-AIDS drugs as well as in serving as new tools for the investigation of HIV cell biology.

G. ACKNOWLEDGMENTS

This investigation was supported by grant AI-25697 from the National Institute of Allergies and Infectious Diseases awarded to K. H. Lee, Y. C. Cheng and A. J. Bodner. The authors thank those collaborators whose names are cited in references 7, 9 and 10.

REFERENCES

1. R. C. Gallo, H. Z. Stericher and S. Broder, "AIDS," Marcel Dekker, New York (1987), p. 1.
2. E. J. DeClercq, Chemotherapeutic Approaches to the Treatment of the Acquired Immune Deficiency Syndrome (AIDS), *J. Med. Chem.* 29:1561 (1986), and literature cited therein.
3. S. Gupta, Therapy of AIDS and AIDS-related Syndromes, *Trends Pharmacol. Sci.* 393 (1986).
4. Y. C. Cheng, G. E. Dutschman, K. F. Bastow, M. G. Sarngadharan and R. Y. C. Ting, Human Immunodeficiency Virus Reverse Transcriptase, *J. Biol. Chem.* 262:2187 (1987).
5. E. DeClercq, ed., "Design of Anti-AIDS Drugs," Elsevier, Amsterdam (1990).
6. M. Nishizawa, T. Yamagishi, G. Nonaka and I. Nishioka, Structure of Gallotannins in *Paeomiae radix, Chem. Pharm. Bull.* 28:2850 (1980).

7. M. Nishizawa, T. Yamagishi, G. E. Dutschman, W. B. Parker, A. J. Bodner, R. E. Kilkuskie, Y. C. Cheng and K. H. Lee, Anti-AIDS Agents 1. Isolation and Characterization of Four New Tetragalloylquinic Acids As a New Class of HIV Reverse Transcriptase Inhibitors from Tannic Acid, *J. Nat. Prod.* 52:762 (1989).

8. T. Tanaka, G. Nonaka and I. Nishioka, Tannins and Related Compounds XLII. Isolation and Characterization of Four New Hydrolyzable Tannins, Terflavins A and B, Tergallagin and Tercatain from the Leaves of *Terminalia catappa L.*, *Chem. Pharm. Bull.* 34:1039 (1986).

9. G. Nonaka, I. Nishioka, M. Nishizawa, T. Yamagishi, Y. Kashiwada, G. E. Dutschman, A. J. Bodner, R. E. Kilkuskie, Y. C. Cheng and K. H. Lee, Anti-AIDS Agents 2. Inhibitory Effects of Tannins on HIV Reverse Transcriptase and Replication in H9 Lymphocyte Cells, *J. Nat. Prod.* 53:587 (1990).

10. Y. Kashiwada, L. Huang, R. E. Kilkuskie, A. J. Bodner and K. H. Lee, *Bioorg. Med. Chem. Lett.*, submitted.

11. W. B. Parker and Y. C. Cheng, Inhibition of DNA Primase by Nucleoside Triphosphates and Their Analinofuranosyl Analogs, *Mol. Pharmacol.* 31:147 (1987).

12. M. C. Starnes and Y. C. Cheng, Cellular Metabolism of 2',3'-Dideoxycytidine, a Compound Active Against Human Immunodeficiency Virus *In Vitro*, *J. Biol Chem.* 262:988 (1987).

13. E. Baril, L. Mitchner and B. Baril, Action of Pancreatic DNase: Requirements for Activation of DNA as a Template-Primer for DNA Polymerase, *Nucleic Acid Res.* 4:2641 (1977).

14. B. C. Baguley, W. A. Denny, G. J. Atwell and B. F. Cain, Potential Antitumor Agents 34. Quantitative Relationships Between DNA Binding and Molecular Structure for 9-Anilinoacridines Substituted in the Anilino Ring, *J. Med. Chem.* 24:170 (1981).

15. B. C. Baguley, W. A. Denny, G. J. Atwell and B. F. Cain, Potential Antitumor Agents 35. Quantitative Relationships Between Antitumor (L-1210) Potency and DNA Binding for 4'-(9-Acridinylamino)methanesulfon-*m*-amisidide Analogues, *J. Med. Chem.* 24:520 (1981).

16. B. C. Baguley and E. M. Falkenhang, The Interaction of Ethidium with Synthetic Double-stranded Polynucleotides at Low Ionic Strength, *Nucleic Acid Res.* 5:161 (1978).

17. Y. Kashiwada, G. Nonaka and I. Nishioka, Tannins and Related Compounds XIII. Rhubarb (4). Isolation and Structures of New Classes of Gallotannins, *Chem. Pharm. Bull.* 32:3461 (1984).

18. G. Nonaka and I. Nishioka, Tannins and Related Compounds X. Rhubarb (2). Isolation and Structures of a Glycerol Gallate, Gallic Acid Glucoside Gallates, Galloylglucoses and Isolindleyin, *Chem. Pharm. Bull.* 31:1652 (1983).

19. G. Nonaka, I. Nishioka, T. Nagasawa and H. Oura, Tannins and Related Compounds I. Rhubarb (1), *Chem. Pharm. Bull.* 29:2862 (1981).

20. M. Nishizawa, T. Yamagishi, G. Nonaka and I. Nishioka, Tannins and Related Compounds Part 9. Isolation and Characterization of Polygalloylglucoses from Turkish Galls (*Quercus infectoria*), *J. Chem. Soc. Perkin Trans. 1* 961 (1983).

21. M. Nishizawa, T. Yamagishi, G. Nonaka and I. Nishioka, Tannins and Related Compounds Part 5. Isolation and Characterization of Polygalloylglucoses from Chinese Gallotannin, *J. Chem. Soc. Perkin Trans. 1* 2963 (1982).

22. H. Nishimura, G. Nonaka and I. Nishioka, Seven Quinic Acid Gallates From *Quercus stenophylla*, *Phytochem.* 23:2621 (1984).

23. K. Ishimaru, G. Nonaka and I. Nishioka, Gallic Acid Esters of *proto-* Quercitol, Quinic Acid and (-)-Shikimic Acid from *Quercus mongolica* and *Q. myrsinaefolia*, *Phytochem.* 25:1501 (1987).

24. G. Nonaka, M. Ageta and I. Nishioka, Tannins and Related Compounds XXV. A New Class of Gallotannins Possessing a (-)-Shikimic Acid Core from *Castanopsis cuspidata* var. *sieboldii* NAKAI (1), *Chem. Pharm. Bull.* 36:96 (1985).

25. G. Nonaka, unpublished data.

26. G. Nonaka, K. Ishimaru, T. Tanaka and I. Nishioka, Tannins and Related Compounds XVII. Galloylhamameloses from *Castanea crenata* L. and *Sanguisorba officinalis* L., *Chem. Pharm. Bull.*, 32:483 (1984).

27. T. Tanaka, G. Nonaka and I. Nishioka, Tannins and Related Compounds Part 28. Revision of the Structures of Sanguiins H-6, H-2, and H-3 and Isolation and

Characterization of Sanguiin H-11, a Novel Tetrameric Hydrolyzable Tannin, and Seven Related Tannins from *Sanguisorba officinalis*, *J. Chem. Res.* (S) 176 (1985); (M) 2001 (1985).

28. G. Nonaka, R. Sakai and I. Nishioka, Hydrolyzable Tannins and Proanthocyanidins from Green Tea, *Phytochem.* 23:1753 (1984).

29. H. Feng, G. Nonaka and I. Nishioka, Hydrolyzable Tannins and Related Compounds from *Castanea mollissima*, *Phytochem.* 27:1185 (1988).

30. T. Tanaka, G. Nonaka and I. Nishioka, Punicafolin, an Ellagitannin from the Leaves of *Punica granatum*, *Phytochem.* 24:2075 (1985).

31. G. Nonaka, M. Harada and I. Nishioka, Eugeniin, a New Ellagitannin from Cloves, *Chem. Pharm. Bull.* 28:685 (1980).

32. G. Nonaka, S. Morimoto and I. Nishioka, Elaeocarpusin, a Prototype of Geraniin from *Geranium thunbergii*, *Chem. Pharm. Bull.* 34:941 (1986).

33. T. Tanaka, G. Nonaka, I. Nishioka, K. Miyahara and T. Kawasaki, Tannins and Related Compounds Part 37. Isolation and Structure Elucidation of Elaeocarpusin, a Novel Ellagitannin from *Elaeocarpus sylvestris* var. *Ellipticus*, *J. Chem. Soc. Perkin Trans. 1* 369 (1986).

34. T. Tanaka, G. Nonaka and I. Nishioka, Tannins and Related Compounds XL. Revision of the Structure of Punicalin and Punicalagin, and Isolation and Characterization of 2-*O*-Galloylpunicalin, *Chem. Pharm. Bull.* 34:650 (1986).

35. R. Saijo, G. Nonaka and I. Nishioka, Tannins and Related Compounds LXXXIV. Isolation and Characterization of Five New Hydrolyzable Tannins from the Bark of *Mallotus japonicus*, *Chem. Pharm. Bull.* 37:2063 (1989).

36. T. C. Lin, G. Nonaka, I. Nishioka and F. C. Ho, Tannins and Related Compounds CII. Structures of Terchebulin, an Ellagitannin Having a Novel Tetraphenylcarboxylic Acid (Terchebulic Acid) Moiety, and Biogenetically Related Tannins from *Terminalia chebula* RETZ, *Chem. Pharm. Bull.* 38:3004 (1990).

37. G. Nonaka, S. Nakayama and I. Nishioka, Tannins and Related Compounds LXXXIII. Isolation and Structures of Hydrolyzable Tannins, Phillyraeoidins A-E from *Quercus phillyraeoides*, *Chem. Pharm. Bull.* 37:2030 (1989).

38. G. Nonaka, T. Sakai, T. Tanaka, K. Mihashi and I. Nishioka, Tannins and Related Compounds XCVII. Structure Revision of *C*-Glycosidic Ellagitannins, Castalagin, Vescalagin, Casuarinin and Stachyurin, and Related Hydrolyzable Tannins, *Chem. Pharm. Bull.* 38:2151 (1990).

39. T. Tanaka, G. Nonaka and I. Nishioka, Tannins and Related Compounds XLI. Isolation and Characterization of Novel Ellagitannins, Punicacorteins A, B, C, and D, and Punigluconin, from the Bark of *Punica granatum* L., *Chem. Pharm. Bull.* 34:656 (1986).

40. G. Nonaka, K. Ishimaru, R. Azuma, M. Ishimatsu and I. Nishioka, Tannins and Related Compounds LXXXV. Structures of Novel C-Glycosidic Ellagitannins, Grandinin and Pterocarinins A and B, *Chem. Pharm. Bull.* 37:2071 (1989).

41. M. Ageta, G. Nonaka and I. Nishioka, Tannins and Related Compounds LXVII. Isolation and Characterization of Castanopsinins A-H, Novel Ellagitannins Containing a Triterpenoid Glycosidic Core from *Castanopsis cuspidata* var. *sievoldii* NAKAI. (3), *Chem. Pharm. Bull.* 36:1646 (1988).

42. S. Morimoto, G. Nonaka and I. Nishioka, Tannins and Related Compounds XXXVIII. Isolation and Characterization of Flavan-3-ol Glucoside and Procyanidin Oligomers from Cassia Bark (*Cinnamomum cassia* BLUME), *Chem. Pharm. Bull.* 34:633 (1986).

43. Y. Kashiwada, G. Nonaka and I. Nishioka, Tannins and Related Compounds XLVIII. Rhubarb (7). Isolation and Characterization of New Dimeric and Trimeric Procyanidins, *Chem. Pharm. Bull.* 34:4083 (1986).

44. G. Nonaka, S. Morimoto and I. Nishioka, Tannins and Related Compounds Part 13. Isolation and Structures of Trimeric, Tetrameric, and Pentameric Proanthicyanidins from Cinnamon, *J. Chem. Soc. Perkin Trans. 1* 2139 (1983).

45. K. Ishimaru, G. Nonaka and I. Nishioka, Tannins and Related Compounds LV. Isolation and Characterization of Acutissimins A and B, Novel Tannins from *Quercus* and *Castanea* Species, *Chem. Pharm. Bull.* 35:602 (1987).

46. G. Nonaka, K. Ishimaru, K. Mihashi, Y. Iwase, M. Ageta and I. Nishioka, Tannins and Related Compounds LXIII. Isolation and Characterization of Mongolicains A and B, Novel Tannins from *Quercus* and *Castanopsis* Species, *Chem. Pharm. Bull.* 36:857 (1988).

47. T. Tanaka, S. Morimoto, G. Nonaka, I. Nishioka, T. Yokozawa, H. Y. Chung and H. Oura, Magnesium and Ammonium - Potassium Lithospermates B, the Active Principles Having a Uremia-preventive Effect from *Salvia miltiorrhiza*, *Chem. Pharm. Bull.* 37:340 (1989).
48. J. P. Brown, A Review of the Genetic Effects of Naturally Occurring Flavonoids, Anthraquinones and Related Compounds, *Mutation Res.* 75:243 (1980).
49. I. Nishioka, Chemistry and Biological Activities of Tannins, *Oriental Healing Arts Int'l. Bull.* 11:9 (1986).
50. T. Okuda, T. Yoshida and T. Hatano, Ellagitannins As Active Constituents of Medicinal Plants, *Planta Medica* 55:117 (1989).

MODE OF ACTION OF HYPERICIN AS AN ANTIRETROVIRAL AGENT AND OTHER RELEVANT FINDINGS

Daniel Meruelo[1], Steven Degar[1], Nuria Amari[1], Yehuda Mazur[2], David Lavie[2], Brandi Levin[1], and Gad Lavie[1]

From the [1]Department of Pathology and Kaplan Cancer Center, New York University Medical Center, New York, N. Y. 10016, and [2]Department of Organic Chemistry, The Weizmann Institute of Science, Rehovot 76 100, Israel

INTRODUCTION

Hypericin (Hy), 4,5,7,4',5',7'-hexahydroxy 2,2'dimethyl-meso naphthodianthrone, is a naturally occurring, polycyclic aromatic dianthraquinone (Figure 1). The molecule assumes a planar configuration and exhibits red fluorescence with an excitation peak at 495nm and emission peaks at 595nm and 645nm. It is a lipophilic compound soluble in polar organic solvents such as methanol, ethanol, acetone and DMSO. Hypericin can be extracted from plants of the *Hypericum* genus, commonly known as Saint Johnswort. Hypericin usually co-purifies with a congener, pseudohypericin, with the two forming up to 0.1% of the total dry weight of the plant. Pseudohypericin differs from hypericin by a single hydroxyl group substitution in one methyl group. Hypericin is also found in the prosthetic group of the photoreceptor of the photophobic ciliated protozoa *Stentor ceruleus* (1). The compound can also be synthesized from emodin by an alkaline condensation via a method described by Spitzner (2).

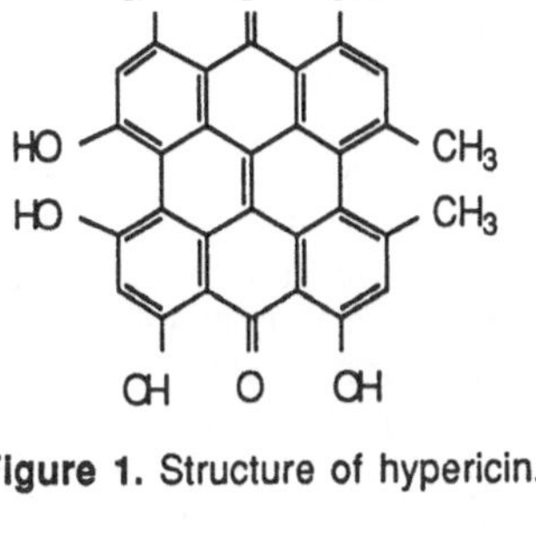

Figure 1. Structure of hypericin.

Natural Products as Antiviral Agents, Edited by C.K. Chu and H.G. Cutler, Plenum Press, New York, 1992

We have reported that hypericin has potent antiretroviral activities *in vivo* and *in vitro* (3,4). In BALB/c mice infected with Friend Virus (FV), the administration of a single dose of hypericin shortly following inoculation with the virus results in significant reduction in FV-induced splenomegaly. Hypericin also inhibits retroviral associated reverse transcriptase (RT) activity from murine Radiation leukemia virus (RadLV), FV, LPBM5 murine immunodeficiency virus, and human immunodeficiency virus (HIV) *in vitro*. The compound, however, has no inhibitory effect on the enzymatic activity of commercially purified preparations of RT (3,4). Hypericin has been shown to inhibit RT activity from HIV-infected peripheral blood lymphocyte cultures (4,5). The antiretroviral properties of hypericin have also been demonstrated against Equine Infectious Anemia Virus (EIAV), a lentivirus closely related to HIV (6). Hypericin has also been shown to inactivate murine cytomegalovirus, Sindbis virus (7-9), herpes simplex virus, influenza virus A, and Moloney murine leukemia virus (Mo-MLV); as well as simian immunodeficiency virus and human cytomegalovirus (M. Nasr, personal communication). It has been suggested that the compound may be active against all enveloped viruses (9). Hypericin is currently being studied in phase I clinical trials for the treatment of the acquired immune deficiency syndrome (AIDS) and AIDS related complex (ARC).

Following the initial description of the antiretroviral activity of hypericin by our group, a large number of additional quinone group-containing organic compounds have been analyzed for antiretroviral and antiviral activity (Meruelo and Lavie, unpublished; 5). While many of the compounds showed some degree of antiretroviral activity, none were reported to surpass the activity exhibited by hypericin (Meruelo and Lavie, unpublished; 5).

RESULTS

I. Scope of action

In pursuit of novel antiretroviral therapies for the treatment of AIDS, various stages in the retroviral life cycle have been identified as potential targets for therapeutic intervention. In contrast to the many antiretroviral compounds which target intracellular phases of the retroviral replicative cycle, the activity of hypericin is primarily directed against the retroviral particle. In a manner related to the direct inhibitory effects on retroviruses, hypericin can inhibit the production of infectious retroviral particles, and interfere with *de novo* infection of hypericin treated cells (3,4,10).

1. Effects on virus assembly or budding into the extracellular medium

Since the reverse transcriptase enzyme is a component of retroviruses, the determination of levels of RT activity in the culture supernatant of cells infected with retroviruses, can be used to evaluate the release of infectious virus particles by cells and the antiviral activity of HY. Incubation of virus producing infected lymphoblastoid cells with hypericin for even very short periods of time, leads to a drastic reduction of detectable RT activity from the growth media (3). There are two possible reasons for this reduction of activity. One is a direct effect on the virus, a possibility suggested by the rapid action of the compounds. A second possibility is an effect of the compounds on assembly and/or budding of normal virions from infected cells.

Exposure of virus producing cells for 30 minutes to hypericin at various doses (with removal of cell-free compound by extensive washing of the cells) diminishes, for at least 48 hours, subsequent production by the cells of virus particles containing measurable RT activity (4). Inhibition is complete at a concentration of 10 µg/ml (4). A similar phenomenon has been observed with cells infected with and producing murine Cytomegalovirus (MCMV) (7,8,11). In addition to its direct effects on the viral particles, hypericin can also remain bound to the cell membrane (as determined by fluorescent microscope analysis) and interfere with the production of virus particles from which reverse transcriptase activity can be retrieved. We have shown that hypericin does not affect steady state viral mRNA levels in infected cells (3), nor cleavage of polyproteins by the retroviral encoded protease (S. Degar, unpublished). Translation and transport of viral proteins or glycoproteins to the cell membrane is also unaffected (3).

2.   Pretreatment of cells with hypericin reduces the ability of viruses to establish productive infections

The effect of hypericin on the infectivity of tissue culture derived RadLV was examined after pretreatment of uninfected RL-12 cells with 1 μg/ml of hypericin, for various periods of time (Figure 2A). Little infectivity was noted when cells were pretreated 1 hour before infection. Reductions in infectivity were noted when assayed 4 and 7 days after infection, if pretreatment was for 4 hours before infection. Even pretreatments 64 hours before infection had a temporary effect 4 days post-infection.

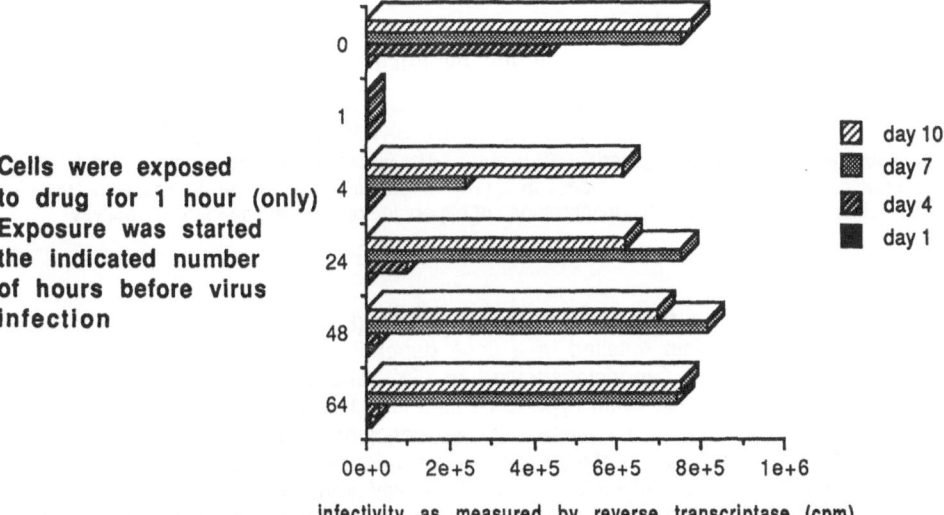

Cells were exposed to drug for 1 hour (only) Exposure was started the indicated number of hours before virus infection

infectivity as measured by reverse transcriptase (cpm)

Figure 2A. The effect of hypericin on the infectivity of tissue culture derived RadLV was examined after pretreatment of uninfected RL-12 cells with 1 μg/ml of hypericin, for various periods of time. Little infectivity was noted when virions were pretreated 1 hour before infection. Reductions in infectivity were noted when analyzed after 4 and 7 days, if pretreatment was for 4 hours before infection. Even pretreatments 64 hours before infection had a temporary effect 4 days post-infection.

3.   Direct effects on virion particles

(a) Inability to release reverse transcriptase activity from retroviral cores after hypericin treatment

To detect RT activity from retroviral particles *in vitro*, the virions must be disrupted with a non-ionic detergent to release their enzymatic activity (12,13). We compared the ability of hypericin to inhibit RT activity from intact *versus* pre-disrupted HIV. Increasing concentrations of hypericin were added to HIV and incubated. This was followed by the addition of a non-ionic detergent. Alternatively, HIV was first disrupted with detergent, and then subjected to the same concentrations of hypericin.

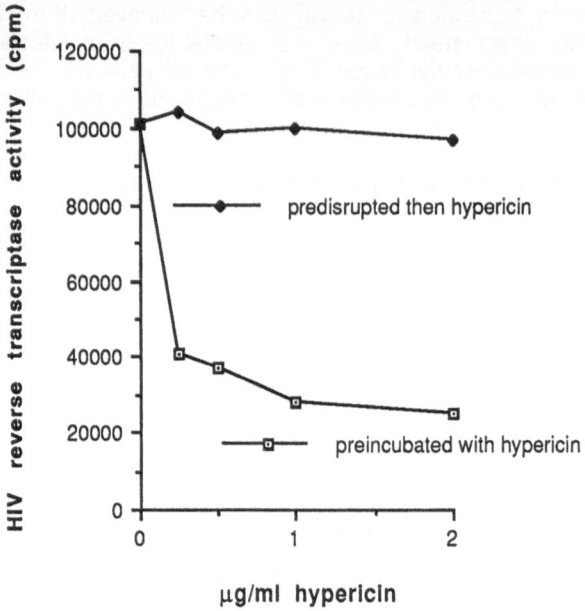

Figure 2B. Hypericin inhibits the release of RT activity from pre-treated HIV. Purified HIV (Pharmacia) was diluted with PBS to a concentration of 5×10^6 TCID$_{50}$/ml. Hypericin was added to the HIV preparations and incubated for 30 minutes. The virus was then treated with 0.5% Triton X-100 and analyzed for RT activity as described (13). In parallel, HIV was pre-treated with Triton X-100, adjusted to the same final concentrations of hypericin and then assayed for RT activity.

Both preparations were then assayed for RT activity (Figure 2B). The pre-incubation of intact HIV with hypericin resulted in a dose dependent decrease in RT activity. HIV treated with 2.0 μg/ml hypericin showed a 75% reduction in activity compared to untreated HIV. In contrast, when hypericin was added to pre-disrupted HIV, no inhibition of RT activity occurred. From these data we conclude that hypericin is not a direct inhibitor of RT activity, but rather, prevents the release of RT activity from hypericin treated virions. The partial inhibition of RT activity from pre-treated HIV may reflect the presence of disrupted HIV particles which may be present in purified preparations of retroviruses (12). A complete inhibition of RT activity results when hypericin-treated HIV is pelleted from cell culture supernatant before the RT assay (4).

Hypericin intercalates into cell membranes. In addition, hypericin binds to retroviral particles, probably by associating with the membrane-derived lipid envelope. In membranes the most plausible targets for hypericin binding are phospholipids such as phosphatidyl choline. The evidence for such binding of hypericin to phosphatidyl choline comes from two sources. First, liposomes composed exclusively of phosphatidyl choline have been found to bind and retain hypericin. In addition, hypericin bound to phosphatidyl choline exhibits optical activity (as discussed later). Since neither hypericin nor phosphatidyl choline are chiral molecules, the optical activity indicates a close interaction between the two. Virion-associated hypericin may stabilize HIV structure and prevent the non-ionic detergent disruption and release of RT activity. Such a stabilization could also prevent *de novo* infection of cells by preventing reverse transcription of the retroviral genome following attachment and entry of the virus into the host cell cytoplasm.

Such a stabilizing effect has been shown for WIN-51711 (14-16) in rhinoviruses. This agent selectively inhibits viral replication *in vivo* and *in vitro* by binding to hydrophobic domains of the viral capsid proteins, stabilizing the capsid structure and thereby preventing the uncoating process. Hydrophobic binding domains may also be present on the HIV major capsid protein (17). However, using tritium labeled hypericin, we have not been able to detect binding of hypericin to the major capsid protein (p24) of HIV. Nevertheless, and in support of the postulated indirect interaction between hypericin and the major capsid protein, Western blots of HIV inactivated with hypericin show a dose dependent shift in the mobility of p24 and the *gag* precursor polyproteins (Figure 6). Thus, while the binding of hypericin to p24 cannot be demonstrated with radiolabeled hypericin, the mobility shifts on Western blots suggest that p24 may be crosslinked or chemically altered (through an active oxygen species mediated reaction).

In order to test the effects of hypericin on the postulated rigidity of the viral capsid, we analyzed the capacity of hypericin-treated virions to bind and enter target cells, and establish a productive infection.

(b) Decreased infectivity of virus preincubated with hypericin

To determine if hypericin has a direct inhibitory effect on the replication of HIV, increasing concentrations of hypericin were added to supernatants containing HIV. These

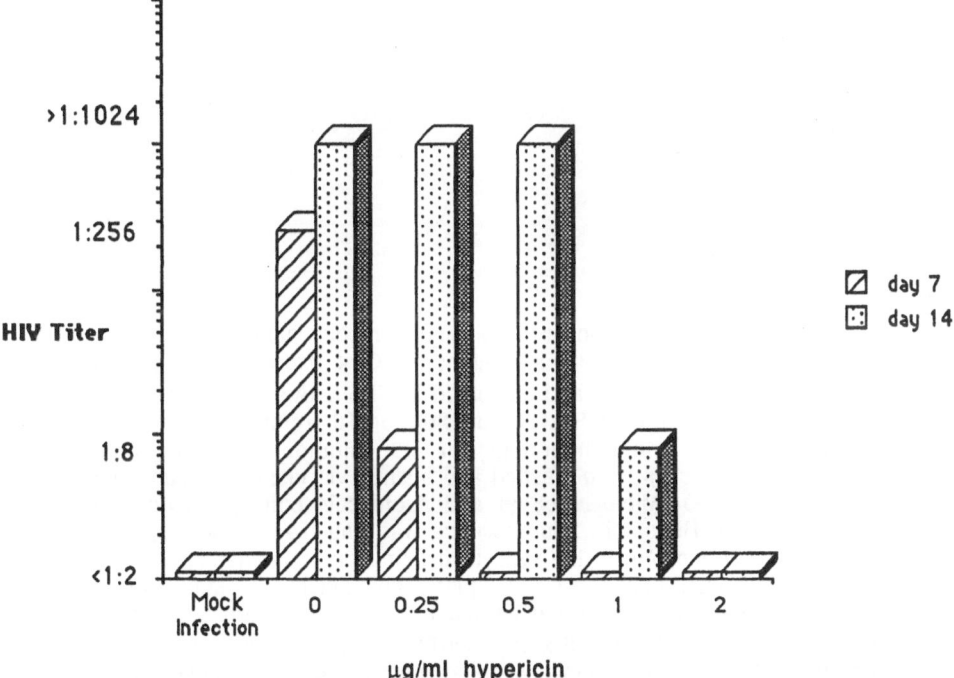

Figure 3. Incubation of HIV with hypericin inhibits infectivity and blocks the production of virus through 14 days in culture: Supernatants containing 10^5 tissue culture infective doses ($TCID_{50}$) of HIV/ml were incubated with synthetic hypericin for 30 minutes at room temperature. 10^7 CEM cells were pelleted and resuspended in the HIV supernatants with polybrene as described (33). The cells were split 1:10 with fresh media (RPMI-1640 with 10% FCS) 18 hours later. The cells were returned to culture and split twice a week with fresh medium. Cell culture supernatants were monitored for HIV production on days 7 and 14 by an enzyme linked immunoadsorbant assay (ELISA) using rabbit polyclonal antisera reactive with recombinant HIV p55. HIV titer was determined by serial dilution of the cell culture supernatants. Samples with twice the average OD_{490} of the negative control were considered positive. In all experiments presented, lyophilized hypericin was dissolved in ethanol and sonicated; small quantities were added to the samples to reach the desired final concentrations. Ethanol was added to all untreated samples as a control. The Mock infection contains heat inactivated HIV (56°C, 30 minutes).

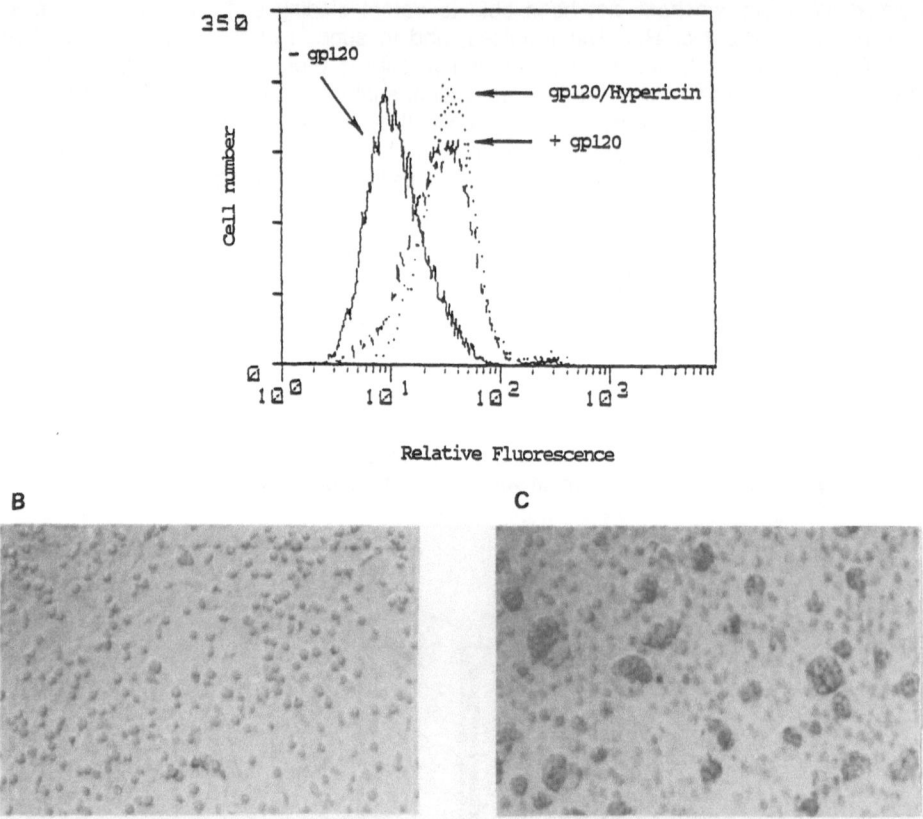

Figure 4. Hypericin does not inhibit the binding of recombinant gp120 to CD4 positive cells nor does it inhibit syncytium formation: (A) CD4 expressing SupT1 cells (2 x 10^6/sample), were resuspended and incubated in PBS/BSA-azide (- gp120); 10 µg/ml recombinant gp120 (MicroGeneSys) (+ gp120); or 10 µg/ml gp120 that had been pre-treated with 1.0 µg/ml hypericin for 30 minutes at room temperature (gp120/hypericin). The cells were washed and stained via indirect immunofluorescence using rabbit anti-gp160 (MicroGeneSys) and fluorescein conjugated goat anti-rabbit IgG (Jackson ImmunoResearch). Fluorescence was analyzed by flow cytometry on a FACScan (Becton Dickinson). In the syncytium formation experiments, HIV gp120/gp41 expressing K1 cells were plated in six well plates and cultured in Hams-F10 media with 10% FCS until confluent. Various concentrations of hypericin were added to the wells, 30 minutes before the addition of 5 x 10^5 SupT1 cells (SupT1 cells added in less than 200 µl of PBS). Syncytia were qualitatively scored 24 hours later and photographed at 100X magnification. The photographs show syncytium formation in the presence of (B) 8.0 µg/ml dextran sulfate (Pharmacia) or (C) 2.5 µg/ml hypericin. No syncytia were observed when the parental K1 cells that lack the HIV gp120/gp41 envelope proteins were co-cultured with SupT1 cells. Qualitatively there was no difference in the syncytia observed in the hypericin treated and untreated assays.

preparations were then used to infect CD4 positive cells and infection was followed by assaying the cell culture supernatants for HIV production (Figure 3). Hypericin-treated HIV has a diminished ability to infect CD4 positive cells as shown by a hypericin dose-dependent decrease in HIV titer. Seven days after infection with HIV pre-exposed to 0.25 µg/ml (0.5 µM) hypericin, HIV production was reduced by more than 95%. HIV treated with 2.0 µg/ml hypericin showed no evidence of infecting the cells after two weeks in culture. Hypericin in this dose range was non-toxic to the cells, which remained viable (data not shown). Cells which became infected with HIV exhibited syncytium formation and a loss in viability.

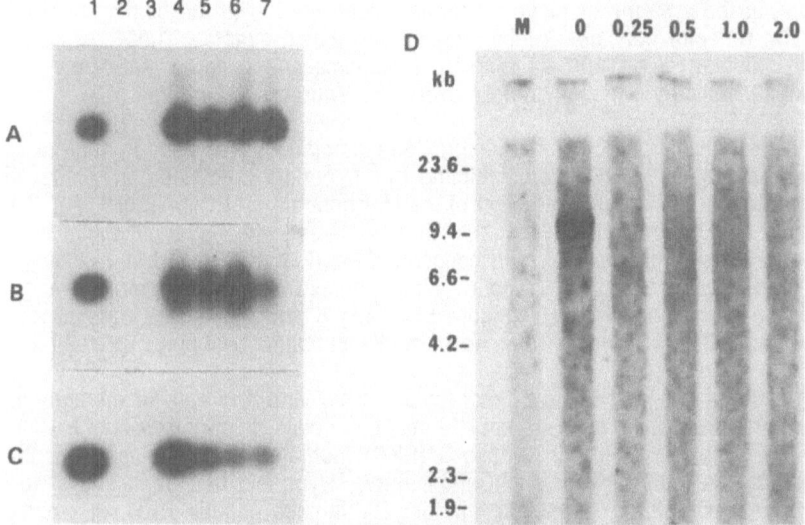

Figure 5. Polymerase chain reaction (PCR) analysis of (**A**) RNA from HY-treated HIV, (**B**) RNA from cell associated, HY-treated HIV (obtained 15 minutes post virus addition), (**C**) HIV cDNA synthesized 18 hours after infection with HY-treated HIV. Total nucleic acid preparations were subject to 25-30 PCR cycles and subject to Southern blot analysis using the SK19 probe (Perkin-Elmer Cetus HIV PCR kit). RNA samples were treated with RNAse free DNAse, then reverse transcribed (RT). Lane1= positive control (HIV DNA). Lane 2= negative control (no RT of DNAse-treated positive control or exogenous DNA). Lane 3= Mock infection (Fresh RPMI medium). Lane 4= Untreated HIV. Lane 5=HIV/0.5 μg/ml HY. Lane 6=HIV/1.0 μg/ml HY. Lane7= HIV/2.0 μg/ml HY. (**D**) Hirt extraction samples from cells infected with untreated or HY-treated HIV. Samples run on a 0.8 % agarose gel, blotted to nitrocellulose, and hybridized with a nick translated HIV probe isolated from pBH10 (Provided by the AIDS reference and reagent program). M= mock infection with heat inactivated HIV (56°C, 30 minutes).

The first event in the infection of a CD4 positive cell by HIV is the binding of the HIV envelope protein, gp120, to the CD4 receptor (19). We tested the ability of hypericin to block this interaction. The binding of recombinant HIV gp120 to cell surface CD4 can be detected by flow cytometry following staining with antisera to gp120 and a fluoresceinated antibody to immunoglobulin. Compounds such as aurintricarboxylic acid that interact with CD4 (20), and N-carbomethoxycarbonyl-propyl-phenylalanyl benzyl esters (CPFs) that interact with gp120 (22), inhibit the binding of recombinant HIV gp120 to CD4 positive cells (18,21 respectively). When hypericin is added to the gp120 binding assay, no inhibitory effect on binding occurs (Figure 4A).

In HIV particles, the gp120/gp41 envelope protein is embedded in the cell membrane-derived lipid envelope. These envelope proteins are required for the binding of HIV to a target cell, and for fusion of the viral and cell membranes, allowing the entry of the retroviral capsid into the host cell cytoplasm (22). To test the possibility that membrane associated gp120 binding and subsequent membrane fusion events may be inhibited by hypericin treatment, the effects of hypericin on syncytium formation was investigated. In the system employed (23), K1 cells that constitutively express the HIV envelope proteins gp120 and gp41 (K1/env cells), are co-cultured with CD4 positive SupT1 cells. The SupT1 cells bind and fuse with the K1/env cells, and the resulting heterokaryons continue this process, eventually forming large multinucleate syncytia. We evaluated the ability of hypericin to interfere with syncytia formation *in vitro*. Compounds such as dextran sulfate, which block the binding of HIV to CD4 positive cells (24), also block the formation of syncytia (Figure 4B). No inhibitory effect was observed when hypericin was added to the syncytia formation assay (Figure 4C). Therefore, in the systems employed, hypericin does not inhibit the binding of HIV gp120 to the CD4 receptor nor does it inhibit membrane fusion events. These results suggest that the inhibition of gp120/CD4 binding and membrane fusion events may not be a major mechanism by which hypericin exerts its antiretroviral effect.

Next polymerase chain reaction (PCR) methodology was used to analyze whether entry and uncoating of HIV were inhibited by hypericin pretreatement of the virions. Hypericin does not inhibit the detection of the HIV genome by PCR (Figure 5A). Although analysis of HIV genomic RNA suggests that treatment with hypericin might inhibit the binding of HIV to target cells at high concentrations (Figure 5B), efficient binding of HIV is observed (Figure 5B Lane 5 and 6) with hypericin doses that result in a dramatic inhibition in HIV cDNA synthesis (Figure 5C, Lane 5 and 6). These data suggest that events after retroviral binding may be primarily inhibited by hypericin treatment.

HIV enters the host cell by direct fusion of the viral and cellular membranes (25). There are also data which suggest that HIV may enter cells via endocytosis of HIV bound to the CD4 receptor (26). Protein Kinase-C (PKC) mediated phosphorylation of the CD4 receptor may be important for such virion entry. Hypericin has been shown to inhibit protein Kinase C (27). It is unclear whether HY inhibits cellular PKC, or whether this may be significant to the antiviral effects presented.

Following entry into the host cell cytoplasm, the retroviral capsid uncoats, permitting reverse transcription of the retroviral genomic RNA to cDNA (24). To determine whether hypericin-treated HIV reverse transcribes its genome, low molecular weight DNA was isolated from cells 18 hours after infection, using the Hirt extraction procedure (28). The DNAs were separated by agarose gel electrophoresis, blotted to a nitrocellulose filter, hybridized with a radiolabelled HIV probe and then examined via autoradiography (Figure 5D). Full length, linear HIV DNA was detected in the cells infected with untreated virus but was undetectable in the Hirt extraction samples obtained from cells infected with hypericin-treated HIV. These results again suggest that HIV replication is blocked prior to the reverse transcription of the HIV genome. Therefore, hypericin may interfere with reverse transcription of the viral genome directly, or indirectly by inhibiting retroviral uncoating.

The degree of uncoating required for reverse transcription of the retroviral genome is unclear. It has been shown that intact capsids isolated from EIAV possess endogenous RT activity; however, low concentrations of non-ionic detergent enhance this activity (S. Oroszlan and M. M. Roberts, personal communication; 29-30). Incubation of these capsids in vitro results in the retroviral protease-induced cleavage of the nucleocapsid protein (p11). Following p11 cleavage and reverse transcription, the resulting DNA containing capsids have altered morphologies. Following HIV attachment and entry, morphologically intact capsids have not been observed in the cell cytoplasm (31). These data support the hypothesis that conformational changes in the capsid structure are likely to occur following entry, which may be important for reverse transcription. It is unlikely that the capsid completely disintegrates, as full length retroviral DNA remains associated in a ribonucleoprotein complex that can be precipitated with some, but not all, anti-capsid sera (32).

Other retroviral components may be important for the uncoating process. Mutations in the p12 and p15 proteins of Mo-MLV cause a block in infectivity before the reverse transcription of the retroviral genome (33). It has been postulated that the uncoating process is inhibited in these mutants. Hypericin could interact with such proteins and thereby inhibit uncoating.

Western blot analysis of HY-treated HIV reveals dramatic changes in the mobility of the HIV major capsid protein, p24, and the p24 containing *gag* precursor, p55. Recombinant p24 also shifts up in apparent molecular weight following hypericin-treatment. The presence of immunoreactive material with an apparent molecular weight of 48 kilodaltons and above, suggest that cross-linking of recombinant p24 molecules may also occur. These changes require hypericin incubations in the presence of visible light. Collectively, these data suggest that hypericin may mediate the photochemical alteration of the HIV capsid. Such alterations may inhibit the detergent release of RT activity from treated virions and block infectivity by inhibiting the uncoating process.

4.   *In summary*

Although experiments targeting three different points in the retroviral replication cycle have been described (budding, direct effects on infectious particles, and pretreatment of cells before infection), a common mechanism of action could explain all the data. As reverse transcription is inhibited by direct treatment of the virus with hypericin, it is possible that virus particles budding or entering cells may be rendered incapable of reverse transcription if they come in contact with hypericin localized within the cell membrane.

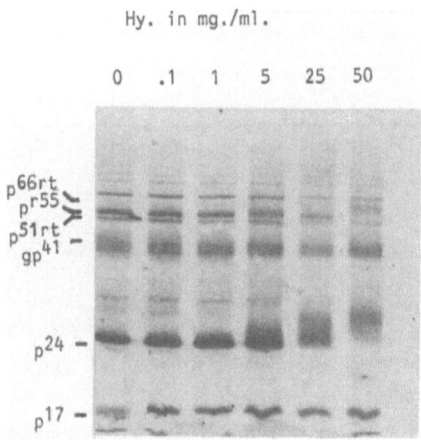

Figure 6. Western blot analysis of hypericin treated HIV. Purified HIV was treated with increasing doses of hypericin for 30 minutes, then disrupted in SDS-PAGE loading buffer and subject to Western blot analysis using antisera obtained from HIV positive donors (Provided by A. M. Prince)

II.   *The role of hypericin's photodynamic properties in its antiviral activities*

It is well documented that hypericin, a dark purple compound with red fluorescent properties, induces photodynamic reactions in biological systems (34). In the presence of light (and possibly other sources of energy) this compound excites oxygen to its singlet state and is capable of generating superoxide radicals which can lead to the oxidation of tryptophan imidazole groups in proteins and to oxidation of fatty acids in biological systems (3,4). We have previously published (10,35) that generation of excited singlet oxygen may play a significant role in the antiviral activity of hypericin *in vitro*. Pre-exposure of retroviruses to quenchers of singlet oxygen reactions, such as sodium azide, crocetin or β-carotene interferes with the anti-retroviral activity of hypericin. The inhibition of reverse transcriptase activity which occurs when the intact virus is exposed to hypericin, is markedly reduced if the singlet oxygen quencher is present prior to the administration of hypericin (Figure 7A). However, if hypericin is allowed to come in contact with virus for 30 minutes at 4°C and the singlet oxygen

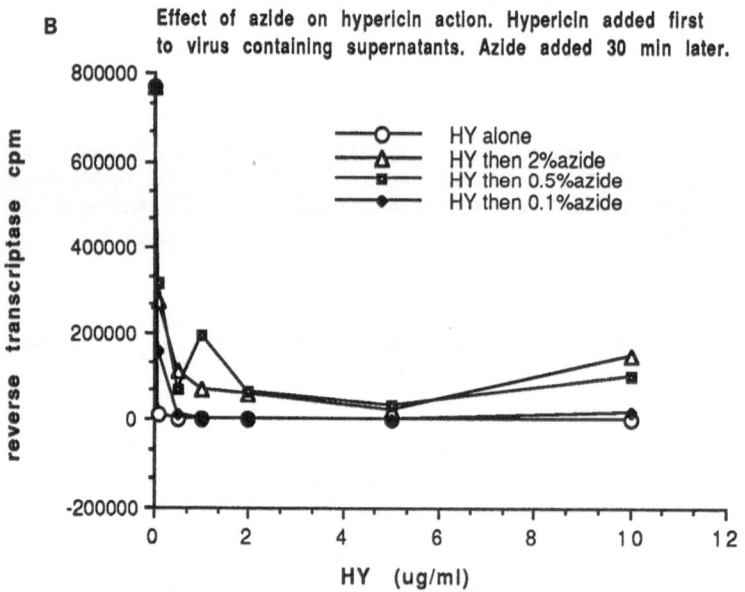

Figure 7. Effect of sodium azide on the hypericin-mediated reduction in virus-derived reverse transcriptase activity. RadLV derived from an AQR strain murine T cell leukemia cell line was as used in this experiment. Sodium azide at final concentrations of 0.1%, 0.5% and 2% was used. (A) Treatment with NaN_3 for 10 minutes and then with HY for 30 minutes. (B) Treatment of virus with HY for 30 minutes and with NaN_3 for 30 minutes subsequently. Sodium azide alone at concentrations of above 1% was inhibitory to the RT assay.

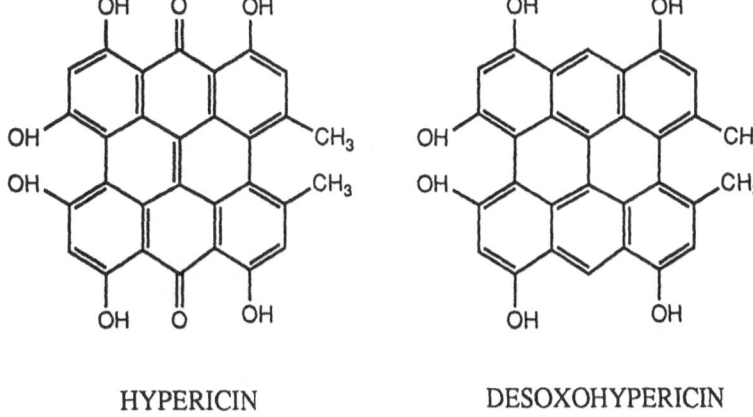

HYPERICIN   DESOXOHYPERICIN

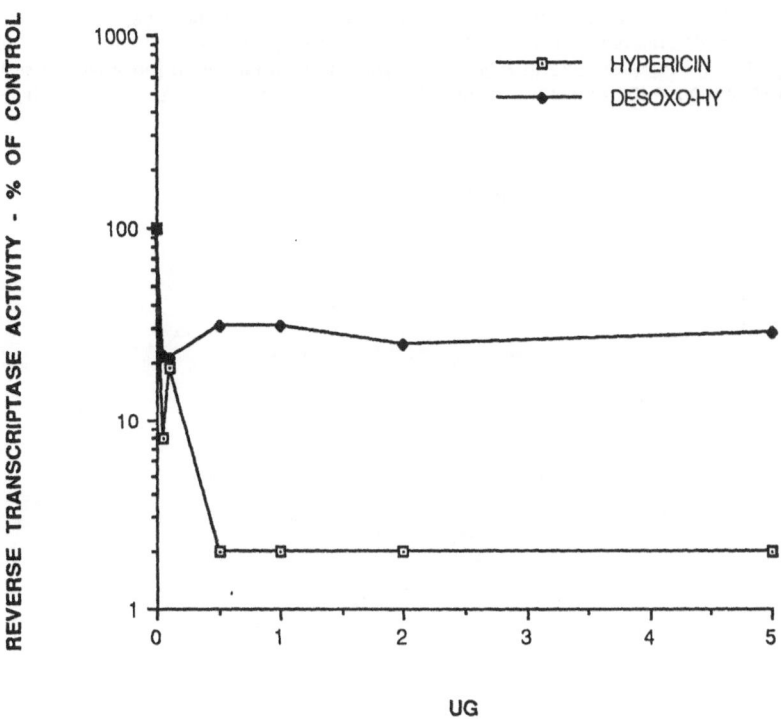

Figure 8. Comparison of the effects of hypericin and the quinone deficient analog, desoxohypericin on reverse transcriptase activity of murine Radiation Leukemia virus.

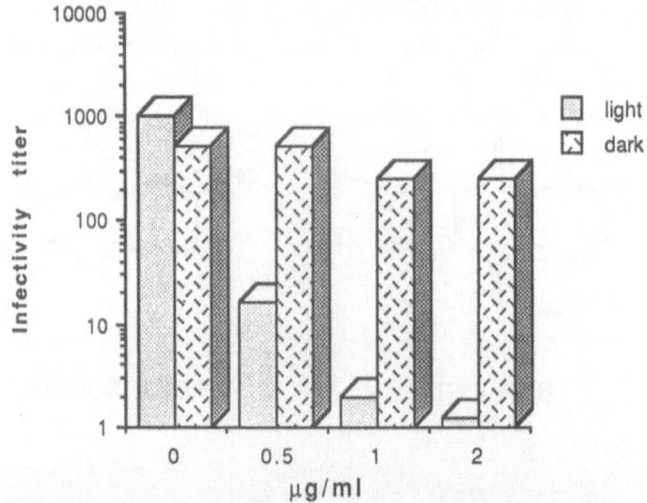

Effect of light on hypericin's ability to inactivate HIV in a direct inactivation assay.

Figure 9. Effect of light on the anti-HIV activity of hypericin. Hypericin was added to HIV in the presence or absence of light. Infection of CEM cells, and determination of HIV titers were performed as described in Figure 3. Hypericin incubations in the absence of light significantly reduce the antiviral activity of hypericin against the human immunodeficiency virus.

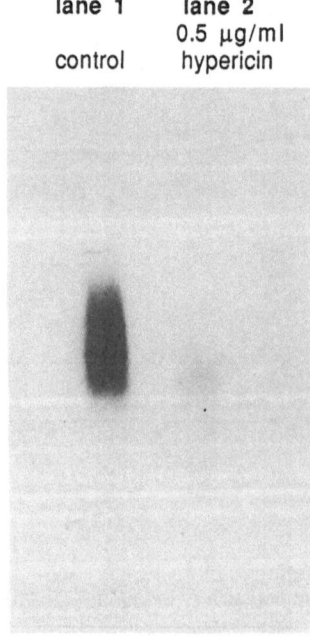

Figure 10. Extraction of RNA using guanidinium hydrochloride from virions pretreated with hypericin prior to disruption is markedly inhibited.

quencher is added subsequently, the inhibition of RT activity by hypericin remains unaffected (Figure 7B) suggesting that the hypericin effect is irreversible. We have also seen reduced antiretroviral activity in hypericin analogs, such as desoxohypericin, devoid of the quinone groups which appear to be involved in singlet oxygen generation (for example, see Figure 8 and reference 35).

Data on Figure 9 demonstrate the role that light (hence, probably generation of singlet oxygen) plays in the antiviral activity of hypericin. These findings are similar to those recently published by others (11,36). Such observations strongly suggest that the changes in the virus which are caused by hypericin may be associated with excitation of molecular oxygen to the singlet excited state by the compound. Superoxide radicals may also be involved in eliciting what appears to be increased resistance of viral particles to lysis. This can also be inferred from the fact that it is much more difficult to release RT activity from virions treated with Triton-X-100, and RNA from guanidinium hydrochloride treated virions after their exposure to hypericin (for example, see Figures 1,2 and 10).

Photodynamic reactions have been shown to occur in the dark. It is possible that such dark reactions may explain the apparent *in vivo* antiretroviral activity we observe with hypericin in the Friend virus (FV) system. In addition, Hudson *et al.* (7) working with murine cytomegalovirus and Sindbis virus have shown that while light contributes to the antiviral activity of hypericin, this compound's activity can also be measured in the dark.

III. Other effects of hypericin relevant to its antiretroviral activity

1. In-vitro toxicity of hypericin

When grazing animals, such as cattle, ingest excessive amounts of *Hypericum* plants which contain hypericin, these compounds reach the skin and induce a sunlight-mediated reaction which results in erythema and swelling. Such toxicities, although extensively documented in the literature for animals, have not been reported for humans, possibly because the quantities of plant-materials that humans have ingested have usually been much smaller. Nevertheless, the photodynamic properties of hypericin probably account for its toxicity to tissue culture grown cells (3), as it does for other cellular organisms (37). We have previously reported that this toxicity of hypericin interfered with our ability to distinguish certain antiviral activities from effects due to cellular toxicity, particularly when human cells were used (3). To address this issue we set out to evaluate the maximal hypericin dose tolerated by murine and human lymphoblastoid cells grown in culture. We also attempted to impart resistance to hypericin sensitive cells.

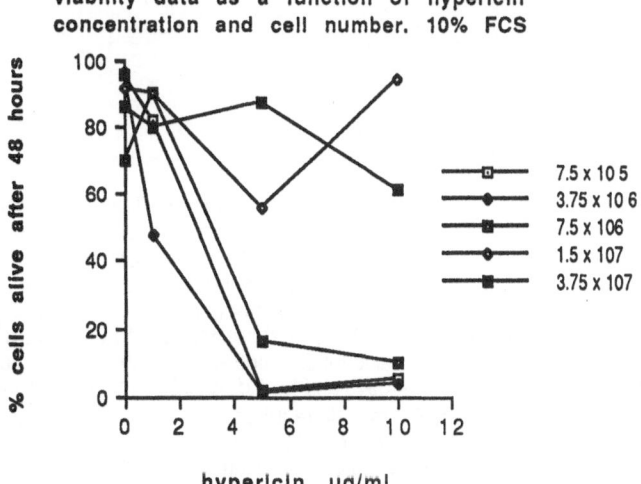

Figure 11. AQR cells were cultured at various initial cell concentrations with hypericin for 48 hours and their viabilities determined by trypan blue exclusion assay.

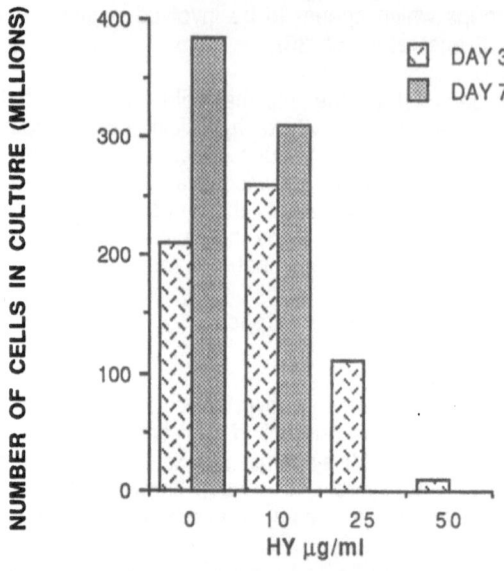

HY EFFECTS ON CELL PROLIFERATION OF MOUSE B10.T(6R) CELLS.

Figure 12. B10.T(6R)cells were incubated for either 3 or 7 days in the presence of various concentrations of HY. Numbers of viable cells were determined by trypan blue exclusion assay.

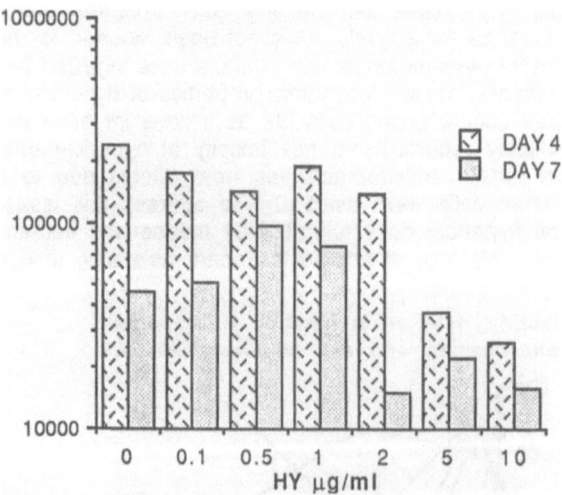

Figure 13. Cells of the H9 human lymphoblastoid cell line were brought to a concentration of 2×10^5/ml and plated in 25 cm^2 tissue culture flasks (Corning), 10 ml/flask (total of 2×10^6 cells) in medium RPMI-1640 (GIBCO) supplemented with 10% fetal calf serum (GIBCO). Synthetic hypericin was solubilized in 2% aqueous benzyl alcohol to a stock solution of 2 mg/ml. Dilutions were then made in 2% benzyl alcohol and the compound added to the culture media to yield final concentrations of 10, 5, 2, 1, 0.5 and 0.1 μg/ml of hypericin (in final benzyl alcohol concentration of 0.01%). The cultures were maintained in a 5% CO_2 incubator for 4 days and for 7 days, after which time 1 ml samples of each flask were transferred to 12x75 mm. culture tubes. One μCi of ^{3}H-thymidine was then added to each tube, incubated for 4 hours, the cells then harvested onto GF/C glass microfibre filters (Whatman) and counted in a scintillation counter for radiolabeled thymidine incorporation.

We exposed cells of the mouse thymoma derived cell lines AQR or B10.T(6R) or of the human lymphoblastoid H9 and CEM cell lines to varying concentrations of hypericin for various lengths of time and determined their viabilities either by trypan blue exclusion or by thymidine incorporation assays. Figures 11 and 12 show the effects of hypericin on mouse lymphoblastoid cell lines. The Figures show that loss of cell viabilities correlates directly with the length of the incubation time and the dose of hypericin, and correlates inversely with the initial numbers of cells cultured. Figure 13 shows the results of hypericin incubation with human, H9, cells. Cells grown for 4 days in the presence of hypericin concentrations of up to 2 µg/ml remained essentially healthy. Cell proliferation declined by seven days in all the cultures, but more substantially for cells grown at hypericin concentrations higher than 1 µg/ml.

2. Development of resistance to hypericin-mediated toxicity

In an attempt to render cells more resistant to the toxicity associated with hypericin, while retaining their usefulness in antiviral assays, we have adapted cells gradually and progressively to growth in increasing concentrations of hypericin continuously and for time periods of several months. Preliminary results suggest that over time cells can develop some resistance to hypericin's toxic activity. Thus, it is possible to adapt murine AQR lymphoblastoid cells to grow continuously in medium containing 10 µg/ml of hypericin.

We conducted preliminary studies to characterize the cellular alterations that mediate this adaptation to growth in the presence of high concentrations of hypericin. It appears that after long term exposure of cultures to high concentrations of hypericin, cell lines either take up less hypericin, or alternatively after taking up the compound, develop an increased capacity to exclude it. This can be readily demonstrated by reduced cell-associated hypericin fluorescence intensity in cell sorter analyses (Figure 14). The mechanism by which such reductions might occur resembles the expulsion of many chemotherapeutic drugs from cells (38). It may involve increased activity of cellular exclusion pumps due to induction of higher levels of expression and/or production of multidrug resistance (mdr) gene products, which are known to encode for such transport proteins (39-42). Preliminary results shown in Figure 9 are concordant with this possibility. In these experiments approximately 5-10 fold reduction in cell-associated fluorescence intensity have been observed. Further studies are underway to quantitate levels of mdr-mRNA expression in cultures resistant to the toxic effects of hypericin.

Our initial studies suggest that verapamil, a drug known to counteract the effects of increased mdr-encoded glycoprotein expression (43-45), does not reverse hypericin-induced levels of these glycoproteins (Figure 15). However, other agents might prove efficacious at reversing hypericin-induced increases in expression of mdr encoded glycoproteins.

During these studies it was noted that virus produced by the cell lines which developed relative resistance to the toxicity of hypericin continued to be inactive as measured by our reverse transcriptase enzymatic assay, even after months of exposure of the cells to hypericin. (For such an assay hypericin was removed from the media and the cells cultured overnight, so that the only hypericin present would presumably be intracellular.) However, examination of the virions by electron microscopy showed no increased number of virions with morphologically altered capsids in the hypericin treated cells versus controls. This finding differs from results we obtained earlier (4) with cells freshly treated with hypericin (i.e., not cultured continuously for months in media containing hypericin). Given the low concentrations at which hypericin is effective in inactivating certain retroviruses, it is possible that even the lower levels of hypericin remaining in the cells are sufficient for the partial effects observed on viruses. Furthermore, it remains unclear whether the possible development of reduced uptake of hypericin seen in our in vitro studies will have any impact on the blockage of virus replication which hypericin might achieve in vivo. Another consideration relevant to our observations are the findings of Gollapudi and Gupta (41), which

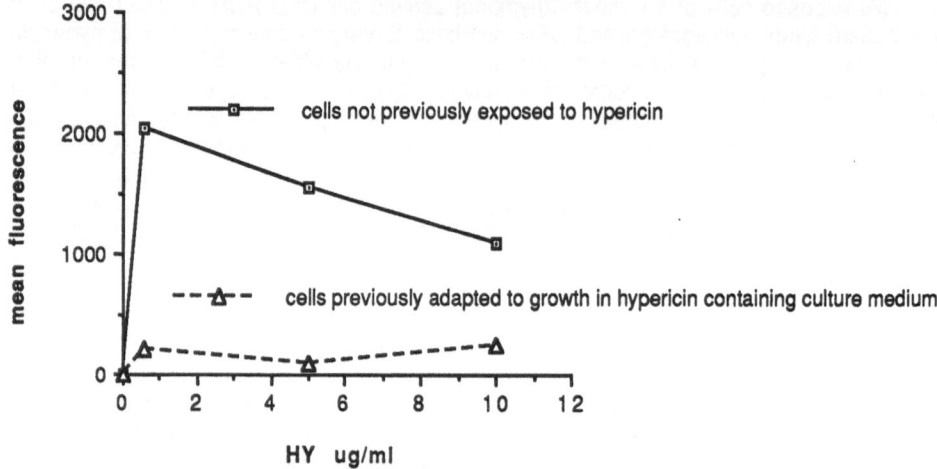

Figure 14. Reduced uptake of hypericin by cells previously exposed continuously to this compound *in vitro*. Hypericin was removed from the growth medium 24 hours before the beginning of the experiment. These cells, as well as cells not previously exposed to the compound were then pulsed with hypericin for a period of 30 minutes, washed three times, and analyzed for fluorescence uptake using a FACS analyzer. Reduced uptake of hypericin can be seen in the cultures which were previously adapted to growth in the presence of hypericin.

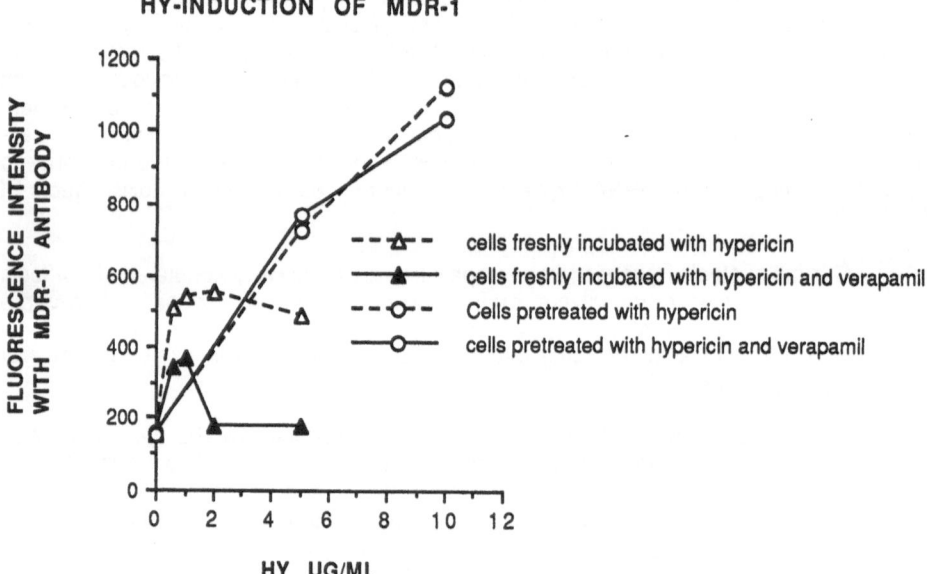

Figure 15. Cells freshly incubated with various concentrations of hypericin appear to increase their expression of mdr glycoproteins after overnight exposure to the compound. The increased expression of the glycoproteins seems to be reduced by treatment with verapamil (5 µg). Verapamil has been reported to reverse the induction of mdr expression by other therapeutic drugs, possibly increasing the time interval during which the drug therapy may continue to be effective. By contrast, cells adapted to growth in hypericin show higher levels of mdr glycoprotein expression which is not reversed by verapamil.

indicate that HIV-1 can induce expression of P-glycoproteins. Thus, both H-9 and U937 cells, upon infection with HIV, expressed increased levels of P-glycoprotein and accumulated significantly less AZT and daunorubicin as compared to uninfected cells.

3. Possible effects of hypericin on the immune system.

Recently, hypericin has been reported to inhibit protein kinase C (21). The use of hypericin in immunocompromised individuals afflicted with retroviral diseases has been discussed. Given the important role that PKC can play in a variety of biological responses, particularly immunological ones, we decided to investigate the effects of hypericin on cell surface expression of molecules important to the immune system, like CD4, which is crucial to cellular interactions via class II molecules.

Antigen-induced activation of the T lymphocyte requires binding of the antigen receptor to the combination of antigen and self-histocompatibility molecules on the surface of antigen-presenting cells. This complex interaction is followed by a series of finely orchestrated events which render the lymphocyte immunologically functional. The enzyme protein kinase C (PKC) plays a crucial role in signal transduction and proliferation initiated by the binding of antigen to the T-cell receptor complex. This membrane associated enzyme is a Ca^{++} and phospholipid-dependent protein, activated by diacylglycerol produced by the signal-induced turnover of inositol phospholipid (46-49).

Phorbol esters such as phorbol 12-myristate 13-acetate (PMA) intercalate into the cell membrane and directly activate protein kinase C, bypassing the diacylglycerol requirement. PMA has been used by several laboratories to study the physiological response of the T cell after stimulation and as a tool in the study of the differentiation block in leukemic cells as well as a means to induce differentiation of these cells. It has been shown that PMA down modulates CD4 on human cells (50, 51) and L3T4 on murine lymphocytes (52). Other observations include a loss of CD4 and CD8 antigen expression in $CD4^+$ $CD8^+$ cells with a concomitant increase of CD5 and a transient down regulation of CD3 (53). The effect of PMA on CD8 expression on human or murine T cells remains controversial (50,52,53).

An important postulate put forth to explain the modulation by PMA of cell surface antigens has been the obligatory role of PKC. In our studies hypericin was not able to reverse the effects of PMA, but if used independently or in conjunction with PMA, it was capable of lowering the cell surface expression of CD4 and CD8 on T-cell leukemias (Figure 16).

This finding could have virologic and immunologic implications since CD4 is thought to play an important role in certain immune responses as well as in HIV attachment to cells during infection. (Of equal importance is the question, not explored here, of whether hypericin, in addition to reducing levels of cell surface expression of particular receptors, might not also alter their affinities for the appropriate ligands as well.) While reduced CD4 expression may help decrease viral infection of cells in which virus entry requires binding of the virus to this cell surface molecule, preliminary evidence suggests that hypericin may also impair at least some immunological functions in mice (see below). In this respect, very preliminary results from our laboratory indicate that hypericin might have some usefulness in the treatment of certain autoimmune conditions. Given recent reports that AIDS might be associated with autoimmune components, hypericin's effects on the immune system in AIDS patients are difficult to predict (60-62).

The effects of hypericin on CD4 may be the result of interactions of this compound with cell membranes. Membrane mediated activities of the immune system involve, among others, the ability of cytotoxic T-lymphocytes (CTL) to lyse target cells.

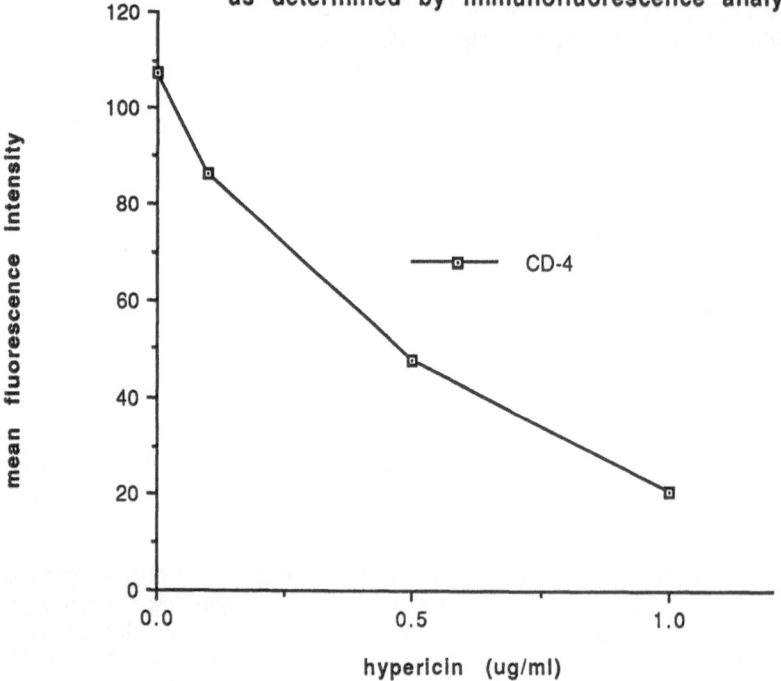

Incubation of human Supt-1 cells with hypericin results in a net decrease of CD-4 expression as determined by immunofluorescence analysis.

mean fluorescence Intensity

CD-4

hypericin (ug/ml)

Figure 16. Human SupT-1, CD4+ cells were exposed to hypericin for a period of 24 hours. The compound was then removed and the cells stained with fluorescein labeled anti-human CD4 (Leu3a) monoclonal antibodies and analyzed using a FACS analyzer. As shown, exposure of human CD4 positive cells to hypericin resulted in substantial decreases in expression of this cell surface antigen.

In order to analyze the effects of hypericin on the activity of CTL, cytotoxic T lymphocytes were generated *in vitro* by sensitization of spleen cells derived from C3H/DiSn mice (H-2^k) with stimulator spleen cells from (C3HxBALB/c)F1 mice (H-2^{kd}), irradiated with 2000 rad and propagated in culture with recombinant IL-2 (10 ng/ml) for four or eight months. The cytotoxic activity of CTL cells was assessed against P-815 murine mastocytoma target cells which express class I H-2^d, but not class II antigens, using the [51]Chromium release assay. Hypericin was added at the onset of the lytic reaction at concentrations between 0.2-10 µg/ml. The results, shown in Figure 17 indicate that hypericin interferes with the cytotoxic activity of CTL cells, and that since the effector cells were presensitized before the addition of hypericin the inhibitory effect shown in Figure 17 pertains to the lytic reaction and not to the sensitization phase.

The lytic activity of murine CTLs is associated with the concomitant release of granules containing numerous proteinases (granzymes) (54-55) and Ca^{++}-dependent pore forming proteins (perforins) (56-57). We examined the effect of hypericin on granzyme release into the reaction supernatant by examining the supernatant of the cytotoxicity assay for enzymatic hydrolysis assay of N-benzyloxycarbonyl-L-lysine thiobenzyl ester (BLT) (58). The results (Figure 18) show that, while hypericin markedly inhibited chromium release from target cells, it had no effect on secretion of granzyme-containing granules into the reaction medium.

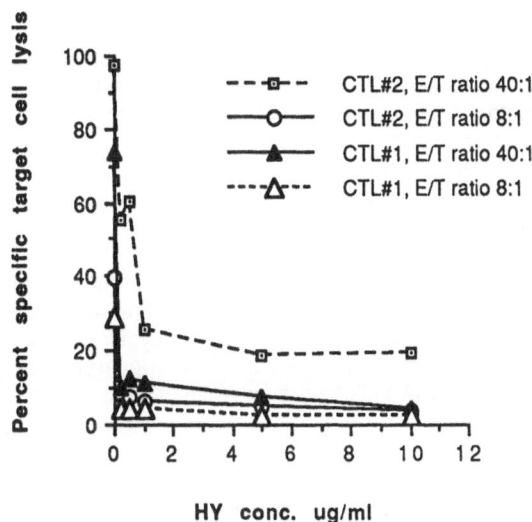

Figure 17. The effects of hypericin on the cytolytic activity of murine cytotoxic T-lymphocytes sensitized against MHC Class I antigens.

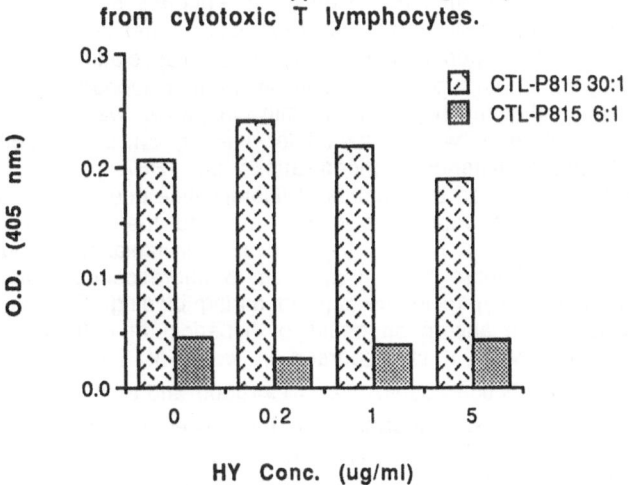

Figure 18. The effect of hypericin on the secretion of BLT-proteases from cytoplasmic granules during a CTL mediated cytotoxicity reaction. Hypericin at concentrations of 0.2, 1 and 5 μg/ml was administered at the onset of a cytotoxicity reaction. Two effector/ target cell ratios of 30:1 and 6:1 have been analyzed. Samples of the supernatant were collected after 5 hours of incubation and analyzed for hydrolysis of BLT as described in (58).

4. The interaction of hypericin with serum albumin and other proteins

It has been reported by others (5) that hypericin can inhibit HIV reverse transcriptase activity directly, provided that the assay conditions are devoid of serum albumin. This finding differs from our results, obtained when hypericin's action on purified reverse transcriptase was measured in the presence of serum. The findings of direct inhibitory activity of hypericin on reverse transcriptase activity can, however, be readily explained by the affinity that hypericin displays for many proteins. From the studies described below it is surmised that, while in the absence of serum proteins hypericin's binding might be important and lead to inactivation of certain enzymes like reverse transcriptase, the reported effect on such enzymes is not the primary nor a significant mode by which hypericin exerts its antiretroviral activity in tissue culture media or other biological fluids, in which serum is present.

The evidence for binding of hypericin to human serum albumin (HSA) is obtained from circular dichroism. Hypericin itself is not a chiral molecule and is devoid of optical activity. A circular dichroism spectral analysis of the optical rotation of hypericin bound to HSA is shown in Figure 21. The appearance of optical activity after the addition of hypericin to HSA indicates a close interaction between the protein and the polycyclic dione.

In its monomeric form hypericin exhibits red fluorescence with emission maxima at 595 and 645 nm. However, in aqueous media the compound is entirely devoid of fluorescence. (59). In aqueous media hypericin occurs as high molecular weight loosely associated aggregates which do not cross through semipermeable membranes with molecular weight cut off values of up to 100,000 daltons. These polymers were found to be dissociable in polar organic solvents as determined by shifts in absorption spectra and acquisition of fluorescence properties which occur when a polar solvent such as ethanol is applied to aqueous hypericin at a volume which exceeds 30%. Thus, the loss of fluorescence in water is associated with significant polymerization or aggregation of hypericin.

The addition of human serum albumin to aqueous suspensions of hypericin aggregates results in resumed fluorescent properties. Fluorescence is characteristic to the monomeric form of hypericin and occurs after dissociation of the aggregates, suggesting that hypericin binds to serum albumin. Further support of this notion comes from the observation that the hypericin-albumin complexes were found to transverse semipermeable membranes with a molecular weight cut off of 100,000 daltons. Titration of increasing hypericin concentrations into HSA results in increase in hypericin fluorescence followed by a phase of fluorescence quenching. A Scatchard plot analysis extrapolating the bound/free molar ratios to the 0 abscissa shows that maximum fluorescence occurs at a 1:1 hypericin/albumin molar ratio and diminishes at higher HY/HSA ratios (Figure 19). In addition to fluorescence, additional spectral absorption properties of hypericin are altered during binding. For example, the two major visible light region absorption peaks of hypericin at 590nm and 550nm with extinction coefficients when it is dispersed in water of: $\varepsilon=6000$ and $\varepsilon=10,000$, respectively, acquire extinction coefficients of $\varepsilon=46000$ and $\varepsilon=22,000$ when associated with albumin. These spectroscopic and fluorescence changes facilitate the quantitation of hypericin in biological fluids. It is important to note that the binding of hypericin to serum albumin, does not diminish hypericin's antiviral activity (if anything, it enhances it), as shown in Figure 20. There are several possible explanations for this. First, exposure of hypericin-albumin complexes to murine lymphoblastoid T cell lines *in vitro* results in the transfer of hypericin to the cell membrane. This type of behavior may be an important tool in the transport and delivery of hypericin *in vivo* to target cells and virions. Second, serum albumin seems to help hypericin remain in its most biologically active state, the monomeric form.

In addition to its binding to serum albumin, we have found that hypericin binds to lipoproteins. Agarose gel electrophoresis of serum treated with hypericin reveals the migration of the red fluorescent compound also with low density lipoproteins (LDL) and

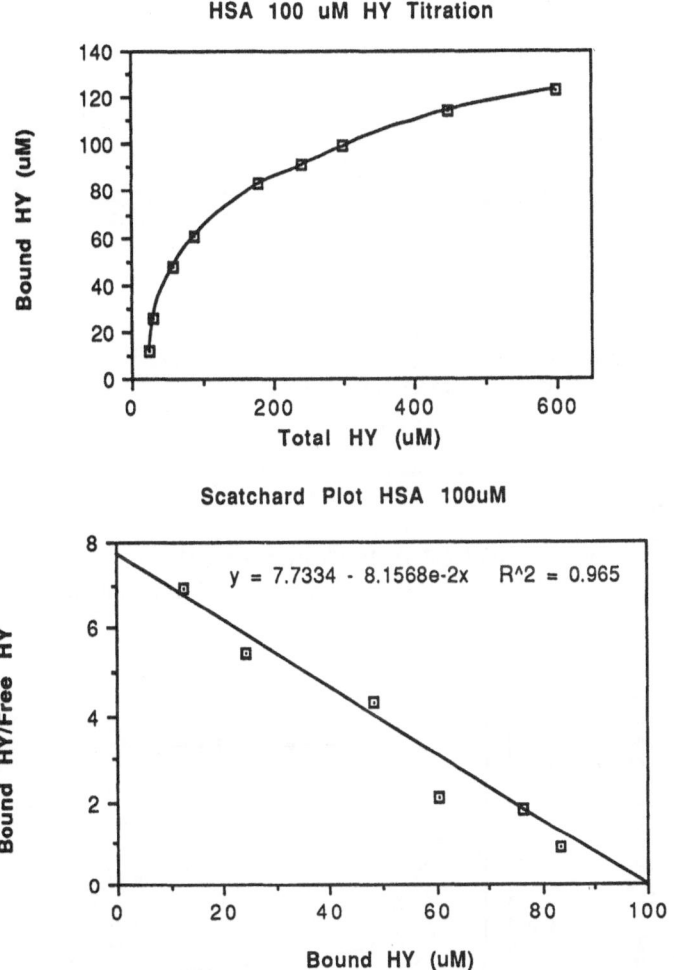

Figures 19. Scatchard plot analysis of the binding of hypericin to human serum albumin. Extrapolation of the bound/free molar ratio to the 0 abscissa intercept at 94.8 μM indicates essentially a 1:1 HY:HSA binding ratio.

A

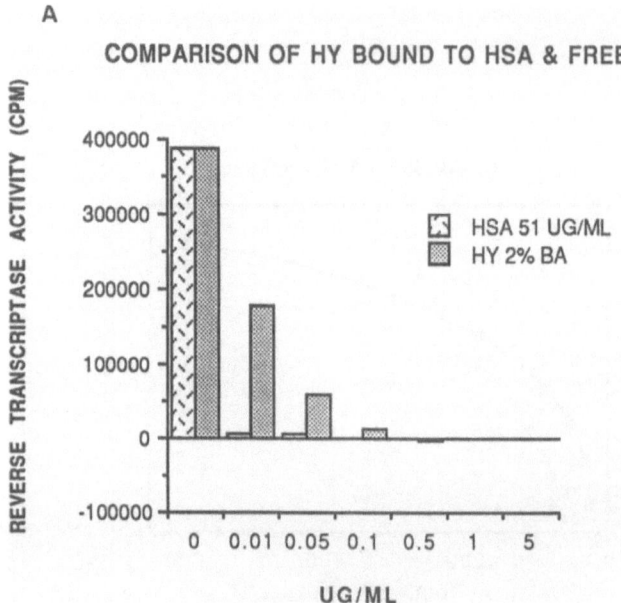

COMPARISON OF HY BOUND TO HSA & FREE

B

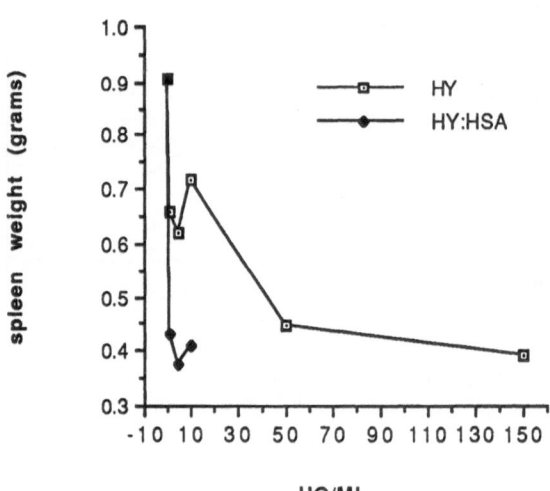

**COMPARISON OF IN VIVO ACTIVITY OF HY
VERSUS HY:HSA IN FV-INDUCED SPLENOMEGALY**

Figure 20. Comparisons of the antiviral activity of hypericin and albumin-bound hypericin. A. Analysis of the effect of HY-albumin on direct inactivation of murine RadLV *in vitro*, as measured by release of reverse transcriptase enzymatic activity. B. Effect of HY-albumin on retrovirus-induced disease *in-vivo*, as measured by the effect of the complexes on Friend-virus induced splenomegaly in BALB/c mice.

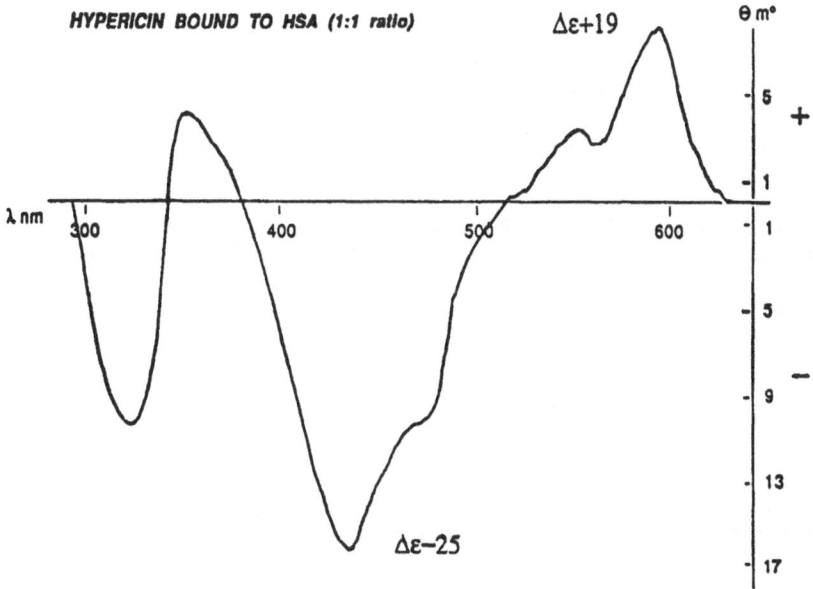

Figure 21. Circular dichroism spectrum of hypericin after the addition of the compound to HSA. Δε = amplitude of Cotton effect. Θ = angle of rotation expressed in millidegrees.

with high density lipoproteins (HDL). However, the large molar ratios of binding of hypericin to LDL and to HDL compared to HSA seem to suggest that in the case of lipoproteins binding may occur with the phospholipid shell of lipoprotein particles. The evidence for such binding of hypericin to phosphatidyl choline comes from analyses of the optical activity of hypericin in phosphatidyl choline. A circular dichroism spectral analysis of the optical rotation of hypericin bound to phosphatidyl choline is shown in Figure 22. Since neither hypericin nor phosphatidyl choline are chiral molecules the optical activity indicates a close interaction between the two.

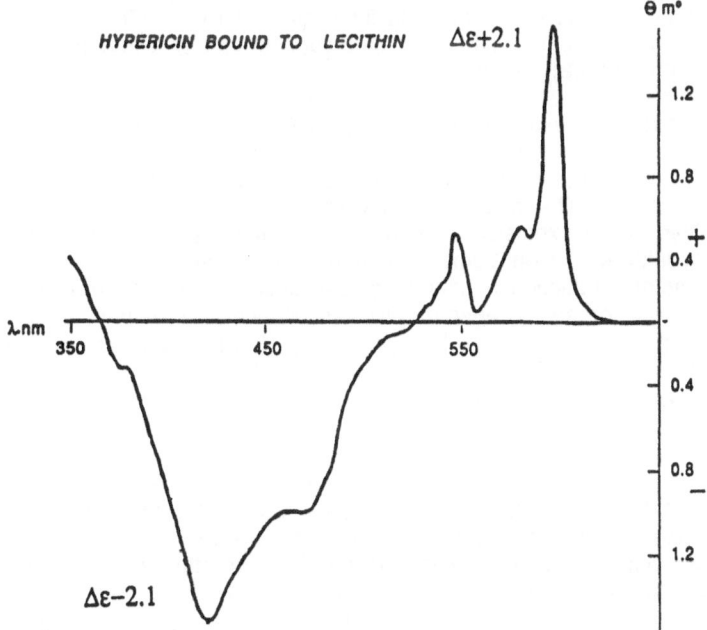

Figure 22. Circular dichroism spectrum of hypericin which is induced upon binding to phosphatidyl choline. Δε = amplitude of Cotton effect. Θ = angle of rotation expressed in millidegrees.

DISCUSSION

Hypericin, a polycylic aromatic compound with a naphthodianthrone skeleton originally isolated from the common herb Saint Johnswort, inhibits the replication of many viruses and retroviruses, *in vivo* and *in vitro*. Of particular importance is that hypericin has been shown to inactivate HIV *in vitro*. Incubation of HIV with non-cytotoxic concentrations of hypericin renders the virus non-infectious. Hypericin does not inhibit the binding of HIV gp120 to the CD4 receptor nor does it inhibit membrane fusion events. However, the quantity of reverse transcribed HIV DNA is dramatically reduced in cells infected with hypericin-treated HIV. Available Data suggest that hypericin inhibits the non-ionic detergent disruption and release of reverse transcriptase activity from pre-treated HIV, and further that hypericin stabilizes HIV structure to block infectivity by inhibiting the uncoating process. Hypericin may interact with putative hydrophobic binding domains on the HIV major capsid protein, or may mediate the photodynamic oxidation of retroviral components.

Retroviral uncoating has been a poorly understood phenomenon, but recent work indicates that it may be a complex step involving proteolytic cleavage of the nucleocapsid protein and conformational changes of the capsid structure. Compounds like hypericin that appear to inhibit uncoating, may serve as valuable tools for further elucidation of this process.

It has been reported by others (5) that hypericin inhibits reverse transcriptase activity directly. However, the addition of BSA abolished this inhibitory effect. These data suggest that non-specific interaction of hypericin with RT may occur in the absence of other proteins. It is unlikely that the direct inhibition of RT in the absence of BSA is relevant to the antiretroviral effects presented. In our studies, HIV is inactivated in the presence of 10% fetal calf serum. Also, the addition of BSA does not inhibit the photochemical alteration in mobility of the HIV major capsid protein (unpublished data).

Additional findings that may be of significance to understanding the mode of action of hypericin as an antiretroviral agent relate to the effects of light. Hypericin excites oxygen to its singlet state. These reactions may lead to stabilization of the capsid, and result in virions which are inactive because they cannot uncoat. However, sources of energy other than light may also allow the excitation of oxygen by hypericin. *In vivo* injection of hypericin into mice, following inoculation of FV, is effective in delaying the onset of FV-induced disease (3,4). Such data suggest that light may not be required for hypericin's antiretroviral activity.

After long term exposure of cultures to high concentrations of hypericin, the cells either take up less hypericin, or develop an increased capacity to exclude it. This can be demonstrated by a reduction in cell-associated hypericin fluorescence intensity in cell sorter analyses. The mechanism by which such reductions occur might be related to that involved in the expulsion of many chemotherapeutic drugs by cells. Increased activity of cellular exclusion pumps might be due to induction of higher levels of expression and/or production of multidrug resistance (*mdr*) gene products, which are known to encode for such transport proteins. Preliminary results presented in this manuscript are concordant with this possibility. Further studies are underway to quantitate levels of *mdr*-mRNA expression in resistant cultures and to determine more precisely the potential role of *mdr* encoded glycoproteins in transport of hypericin.

While the development of cellular resistance to hypericin via reduced drug uptake may have clinical implications, if it occurs in humans, there are, at the present, no correlations that would allow extrapolations from these findings obtained with a murine cell line *in vitro* to humans. Furthermore, efforts are underway by many clinicians and scientists to deal with this problem for other drugs, and if a similar situation develops for hypericin, knowledge gained from such studies may be helpful. Our initial experiments suggest that verapamil, a drug known to counteract the effects of increased *mdr*-encoded glycoprotein expression, does not reverse hypericin-induced levels of these glycoproteins. However, other agents might prove efficacious in reversing hypericin-induced increases in expression of *mdr* encoded glycoproteins.

A third novel observation is that in our studies hypericin was capable of lowering the cell surface expression of CD4 and CD8 on T-cell leukemias. At the moment these studies have been limited to *in vitro* systems. Thus, we have no evidence that this phenomenon also occurs *in vivo*. If this observation proves relevant *in vivo*, it could have virologic and immunologic implications since CD4 plays an important role in certain immune responses as well as in HIV attachment to cells during infection. Not explored here, but of equal importance is the question of whether hypericin might in fact not only reduce levels of cell surface expression of particular receptors but their affinities for the appropriate ligands as well. While reduced CD4 expression may help decrease viral infection of cells in which virus entry requires binding of the virus to this cell surface molecule, preliminary evidence suggests that hypericin may also impair at least some immunological functions in mice. This effect of hypericin may have applicability to the treatment of autoimmune diseases.

The effects of hypericin on CD4 may be the result of interactions of this compound with cell membranes. The interaction with membranes might occur because of affinity of hypericin to the glyceryl moiety of phospholipids, as it was found to bind readily to liposomes composed of phosphatidyl choline (Mazur Y., unpublished).

Membrane mediated activities of the immune system involve, among others, the activity of cytotoxic T-lymphocytes (CTL) and their ability to lyse cells which form the target to specific sensitization by clones of this cell lineage. Our preliminary results indicate that hypericin interferes with the cytotoxic activity of CTL cells at the lytic stage of the cytotoxicity reaction and not the sensitization phase. The lytic activity of murine CTLs is associated with the concomitant release of granules containing numerous proteinases (granzymes). However, hypericin does not seem to affect the levels of secretion of granzyme-containing granules. Thus, it is likely that hypericin affects the interaction of CTL and target cells directly.

In conclusion, our studies of the past few years and those more recently published by several laboratories have clearly established the antiviral activities of hypericin and related compounds. Our studies have brought to light the potential mode of action of these agents. They have also begun to uncover potential interactions of hypericin with cell membranes; possible effects of hypericin on levels and affinities of cell surface receptors like CD4, and begun to explore the effects of hypericin on the immune system. Two critical issues, which will hopefully be addressed in the near future, are whether the antiviral activities of hypericin can be harnessed to serve a useful role in patients, and whether any of the other observed effects of hypericin, such as *in vitro* toxicities, development of resistance, modulation of surface receptors, etc., prove to be either manageable or unimportant (or in some situations even desirable).

REFERENCES

1.   Walker E.B., T.Y. Lee, and P.S. Song . 1979. Spectroscopic characterization of the Stentor photoreceptors. Biochim. Biophys. Acta 587:129-144.

2 .   Spitzner D. 1977. Synthesis of protohypericin from emodin. Angew. Chem. Int. Ed. Engl. 16: 46.

3.   Meruelo D., Lavie G., and Lavie D. 1988. Therapeutic Agents with Dramatic Antiretroviral Activity and Little Toxicity at Effective Doses: Aromatic Polycyclic Diones Hypericin and Pseudohypericin. Proc. Natl. Acad. Sci. USA 85:5230-5234.

4 .   Lavie, G., Valentine, F., Levin, B., Mazur, Y., Gallo, G., Lavie, D., Weiner, D., Meruelo, D. 1989. Studies of the mechanisms of action of the antiretroviral agents hypericin and pseudohypericin. Proc. Natl. Acad. Sci. USA 86: 5963-5967.

5.   Schinazi R.F., Chu C.K., Babu J.R., Oswald B.J., Saalmann V., Cannon, D.L., Eriksson B.F.H., and Nasr M. 1990. Anthraquinones as a new class of antiviral agents against human immunodeficiency virus. Antiviral Res. 13:265-272.

6.   Kraus G.A., Pratt D., Tossberg J., and Carpenter S. 1990. Antiretroviral activity of synthetic hypericin and related analogs. Biochem. Biophys. Res. Commun. 172:149-153.

7.   Hudson, J.B., Lopez-Bazzocchi, I., and Towers, G.H.N. 1991. Antiviral activities of hypericin Antiviral Res. 15; 101-112.

8.   Wood S., Huffman J., Weber N., et al. 1990. Antiviral activity of naturally occurring anthraquinones and anthraquinone derivatives. Planta Medica; Journal of Med. Plant Res. 56: 651-652.

9.   Tang J., Colacino J.M., Larsen S.H., and Spitzer W. 1990. Virucidal activity of hypericin against enveloped and non-enveloped DNA and RNA viruses. Antiviral Res. 13:313-326.

10. Meruelo D., Degar S., Levin B., Lavie D., Mazur Y., and Lavie G. 1991. Inactivation of retroviral particles by hypericin: Possible role of oxidative reactions in the antiretroviral activity [abstract 1-291. HIV Disease: Pathogenesis and Therapy, University of Miami.

11. Lopez-Bazzocchi, I., J.B. Hudson, and G.H.N. Towers. 1991. Antiviral activity of the photoactive plant pigment hypericin. Photochem and Photobiol 54:95-98.

12. Temin, H. M. and S. Mizutani. 1970. RNA-dependent DNA polymerase in virions of Rous sarcoma virus. Nature. 226:1211-1212.

13. Hoffman, A. D., B. Banapour, J. A. Levy. 1985. Characterization of the AIDS associated retrovirus reverse transcriptase and optimal conditions for its detection in virions. Virology. 147:326-335.

14. McSharry, J. J. , L. A. Caliguiri, H. J. Eggers. 1979. Inhibition of uncoating of Poliovirus by Arildone, a new antiviral drug. Virology. 97:307-315.

15. Otto, M. J. et al. 1985. Antimicrob. Agents Chemother. 27:883.

16. Smith, T. J. M.J. Kremer, M. Luo, G. Vriend, E. Arnold, G. Kamer, M.G. Rossmann, M. McKinlay, G.D. Diana, and M.J. Otto. 1986. The site of attachment in Human Rhinovirus-14 for antiviral agents that inhibit uncoating. Science. 233:1286-1293.

17. Rossmann, M. G. 1988. Antiviral agents targeted to interact with viral capsid proteins and a possible application to Human Immunodeficiency Virus. Proc. Natl. Acad. Sci. USA 85:4625-4627.

18. Degar S., G. Lavie, B. Levin, Y. Mazur, D. Lavie, D. Pascual, A. Prince, and D. Meruelo. 1991. Inhibition of HIV infectivity by hypericin: Evidence for a block in capsid uncoating [abstract#1-16]. HIV Disease: Pathogenesis and Therapy, University of Miami.

19. McDougal, J. S., M.S. Kennedy, J.M. Sligh, S.P. Cort, A. Maule, J.K.A. Nicholson. 1986. Binding of HTLV-III/LAV tp T4$^+$ T cells by a complex of the 110K Viral Protein and the T4 Molecule. Science. 231:382-385.

20. Schols, D., M. Baba, R. Pauwels, J. Desmyter, E. De Clercq. 1989. Specific interaction of aurintiricarboxylic acid with the human immunodeficiency virus/CD4 cell receptor Proc. Natl. Acad. Sci. U.S.A. 86:3322-3326.

21. Finberg, R. W., D.C. Diamond, D.B. Mitchell, Y. Rosenstein, G. Soman, T.C., Norman, S.L. Schreiber, and S.J. Burakoff. 1990. Prevention of HIV-1 infection and preservation of CD4 function by the binding of CPFs to gp120. Science. 249:287-291.

22. Kowalski, M., J. Potz, L. Basiripour, T. Dorfman, W.C. Goh, E. Terwilliger, A. Dayton, C. Rosen, W. Haseltine, and J. Sodroski. 1987. Functional regions of the envelope glycoprotein of Human Immunodeficiency virus type 1. Science. 237:1351-1355.

23. Walker, B. D., et al. 1987. Inhibition of Human Immunodeficiency Virus syncytium formation and virus replication by castanospermine. Proc. Natl. Acad. Sci. USA. 84:8120-8124.

24. Mitsuya, H. , D.J. Looney, S. Kuno, R. Ueno, F.W. Staal and S. Broder. 1988. Dextran Sulfate suppression of viruses in the Hiv family: inhibition of virion binding to CD4+ cells. Science. 240:646.

25. Stein, B. S., S. D. Gouda, J. D. Lifson, R. C. Penhallow, K. G. Bensch, and E. G. Engleman. 1987. pH-independent HIV entry into CD4-positive T-cells via envelope fusion to the plasma membrane. Cell 49:659-668.

26. Fields, A.P., D. P. Bednarik, A. Hess, and W.S. May. 1988. Human immunodeficiency virus induces phosphorylation of its cell surface receptor. Nature 333:278-280.

27. Takahashi I, Nakanishi S, Kobayashi E, Nakano H, Suzuki K, and Tamaoki T. 1989. Hypericin and pseudohypericin specifically inhibit protein kinase C: possible relation to their antiretroviral activity. Biochem. Biophys. Res. Commun. 165:1207-1212.

28. Hirt, B. J. 1967. Selective extraction of polyoma DNA from infected mouse cell cultures. J. Mol. Biol. 26:365-371.

29. Roberts, M. M. and S. Oroszlan. 1989. The preparation and biochemical characterization of intact capsids of Equine infectious anemia virus. Biochem. Biophys. Res. Com. 160:486-494.

30. Roberts, M. M. and S. Oroszlan. 1990. In "Retroviral Proteases: Control of Maturation and Morphogenesis." MacMillan Press, London, pp. 131-139.

31. C. Grewe, A. Beck, H. R. Gelderblom. 1990. HIV - Early virus-cell interactions. J. AIDS. 3, 965-974.

32. B. Bowerman, P. O. Brown, J. M. Bishop, H. E. Varmus. 1989. A nucleoprotein complex mediates the integration of retroviral DNA. Genes and Development 3:469.

33. Crawford, S., and Goff, S. 1984. Mutations in gag proteins p12 and p15 of moloney murine leukemia virus block early stages of infection. J. Virology. 42:909-917.

34. Yang, K.C., R.K. Prusti, E.W. Walker, P.S. Song, M. Watanabe, and M. Furuya. 1986. Photodynamic action in *Stentor coeruleus* sensitized by endogenous pigment stentorin. Photochem. Photobiol. 43:305-310.

35. Lavie, G., Y. Mazur, D. Lavie, B. Levin, I. Itach, and D. Meruelo. 1990. Hypericin as an antiretroviral agent. Mode of action and related analogs. In "AIDS: Anti-HIV Agents, Therapies, and Vaccines" St. Georgiev, V. and McGowan, J.J. (eds.), N.Y. Acad. Sci., 616:556-562.

36. Carpenter, S. and G. Kraus. 1991. Photosensitization is required for inactivation of equine infectious anemia virus by hypericin. Photochem. and Photobiol. 53:169-174.

37. Duran, N. and P. S. Song. 1986. Hypericin and its photodynamic action. Photochem. and Photobiol. 43:677-680.

38. Tsuruo, T. 1988. Mechanisms of multidrug resistance and implications for therapy. Jpn. J. Cancer Res. (Gann) 79:285-296.

39. Chaudhary, P.M., and I. B. Roninson. 1991. Expression and activity of P-glycoprotein, a multidrug efflux pump, in human hematopoeitic stem cells. Cell 66:85-94.

40. Lemontt, J.F., M. Azzarie, and P. Gros. 1988. Increased *mdr* gene expression and decreased drug accumulation in multidrug-resistant human melanoma cells. Cancer Res. 48:6348-6353.

41. Gupta, S., and S. Gollapudi. 1990. Human Immunodeficiency virus I-induced expression of P-glycoprotein. Biochem. Biophys. Res. Comm. 171:1002-1007.

42. Van der Bliek, A.M., F. Baas, T. Van der Velde-Koerts, J.L. Biedler, M.B. Meyers, R.F. Ozols, T.C. Hamilton, H. Joenje, and P. Borst. 1988. Genes amplified and overexpressed in human multidrug-resistant cell lines. Cancer Res. 48:5927-5932.

43. Atlas, D., and M. Adler. 1981. α-Adrenergic antagonists as possible calcium channel inhibitors. Proc. Natl. Acad. Sci. USA. 78:1237-1241.

44. NeyFakh, A.A., V.E. Bidneko, and L.B. Chen. 1991. Efflux in *Bacillus subtilitis*: Similarities and dissimilarities with the mammalian system. Proc. Natl. Acad. Sci. USA. 88:74781-4785.

45. Bellamy, W.T., W.S. Dalton, J. M. Kailey, M.C. Gleason, T. M. McCloskey, R.T. Dorr, and D.S. Alberts. 1988. Verapamil reversal of Doxorubicin resistance in multidrug-resistant human myeloma cells and association with drug accumulation and DNA Damage. Cancer Res. 4*:6303-6308.

46. Gardneer, P. 1989. Calcium and T lymphocyte activation. Cell 59:15-20.

47. Crabtree, G. R. 1989. Contingent genetic regulatory events in T lymphocyte activation. Science 243: 355-361.

48. Nishizuka, Y. 1986. Studies and perspectives of protein kinase C. Science 233:305-312.

49. Nishizuka, Y. 1984. The role of protein kinase C in cell surface signal transduction and tumor promotion. Nature London 308:693-698.

50. Ryffel, B., Henning, C. B., and Huberman, E. 1982. Differentiation of human T-lymphoid leukemic cells into cells that have a suppressor phenotype is induced by phorbol 12-myristate 13-acetate. Proc. Natl. Acad. Sci. USA. 79:7336-7340.

51. Solbach, W. 1982. Tumor producing phorbol esters selectively abrogate the expression of the T4 differentiation antigen expressed on normal and malignant (Sezary) T helper lymphocytes. J. Exp. Med. 156:1250-1255.

52. Wang P. T. H., Bigby, M., and Sy, M. 1987. Selective down modulation of L3T4 molecules on murine thymocytes by the tumor promoter phorbol 12-myristate 13-acetate. J. Immunol. 139:2157-2165.

53. Richie, E. R., McEntire, B., Phillips, J., and Allison, J. P. 1988. Altered expression of lymphocyte differentiation antigens on phorbol ester-activated CD4+CD8+ T cells. J. Immunol. 140:4115-4122.

54. Pasternack M.S. & H.N. Eisen .1985. A novel serine esterase expressed by cytotoxic T lymphocytes. Nature 314:743-745.

55. Lobe C.G., Brett Finlay B., Paranchych W. Paetkau V.H. & C. Bleackley. 1986. Novel serine proteases encoded by two cytotoxic T lymphocyte-specific genes. Science 232, 858-861.

56. Dourmashkin R.R., Deteix P., Simone C.B. and P.A. Henkart. 1980. Electron microscopic demonstration of lesions on target cell membranes associated with antibody-dependent cellular cytotoxicity. Clin. Exp. Immunol. 43: 554.

57. Podack E.R. and Dennert G. 1983. Assembly of two types of tubules with putative cytolytic function by cloned natural killer cells. Nature 302:442-445.

58. Vitiello, A., W.R. Heath, L.A. Sherman. 1989. Consequences of self presentation of peptide antigen by cytolytic T lymphocytes. J. Immunol. 143:1512-1517.

59. Liebes, L. Y. Mazur, D. Freeman, D. Lavie, G. Lavie, N. Kudler, S. Mendoza, B. Levin, H. Hochster, and D. Meruelo. 1991. Method for the quantitation of hypericin, an antiviral Agent, in biological fluids by high performance liquid chromatography. Anal. Biochem. 195:77-85.

60. Maddox, J. 1991. AIDS research turned upside down. Nature 353: 297.

61. Kion, T.A., and G. W. Hoffmann. 1991. Anti-HIV and anti-anti-MHC antibodies in alloimmune and autoimmune mice. Science 253:1138-1140.

62. Stott, E.J., Kitchin, P.A., M. Page, B. Flanagan, L.F. Taffs, W.L. Chan, K. H. G. Mills, P. Silvera, and A. Rodgers. 1991. Anti-cell antibody in macaques. Nature 353: 393.

ACTIVITY OF SELECTED AMARYLLIDACEAE CONSTITUENTS AND RELATED SYNTHETIC SUBSTANCES AGAINST MEDICALLY IMPORTANT RNA VIRUSES[1]

Bjarne Gabrielsen,[*,§] Thomas P. Monath,[*] John W. Huggins,[*] Jorma J. Kirsi,[†] Melinda Hollingshead,[†] William M. Shannon,[†] and George R. Pettit[‡]

[*]U.S. Army Medical Research Institute of Infectious Diseases (USAMRIID)
Fort Detrick, Frederick, MD 21702
[†]Southern Research Institute
Birmingham, AL 35255
[‡]Cancer Research Institute and Department of Chemistry
Arizona State University
Tempe, AZ 85287

THE USAMRIID ANTIVIRAL DRUG PROGRAM

RNA viruses comprising the families, *Togaviridae, Bunyaviridae, Arenaviridae* and *Filoviridae* cause hemorrhagic, encephalitic or febrile disease and represent significant health hazards to persons living in or traveling through endemic regions of the world.[2a] These viruses include: a) Japanese encephalitis (JE), yellow fever (YF) and dengue type 1-4 viruses (flaviviruses); b) Rift Valley fever (RVF), sandfly fever (Sicilian, SF), Hantaan (hemorrhagic fever with renal syndrome, HFRS) and Crimean-Congo hemorrhagic fever (CCHF) viruses (bunyaviruses); c) Venezuelan equine encephalitis virus (VEE, alphavirus); d) Lassa (LAS) and Junin (JUN) viruses (arenaviruses); and e) Ebola (EBO) and Marburg (MAR) viruses (filoviruses). In general, these viruses are associated with insect (mosquito, tick) or rodent vectors, occur primarily in Asia, Africa or Central/South America, may require an increased level of biohazard protection, and are often referred to as "exotic" viruses in the context of "tropical" medicine (Figure 1). These diseases occur world-wide but the development of appropriate antiviral chemotherapies is not considered economically feasible by the private sector. The drug discovery program of the U.S. Army Medical Research Institute of Infectious Diseases (USAMRIID) seeks to develop antiviral

[§]Current Address: National Cancer Institute, Frederick Cancer Research and Development Center, P.O. Box B, Building 427, Frederick, MD 21702-1201.

Natural Products as Antiviral Agents, Edited by C.K. Chu and H.G. Cutler, Plenum Press, New York, 1992

Virus (Disease)	Primary Illness / Complications	Geography	In vitro format	Animal Model	Biological Containment
Flavivirus: ss(+)RNA					
Yellow fever, YF	Hemorrhagic fever (high mortality)	S. America, Sub-Saharan Africa	MTT	Mouse, monkey	BL-3
Japanese encephalitis, JE	Encephalitis	Asia, Indian subcontinent	MTT	Mouse, monkey	BL-3
Dengue-4	Febrile Illness/ hemorrhagic fever	Tropical S.E. Asia, Caribbean, Africa Central / South America, Australia	ELISA	Monkey	BL-2
Alphavirus: ss(+)RNA					
Venezuelan equine encephalitis, VEE	Febrile Illness/ encephalitis	Florida, Central / South America	MTT	Mouse, hamster, monkey	BL-3
Bunyavirus: ss(-)RNA					
Rift Valley fever, RVF	Febrile Illness/ hemorrhagic fever/ encephalitis	Sub-Saharan Africa	MTT	Mouse, hamster, monkey	BL-3
Sandfly fever (Sicilian), SF	Febrile Illness	Africa, Europe (Mediterranean), Asia	MTT	(Human Volunteer Model)	
Punta Toro, PT	Febrile Illness (model for RVF)	Far East, USSR, Scandinavia, northern Eurasia	MTT	Mouse, hamster	BL-2
Hantaan (hemorrhagic fever with renal syndrome, HFRS).	Hemorrhagic fever (high mortality)		ELISA	SCID Mouse	BL-3
Crimean-Congo hemorrhagic fever, CCHF	Hemorrhagic fever (high mortality)	Eastern Europe East / West Africa, China	ELISA	Suckling mouse	BL-4
Arenavirus: ss(-)RNA					
Pichinde, PIC	None (model for arenaviruses)		ELISA	Guinea pig, hamster	BL-2
Lassa, LAS	Hemorrhagic fever (high mortality)	West Africa (Nigeria, Liberia, Sierra Leone)	ELISA	Guinea pig, monkey	BL-4
Junin (Argentine hemorrhagic fever), JUN	Hemorrhagic fever (high mortality)/ encephalitis	Argentina	ELISA	Guinea pig, monkey	BL-4
Filovirus: ss(+)RNA					
Ebola, EBO	Hemorrhagic fever (high mortality)	Africa (Zaire, Sudan)	ELISA	SCID Mouse, guinea pig, monkey	BL-4
Marburg, MAR	Hemorrhagic fever (high mortality)	Africa (Zaire, Sudan)	ELISA	SCID Mouse, guinea pig, monkey	BL-4
Poxvirus: (DNA)					
Vaccinia, VV	Vaccine complications		MTT	Mouse (tail pox and encephalitis)	BL-2

Figure 1. Antiviral drug targets in the cooperative antiviral drug development program.

chemotherapeutic agents against several of these viruses. This task is hampered by a limited knowledge of many of the diseases and viruses, the remote location of disease outbreaks with the inherent problems in conducting clinical studies and the requirement for BL3 and BL4 ("spacesuit") laboratories for biocontainment.

The U.S. Army Cooperative Antiviral Drug Development Program includes *in vitro* assays, animal models, preclinical development and clinical trials of promising new antiviral drugs against such exotic infections. The *in vitro* primary screen comprises five viruses: JE, YF, Punta Toro (PT), SF and VEE.[1] Broad-spectrum antiviral drugs, such as ribavirin, 1-(β-D-ribofuranosyl)-1,2,4-triazole-3-carboxamide, have been shown to be highly effective for the prevention or treatment of several bunyaviral and arenaviral infections, such as Lassa fever virus and hemorrhagic fever with renal syndrome (HFRS).[3] Among the bunyaviruses, SF virus causes a self-limiting disease with complete recovery, and was a major source of viral disease among Allied troops stationed in the Mediterranean during World War II. A human volunteer challenge model of SF virus allows for controlled clinical investigation. PT virus serves as a lower biohazard model for RVF virus. S-Adenosylhomocysteine hydrolase inhibitors show promise against Ebola virus (filovirus) infection.[2b] Vaccines are available to control flaviviral infections due to JE, YF and tick-borne encephalitis viruses; however, no vaccine is available to prevent dengue viral infections, the most prevalent flavivirus. With the exception of interferon inducers, there are currently few compounds that appear to be candidates for the treatment of flaviviral and alphaviral diseases.

The *in vitro* antiviral and cytotoxic effects of test compounds are measured by observing inhibition of viral cytopathic effects using an MTT-assay for all viruses except dengue-type 4 virus, which is measured by a general plaque-reduction assay.[1] Antiviral activity and cytotoxicity are reported as: a) viral inhibitory concentration, 50% (IC_{50}), the drug concentration (μg/mL) at which 50% reduction of viral cytopathic effect (CPE) is observed in triplicate test wells; b) cellular toxic concentration, 50% (TC_{50}), the drug concentration (μg/mL) that reduces cell numbers and their metabolic activity by 50%, as compared to the viability of uninfected control cells in duplicate test wells in the MTT assay; and c) the therapeutic or antiviral index, (TI), a single drug concentration measurement of the relative anticellular and antiviral effectiveness of a compound. It is calculated as a ratio of (TC_{50}/IC_{50}). Employing these parameters, the *in vitro* and *in vivo* antiviral properties of certain Amaryllidaceae constituents and related synthetic substances have been evaluated (as have plant, marine, fungal, microbiological and other sources of natural products) in attempts to identify, isolate and characterize new compounds as potential antiviral chemotherapies.

To identify extracts containing traces of such compounds, USAMRIID maintains a semi-quantitative pre-screen protocol wherein three main *in vitro* protocols are evaluated for each extract against YF, PT and VEE viruses as representatives of the respective flavi-, bunya- and alphavirus families. These parameters include the antiviral effect of an extract or compound solution in virus-infected cells and its cytotoxicity, measured in uninfected cells. The general pre-screen procedure, like the primary screen, is an MTT-based photometric assay. It relies on the reduction of the colorless MTT tetrazolium salt, [3-(4,5-dimethylthiazol-2-yl)-2,5-diphenyltetrazolium bromide] to purple-blue formazan by mitochondrial dehydrogenase enzymes in viable cells only. This reduction will not occur in dead cells. The blue formazan is quantified at 570 nm; the optical density reading being directly proportional to the number of viable cells at the end of the test period. Thus, test compound toxicity is determined as a reduction in cell viability in uninfected monolayers. Antiviral properties are measured as protection from or reduction of virus-induced cell destruction (virus cytopathic effect, CPE) in the presence of test compound across a concentration window of 1-1000 μg/mL. Both antiviral effect and cytotoxicity are determined in identical monolayers under identical test conditions. These values are expressed as: a) cellular cytotoxicity (TC_{25}), the compound concentration (µg/mL) that inhibits cellular growth and overall metabolic activity by 25%; b) antiviral inhibitory concentration (IC_{50}), the compound concentration which will prevent 50% of the cells from virus-caused cell destruction (CPE); c) selectivity index, SI, a ratio of (TC_{25}/IC_{50}) indicative of the relative antiviral effectiveness of a test sample. Extracts or compounds with SI values >1 in the initial screen (tested at log dilution) are considered sufficiently active to warrant additional tests and are subjected to further evaluation in the primary screen (tested at 1/3 log dilutions). Compounds active in the primary screen are then tested against additional viruses. Here the term "concentration," as related to crude extracts, does not represent the concentration of active species in the sample.

The pre-screen assay format is ideal for screening large numbers of samples quickly and economically against three viral family representatives. Less than 5 mg of the extract is required per three-virus assay, and five samples may be evaluated on each 96-well plate. Some false positives may occur because of the use of a low SI criteria to avoid missing active leads, or from artificially high optical densities produced by insoluble or highly colored samples. False negatives may also occur as the full-log dilutions (factors of 10) at which the solutions are tested only approximate the antiviral activity; the peak activity could easily occur between two test concentrations. In practice, false negatives are far less common than false positives. Thus, the pre-screen assay is meant simply to serve as an initial marker to identify those extracts that exhibit any antiviral activity at all; further evaluation occurs in the more quantitative primary screen.

Narciclasine, Lycoricidine and Pancratistatin

The Amaryllidaceae family comprises about 110 plant species in 85 genera; they are from tropical to subtropical and are often bulbous. Over 30 species in 11 genera have been used in the primitive treatment of cancer.[4a] Extracts of the genus *Narcissus poeticus* L. were used by the Greek physician Hippocrates of Cos (460-377 B.C.) for the treatment of cancer.[4b] In addition, the use of N. *poeticus* (now known to contain ca. 0.12 g of narciclasine per kg of fresh bulb) in the Middle East and Roman Empire was documented in the first century A.D. by Pliny the Elder.[4b] The *Narcissus* genus (daffodils) comprises 60 species, five of which have been used in cancer treatment. They are a source of a principal compound in this series, narciclasine 1, first isolated by Ceriotti[5] from *Narcissus* bulbs. The biological properties[6] as well as chemistry[7] and synthesis of the Amaryllidaceae alkaloids in general and the narciclasine- and lycorine-type compounds specifically have been extensively reviewed. Narciclasine has strong antimitotic activity and inhibits the growth of *E. coli* and sarcoma 180 tumor cells.[6] It is also a potent and specific inhibitor of peptide bond formation in eukaryotic ribosomes and binds to 60-S ribosomal subunits.[8] A structure/activity study of the inhibition of protein synthesis by narciclasine analogues concluded that *trans*-dihydronarciclasine 11a and isonarciclasine 9a were potent inhibitors similar to 1, while *cis*-dihydronarciclasine 10a and O-methylnarciclasine were nearly inactive.[8] Cytotoxicity studies in HeLa cells revealed a somewhat different order: 1, O-methyl- and *trans*-dihydronarciclasines were highly cytotoxic; *cis*-dihydronarciclasine was much less toxic; and isonarciclasine was least toxic.[6] Considerably less is known concerning the antiviral properties of narciclasines. A water-soluble extract of the bulbs of *Narcissus tazetta* L. (sacred lily), with the lycorine and pseudolycorine alkaloids removed, was found to inhibit RNA-dependent, DNA polymerase (reverse transcriptase) of Avian myeloblastosis virus.[9]

Guided by bioactivity against murine P-388 lymphocytic leukemia, Pettit and co-workers[4b,c] obtained narciclasine, its 7-deoxy analogue lycoricidine 2, and a new alkaloid, pancratistatin 4, from the Hawaiian bulbs of *Pancratium littorale* Jacq. (Amaryllidaceae), a plant that grows on marine sand dunes. (Narciclasine was also obtained from *Narcissus incomparabilis*.) Studies on the chemical reactivity and structural analysis of narciclasine have been reported by Mondon and Krohn.[10] Pancratistatin is currently one of several natural products in various clinical trials as anti-tumor agents.[11] Thus far, the chief source of **4** has been its isolation from Amaryllidaceous plants. The bulbs (45 kg) of *P. littorale* were extracted with methylene chloride-methanol-water; pancratistatin was then concentrated in a butyl alcohol extract of the aqueous phase. Purification was achieved using selective solubility properties, gel permeation chromatography (Sephadex LH-20) and recrystallization from dimethylformamide-methanol-ether to give 6.5 g (0.028%) of pancratistatin.[4b,c] Ten grams of lycoricidine were concurrently isolated with pancratistatin.[4b] Racemic lycoricidine had been previously synthesized by Ohta and Kimoto[12a] while Paulsen and Stubbe[12b,c] synthesized enantiomerically-pure (+)-lycoricidine from D-glucose. The total synthesis of pancratistatin could be achieved by restructuring existing strategies for the 7-deoxy analogues **2** and **5** to account for the 7-hydroxy functionality. Danishefsky and Lee[13] have recently reported a complex synthesis of racemic pancratistatin. Many of the stereochemical obstacles presented by a C-ring containing six stereogenic carbon centers were resolved by a series of vicinal cis functionalization reactions of double bonds. The length of the synthetic pathway (30+ steps) may preclude its use as a large-scale source of **4**. Attempts to synthesize pancratistatin from naturally-occurring narciclasine are underway.[4,13]

The chemistry involved in synthesizing additional quantities of desired analogues for biological testing also yielded interesting results. The synthetic routes described by Ohta and Kimoto[12a] were carried out with modifications for the preparation of 7-deoxypancratistatin **5**, mono-, tetra- and triacetates **6-8** and lycoricidine triacetate **3**, the latter for *in vivo* evaluation as a potential prodrug of lycoricidine **2**. Starting from O-1-tetrahydropyran-protected phenanthridone diol **13**, Ohta and Kimoto synthesized the desired triacetate **3** in seven steps (26% yield). One of the steps was a cis-hydroxylation across the C_3-C_4 double bond, which required an equimolar quantity of highly-toxic osmium tetroxide. Schubert[14] and co-workers carried out the same transformation in three steps with an overall yield of nearly 40% using only catalytic amounts (one mole percent) of osmium tetroxide as regenerated by N-methylmorpholine N-oxide, according to the procedure reported by Van Rheenen and colleagues.[15] The disposal or detoxification of larger amounts of osmium tetroxide was thus minimized.

Table 1. Antiviral (flavivirus/bunyavirus) screening data for narciclasine and 7-deoxy analogues, 1-3.

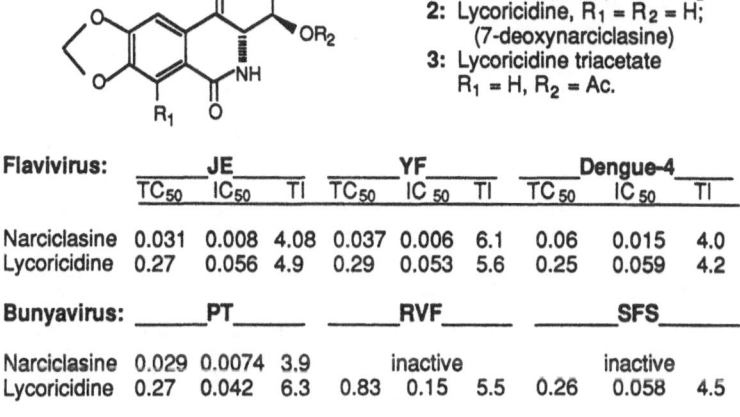

1: Narciclasine, R_1 = OH; R_2 = H
2: Lycoricidine, R_1 = R_2 = H;
   (7-deoxynarciclasine)
3: Lycoricidine triacetate
   R_1 = H, R_2 = Ac.

Flavivirus:	JE			YF			Dengue-4		
	TC_{50}	IC_{50}	TI	TC_{50}	IC_{50}	TI	TC_{50}	IC_{50}	TI
Narciclasine	0.031	0.008	4.08	0.037	0.006	6.1	0.06	0.015	4.0
Lycoricidine	0.27	0.056	4.9	0.29	0.053	5.6	0.25	0.059	4.2

Bunyavirus:	PT			RVF			SFS		
Narciclasine	0.029	0.0074	3.9	inactive			inactive		
Lycoricidine	0.27	0.042	6.3	0.83	0.15	5.5	0.26	0.058	4.5

Narciclasine **1**, lycoricidine **2**, pancratistatin **4** and the synthetic analogue lycoricidine triacetate **3** were furnished by one of us (GRP) and represent the first members of the Amaryllidaceae alkaloids to be evaluated in the USAMRIID RNA-viral screen. They were evaluated *in vitro* against JE, YF, dengue-4, SF, PT, RVF, VEE and HIV-1 viruses. Positive antiviral screening data are provided in Table 1. Antiviral activity for **1** and **2** was observed only against members of two viral families, the Flaviviridae and Bunyaviridae. Both narciclasine and lycoricidine demonstrated activity against the flaviviruses: JE, YF and dengue-4, with therapeutic indices of 4-6; such values indicating a small therapeutic margin between efficacy and cytotoxicity. Only the 7-deoxy analogue, lycoricidine, was active against the bunyaviruses, and yielded similar values for TC_{50} and IC_{50}. Narciclasine was active only against PT virus; inactive against RVF and SF viruses. Triacetate **3** may be considered a prodrug of lycoricidine, hence, it is not surprising that it did not exhibit *in vitro* activity. No antiviral activity was exhibited by any of the compounds in this study against the lentivirus, HIV-1, the alphavirus, VEE, or the DNA-vaccinia virus, VV.[1] This pattern of activity was repeated for all members of this group of Amaryllidaceae constituents and synthetic analogues. Following the antiviral evaluations of **1-3**, pancratistatin **4**, its synthetic 7-deoxy analogue **5**, and acetates **6-8** were obtained or synthesized and evaluated. Their resulting antiviral *in vitro* activity data are presented in Table 2.

An examination of the *in vitro* activity of pancratistatin and its 7-deoxy analogue revealed activity against the three flaviviruses, JE, YF and dengue-4; therapeutic indices were similar (4-8) to those observed for narciclasine and its 7-deoxy analogue, lycoricidine. The presence or absence of the 7-hydroxy (phenolic) substituent in the phenyl A-ring had a greater effect on IC_{50} and TC_{50} values in the pancratistatin than in the unsaturated narciclasine structure. Specifically, the absence of the 7-hydroxy substituent in lycoricidine **2** increased both the TC_{50} and IC_{50} values by factors of 7- to 9-fold over narciclasine **1** in the JE and YF assays. However, in the pancratistatin **4** / 7-deoxypancratistatin **5** series, the absence of the 7-hydroxy substituent in the A-ring resulted in a comparable increase of 22

Table 2. Antiviral (flavivirus/bunyavirus) screening data for pancratistatin **4**, 7-deoxy-pancratistatin **5** and acetates **6-8**.

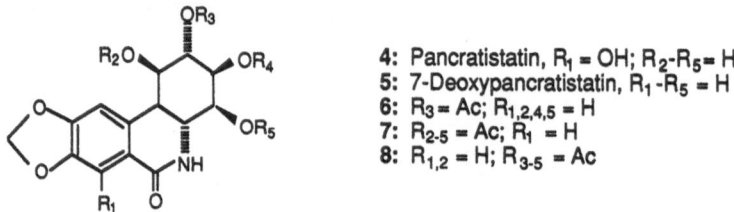

4: Pancratistatin, R_1 = OH; R_2-R_5= H
5: 7-Deoxypancratistatin, R_1 -R_5 = H
6: R_3 = Ac; $R_{1,2,4,5}$ = H
7: R_{2-5} = Ac; R_1 = H
8: $R_{1,2}$ = H; R_{3-5} = Ac

Flavivirus:		JE			YF			Dengue-4		
		TC_{50}	IC_{50}	TI	TC_{50}	IC_{50}	TI	TC_{50}	IC_{50}	TI
Pancratistatin	4	0.092	0.022	4.2	0.079	0.016	4.9	0.5	0.063	8.0
7-Deoxypancrat.	5	2.8	0.48	5.9	2.6	0.4	6.6	2.5	0.67	3.7
Monoacetate	6	10.8	3.3	3.3	21	4.8	4.3	50.0	1.5	33.3
Tetraacetate	7	2020	724	2.8	>1000	262	3.8		untested	
Triacetate	8	16.4	4.5	3.6	8.3	2.2	3.8	100	<5	>20

Bunyavirus:		PT			RVF			SF		
Pancratistatin	4	0.10	<50% inhibition		0.5	0.16	3.1	0.13	inactive	
7-Deoxypancrat.	5	2.9	0.66	4.3	21.5	5.1	4.3	4.5	1.7	2.7
Monoacetate	6	9.6	4.7	2.1	24	5.5	4.4	5.2	inactive	

to 33-fold. Thus, the substituent effect may be more pronounced in the pancratistatin structure containing a saturated polyhydroxy C-ring than in the narciclasine structure containing a C_1-C_{10b} double bond. Activity was also observed against dengue virus; however, it is more difficult to draw quantitative conclusions because the JE/YF MTT assay method differs from the plaque-reduction assay method used to measure activity against dengue virus. Nevertheless, activity against the three representative members of the Flaviviridae was observed for 1, 2, 4 and 5. Antiviral efficacy was also demonstrated by the mono-, tri- and tetraacetates 6-8 of 7-deoxypancratistatin. In general, we observed that esterifying one or more of the hydroxy groups in the C-ring of these alkaloids increased both IC_{50} and TC_{50} values, resulting in equal or less *in vitro* activity as represented by the therapeutic indices. Elevated TI values for 6 and 8 in the dengue virus assay may result from differing methods of evaluating cytotoxicity in the plaque-reduction and MTT assays.

The activity of the pancratistatin analogues against the Bunyaviridae was similar to that observed in the narciclasine series. Pancratistatin 4 was marginally active against RVF, inactive against SF and inhibited the PT virus-induced cytopathic effect by less than 50%. This marginal activity was reminiscent of that of narciclasine. 7-Deoxypancratistatin 5 exhibited *in vitro* activity against the three representative Bunyaviridae members, PT, RVF and SF viruses. Thus, the pattern of increased anti-bunyavirus activity from the 7-deoxy alkaloids was again observed. The method of evaluating cytotoxicity in the RVF plaque-reduction assay may have been responsible for the higher TC_{50} value observed for 5 in that assay. Higher IC_{50} and TC_{50} values were again observed when 7-deoxypancratistatin 5 was compared to 7-deoxynarciclasine 2; this effect of C-ring structure was identical to that observed in the flavivirus assays.

Iso-, *Cis*- and *Trans*-dihydronarciclasines

The observations of *in vitro* flavivirus and bunyavirus antiviral activity by narciclasine, pancratistatin and their 7-deoxy analogues, prompted the synthesis of the isonarciclasines 9, and the *cis*- and *trans*-dihydronarciclasines 10 and 11 containing *cis*- or *trans*-fused B and C rings about the C_{10b}-C_{4a} bond.[1] Catalytic hydrogenation of narciclasine gave a mixture of *cis*- and *trans*-dihydronarciclasines 10a, 11a (obtained via the tetraacetates) and isonarciclasine 9a.[10] Krohn and Mondon[16] also synthesized isonarciclasine by a direct route and established relative configurations at C_2, C_3 and C_4. For our assays, 7-deoxynarciclasine was hydrogenated over PtO_2. Iso-7-deoxynarciclasine 9b crystallized from pyridine-hexane (10% yield) while the *cis*- and *trans*-dihydro-7-deoxynarciclasines were converted to their triacetates, separated and hydrolyzed to yield 10b and 11b in 51% and 8% yields, respectively. Isobe and co-workers synthesized *cis*- and *trans*-dihydro-lycoricidines 10b, 11b by photolysis and hydrolysis of a lactamdiacetate.[17] Most recently, Pettit and co-workers[18] isolated *trans*-dihydronarciclasine 11a by bioassay-directed (P-388 lymphocytic leukemia) isolation from the Chinese medicinal plant *Zephyranthes candida*, wherein it was the principal cytostatic agent. The structures and antiviral activity of these analogues are represented in Table 3 (flaviviruses/bunyaviruses).

The iso-, *cis*- and *trans*-dihydro (and 7-deoxy) narciclasines 9-11 exhibit similar *in vitro* activity against the flaviviruses, JE, YF and dengue-4 to that observed with narciclasine and pancratistatin; the notable exception was the inactivity of iso-7-deoxy-narciclasine 9b against JE virus. The activity generally was modest, therapeutic indices (TC_{50}/IC_{50}) were 2.1-8.5. The high TI value of 18.5 for isonarciclasine 9a against dengue-4 virus may result from an inflated value of TC_{50} (5 µg/mL), when compared with other values (1.5, 0.9, µg/mL) in the MTT assay. Replacement of the 7-hydroxy substituent by hydrogen (i.e., 7-deoxy analogue) again resulted in an increased TC_{50} value (decreased toxicity) within each pair of compounds evaluated: 9a/9b, 10a/10b, 11a/11b. The relative order of cytotoxicity to Vero cells appears to be *trans*-dihydro- > iso- > *cis*-dihydro

Table 3. Antiviral (flavivirus/bunyavirus) screening data for iso-, *cis*- and *trans*-dihydro-narciclasines and 7-deoxy analogues, **9-11**.

9a: Isonarciclasine; R = OH
9b: Iso-7-deoxynarciclasine;
R = H

10a: cis-Dihydronarciclasine;
R_1 = OH; R_2 = H
10b: $R_1 = R_2 = H$

11a: trans-Dihydronarciclasine;
R_1 = OH; R_2 = H
11b: $R_1 = R_2 = H$

Flavivirus:

		JE			YF			Dengue-4		
		TC_{50}	IC_{50}	TI	TC_{50}	IC_{50}	TI	TC_{50}	IC_{50}	TI
Isonarciclasine	**9a**	1.5	0.72	2.1	0.9	0.22	4.1	5.0	0.27	18.5
Iso-7-deoxy-	**9b**	100	inactive		24	5.7	4.2	50	8.5	5.9
Cis-dihydro-	**10a**	4.9	0.96	5.1	5.2	1.3	3.9	>5	2.5	>2.0
Cis-dihydro-7-deoxy-	**10b**	62.5	12.7	4.9	64	9.6	6.6	25	4.4	5.7
Trans-dihydro-	**11a**	0.025	0.004	5.6	0.027	<0.003	>8.5	0.063	0.015	4.2
Trans-dihydro-7-deoxy-	**11b**	0.22	0.039	5.6	0.28	0.037	7.5	2.5	0.5	5.0

Bunyavirus:

		PT			RVF			SF		
		TC_{50}	IC_{50}	TI	TC_{50}	IC_{50}	TI	TC_{50}	IC_{50}	TI
Isonarciclasine	**9a**	1.4	0.28	5.1	25	3.3	7.6	0.72	inactive	
Iso-7-deoxy-	**9b**	26.2	7.2	3.7	50	10	5.0	17.6	inactive	
Cis-dihydro-	**10a**	8.0	2.2	3.6	5.0	1.4	3.6	8.0	<50% inhibition	
Cis-dihydro-7-deoxy-	**10b**	68	14	4.8	not tested			73	25	3.0
Trans-dihydro-	**11a**	0.026	0.008	3.3	not tested			0.027	inactive	
Trans-dihydro-7-deoxy-	**11b**	0.34	0.057	5.9	0.5	0.25	2.0	0.25	inactive	

narciclasines. This order appears to differ somewhat from that observed in HeLa cells[6]: *trans*-dihydro > *cis*-dihydro> iso-narciclasine. The structure/activity study of the protein synthesis inhibitory properties of these alkaloids by Baez and Vasquez[8] revealed that *cis*-dihydronarciclasine is nearly inactive as an inhibitor while iso- and *trans*-dihydro-narciclasines are strong inhibitors. The fact that some antiviral activity was exhibited by the *cis*-dihydro analogues **10a,b** may signify that the narciclasines, in general, exert a specific biological effect on flaviviruses and that the antiviral activity observed may not be simply a result of decreased host cell protein synthesis. In any event, this supposition could be tested further by evaluating these compounds in suitable animal models. The fact that *cis*-dihydronarciclasine was neither cytotoxic to host cells nor an inhibitor of cellular protein synthesis could make **10a** a logical candidate for an *in vivo* study; the only limitation would be its synthetic availability.

The *in vitro* evaluation of **9-11** against bunyaviruses revealed that nearly all of these narciclasine analogues were inactive against SF virus (only **10b** was marginally active). The relative order and magnitudes of TC_{50} and IC_{50} values were similar to those observed against flaviviruses.

Additional Phenanthridone Analogues

During the synthesis of 7-deoxypancratistatin **5**, various intermediates were evaluated for antiviral activity and additional analogues were synthesized. In this manner, the *trans*-1,2-diol **12**, 1-hydroxy analogue **13** and the hitherto-unreported triol **14** were prepared. Compounds **12** and **13** were synthesized as reported by Ohta and Kimoto.[12a] Both were devoid of antiviral activity and neither was cytotoxic to host cells at 100 μg/mL. The *cis*-hydroxylation[15] of **12** by a catalytic amount of osmium tetroxide in t-butanol/aqueous acetone containing N-methylmorpholine-N-oxide gave the 1,2-*trans*, 2,3-*cis*-triol **14**.[1] It also was devoid of antiviral activity and was not cytotoxic at 320 μg/mL.[1] Triol **14** resembles 7-deoxypancratistatin **5** except for the absence of a 4-hydroxy substituent and the epimerization of the 3-hydroxy substituent. If one assumes that the latter has no effect on antiviral efficacy, then one can conclude that the presence of a 4-hydroxy substituent is a prerequisite for the *in vitro* antiviral activity observed with 7-deoxypancratistatin.

12   **13**   **14**

Lycorines and Pretazettine

Lycorine **15**, the most abundant alkaloid of the Amaryllidaceae family, has been found in more than 30 genera. The first isolation of a biologically active Amaryllidaceae constituent was that of lycorine in 1877.[19] Pretazettine **17** was isolated by Furusawa[20] from *Narcissus tazetta*. For this study, lycorine was isolated concurrently with narciclasine from *Narcissus incomparabilis*. Lycorine and pseudolycorine-1-O-β-D-glucosides have been isolated from *Pancratium biflorum*[21] and are examples of the natural occurrence of glucosyloxy alkaloids in Amaryllidaceae. Numerous syntheses of lycorine have been reported, despite a major problem of the stereoselective functionalization of the hydroxylated C-ring.[22]

Lycorines and pretazettine inhibit protein synthesis by preventing peptide bond formation catalyzed by eukaryotic ribosomes. Narciclasine **1** and pretazettine **17** bind to the same area of a yeast ribosomal peptidyl transferase center while lycorine and pseudo-lycorine bind to a different site.[23] All four compounds inhibit endomyocarditis (EMC) viral RNA-directed cell-free polypeptide synthesis.[24] Lycorines and pretazettine inhibit growth of HeLa cells and, in general, exhibit varying degrees of antitumor and antiviral properties.[24]

Lycorines **15-16** and pretazettine **17** have demonstrated antiviral efficacy in cell culture and in animal models. Crude root and leaf extracts of *Clivia miniata* Regel inhibited the cytopathic effect caused by herpes, polio, coxsackie, measles and Semlicki Forest viruses in Vero cells.[25] Lycorine decreased viral protein synthesis in poliovirus-infected HeLa cells possibly at the termination stage.[26] Pretazettine inhibited action of RNA-dependent DNA polymerase (reverse transcriptase) from avian myeloblastosis[9] and from Rauscher leukemia viruses in acutely-infected cells.[27] A structure/activity study of pretazettine, narciclasine and lycorines against Rauscher leukemia virus in NIH/3T3 cells concluded that minimum toxic doses and therapeutic doses were very similar, indicating a very narrow window for prophylaxis and therapy.[28] Furusawa and co-workers[29] isolated a biologically-active compound, narcissidine, identified as 4-methoxypseudolycorine,[30] from *Narcissus tazetta*. Narcissidine is active *in vitro* against JE, vesicular stomatitis and EMC viruses, and moderately active against VV, poliomyelitis 2 and LCM viruses in KB cells.[29a] Furthermore, it prolonged the life span of mice infected with Rauscher leukemia virus[29b]

Table 4. Antiviral (flavivirus/bunyavirus) screening data for lycorines **15**, **16** and pretazettine **17**.

15: Lycorine   16: Pseudolycorine   17: Pretazettine

Flavivirus:		JE			YF			Dengue-4		
		TC_{50}	IC_{50}	TI	TC_{50}	IC_{50}	TI	TC_{50}	IC_{50}	TI
Lycorine	15	2.7	0.33	8.2	2.04	0.28	7.3	2.5	0.24	10.4
Pseudolycorine	16	1.4	0.28	5.0	1.3	0.35	3.7	1.0	0.39	2.6
Pretazettine	17	2.3	0.60	3.8	2.8	0.50	5.6		not tested	

Bunyavirus:		PT			RVF			SF		
Lycorine	15	2.3	0.50	4.6	5.0	0.93	5.4	1.4	inactive	
Pseudolycorine	16	2.3	0.60	3.9	3.8	0.63	6.0	2.5	inactive	
Pretazettine	17	2.3	0.61	3.7	10.0	2.9	3.5	2.5	0.82	3.0

and reduced the mortality of mice infected with several neurotropic RNA viruses such as EMC, JE and LCM.[29b] Administration of 150 µg of narcissidine (subcutaneously, s.c., b.i.d. x 5) 2 hr post-intracranial (i.c.) inoculation of 10,000 LD_{50} of JE virus (Nakayama strain) showed no significant difference in mortality between treated and control mice. However it did prolong survival time.[29a] Narcissidine was inhibitory to the growth of JE virus in brains of mice infected by the intraperitoneal (i.p.) route. *In vitro* antiviral data for lycorine, pseudolycorine and pretazettine are given in Table 4.

Lycorine **15**, pseudolycorine **16** and pretazettine **17** had *in vitro* activity against the flaviviruses, JE, YF and dengue-4 (pretazettine not tested against dengue-4). Values of IC_{50} and TC_{50} agreed within a factor of two. All the compounds were active against the bunyaviruses, PT and RVF, while only pretazettine showed slight activity against SF. Values of IC_{50} and TI_{50} were in excellent agreement against PT virus. As demonstrated by the narciclasine-type alkaloids, the general trend of *in vitro* activity against PT and RVF with less general activity against SF was also observed with the lycorines and pretazettine.

In Vivo Studies

As stated earlier, effective antiviral chemotherapies exist for several bunyaviral and arenaviral infections. Ribavirin has demonstrated efficacy against Junin, Machupo and Lassa viral infections (arenavirus) as well as against SF, PT, RVF, HFRS and CCHF viruses (bunyavirus) in a number of animal models including human.[2a,3] A significant limitation of ribavirin is its inability to penetrate the central nervous system (blood-brain barrier), hence its lack of efficacy against encephalitic viral disease components. No effective antiviral chemotherapies currently exist against alphavirus (VEE) or flavivirus infections though a vaccine is available for protection against VEE infectivity. Regarding the flaviviruses, a live, attenuated (17D) vaccine is available for protection against yellow fever. Candidate drugs against YF must be evaluated, however, in a rhesus monkey model as rodent models do not currently exist. Dengue viruses exist in four serologically different types which generally produce a febrile disease but are capable of causing severe hemorrhagic disease as well. Recent importations (1985) of the larvae of the Asian tiger mosquito (*Aedes albopictus*) into Gulf coast ports is a cause for concern as dengue, JE and LaCrosse encephalitis viruses are vertically transmitted through the eggs of this species, which now inhabits twenty Southern and Midwestern states. No vaccine is currently available to prevent dengue viral infections. Neither ribavirin nor rimantadine had any effect on survival numbers or times in mice inoculated (i.c.) with dengue virus type 2, while ribavirin triacetate exhibited significant increase in survival times.[2a] However, ribavirin was found to be ineffective as a prophylactic treatment of dengue type 1 viral infections in rhesus monkeys.[31] No specific antiviral chemotherapy exists for JE viral infections, although an inactivated vaccine from the Nakayama NIH strain is widely used.[32] A live, attenuated vaccine for JE is currently undergoing clinical evaluation in the People's Republic of China. Prior to the availability of a vaccine, the incidence of JE infection in northern Thailand, for example, exceeded that of poliomyelitis at its peak in the United States. The incidence of encephalitis occurs in about one per 25-500 cases of JEV infection; mortality occurs in 20-50% of the encephalitic cases.[32a] Thus, an optimum antiviral drug should be able to penetrate the central nervous system to be clinically useful.

There are few reports in the literature that describe effective *in vivo* chemotherapy of JE infections. A nuclease-resistant complex of polyriboinosinic-polyribocytidylic acid, poly-L-lysine and carboxymethylcellulose [poly(ICLC)], known to induce interferon, reduced mortality by 50% and increased survival times in four JE virus-infected monkeys when administered therapeutically 8 or 24 hr after intranasal viral challenge.[33a] A tilorone-related interferon inducer, 10-carboxymethyl-9-acridanone (CMA), showed therapeutic and prophylactic effects against JE viral infections in baby hamsters and weanling mice; the degree of protection depended upon the dose and time of administration of CMA in relation to viral challenge.[33b] The sole reports of effective treatment with a non-immunomodulator are those describing the antiviral activity of narcissidine,[29a,b] (characterized as 4-methoxy-pseudolycorine[30]). As noted earlier, narcissidine was isolated from *Narcissus tazetta* and found to be active *in vitro* against JE, EMC and vesicular stomatitis viruses and moderately active against polio 2, vaccinia and lymphocytic choriomeningitis (LCM) viruses. It also prolonged the lives of mice infected with Rauscher leukemia virus.[29b] When administered (s.c.) at 150 μg, twice daily for 5 days to mice inoculated (i.p.) with 10,000 LD_{50} of JE virus (Nakayama strain), no significant differences in mortality were observed between treated and control mice, however, survival times were prolonged.[29a] In our study, narciclasine 1, lycoricidine 2, lycoricidine triacetate 3, pancratistatin 4, 7-deoxypancratistatin 5, monoacetate 6, lycorine 15 and pseudolycorine 16 were evaluated prophylactically in a low-dose JE viral challenge mouse model. The two pancratistatin analogues 4-5 were also evaluated in a high-dose viral challenge model. The experimental details of both murine models are described elsewhere.[1] The low-dose challenge model consisted of C57Bl/6 mice (14-16 g) infected (i.p., 2 hr after the second dose of drug was administered, day 0) with 9 LD_{50} of Beijing-strain JE virus, adequate to produce 90-100% mortality in the diluent controls. Drug was administered once daily for 7 days with the first dose administered on the day (-1) preceding viral challenge. The high-dose (LD_{50}) challenge model differed from the lowdose model with respect to ages of mice, viral challenge, and drug treatment schedule as follows: a) mice were younger, 12-14 g; b) 100 LD_{50} of JE virus was administered s.c. adequate to produce 100% mortality in the diluent controls; c) drugs were administered twice daily (b.i.d.) for 9 days beginning on the day preceding viral challenge (day -1).

Narciclasine 1 was efficacious when administered prophylactically (s.c. in 2% ethanol-saline) in the low-dose challenge model (see Table 5). While it was lethally toxic to 60-80% of the mice at 6 mg/kg/day, narciclasine at 3 mg/kg/day increased their survival to 80-90%, compared to 0% survival in the diluent-treated controls (p = 5.95 x 10^{-5}). At

Table 5. Evaluation of narciclasine 1 and lycoricidine 2 in the murine Japanese encephalitis model (low virus dose challenge).

Treatment	Dose (mg / kg / day)	Uninfected (#dead / total)	ADD[a]	Infected (# dead / total)	ADD (days)	GMTD[b]
Narciclasine	6	4/5	6.0 ± 1.4	5/9	6.4 ± 3.1	15.8
	6	3/5	7.0 ± 5.0	10/10	6.9 ± 2.8	6.3
	3	0/5		2/10	17.5 ± 7.8	25.2
	3	0/5		1/10	15.0 ± 0	26.3
	1	0/5		6/10	15.7 ± 3.4	19.5
	1	0/5		9/10	15.1 ± 1.2	16.0
Lycoricidine	25	4/5	4.8 ± 3.1	4/10	10.5 ± 6.7	16.1
	20	5/5	2.6 ± 1.3	10/10	3.5 ± 2.2	2.8
	12.5	0/5		8/10	14.0 ± 1.8	16.0
	10	0/5		4/10	13.3 ± 1.0	20.7
	6.25	0/5		8/10	13.5 ± 1.5	15.6
	5.0	0/5		9/10	13.4 ± 1.9	14.4
Untreated control		0/5		10/10	14.0 ± 1.8	13.9
Sham-infected		0/10				
2% Alcohol / saline control		0/5		10/10	12.8 ± 1.3	12.7

[a] Average day of death
[b] Geometric mean time to death

1 mg/kg/day, survival was not increased; however, the average day of death (ADD) was significantly prolonged (p = 0.01). Narciclasine was not toxic at 3 or 1 mg/kg/day. Narciclasine was also evaluated in the high dose challenge model.[34] Administration of **1** at 6, 5, 4, 3, 2, 1 and 0.5 mg/kg/day (i.p. in 5% ethanol-saline, b.i.d.) from day -1 through day +7 was ineffective as shown by no increase in survivor numbers or survival times. Thus, the efficacy of narciclasine in the low-dose challenge model was not observed in the high-dose challenge model.

Lycoricidine **2**, when administered (s.c.) in NCI-hydroxypropylcellulose (HPC) in the low-dose challenge model, was toxic at 25 and 20 mg/kg/day and ineffective at 6.25 and 5 mg/kg/day in terms of increased survival number or times. At 10 mg/kg/day, lycoricidine did not produce mortality in the toxicity controls, although mice failed to gain weight. At this dose level, survival rate of virus-infected mice increased significantly to 60% compared to the diluent-treated controls (p = 5.4 x 10^{-3}); however, the ADD of the non-surviving animals was not prolonged. Thus, the efficacy exhibited by lycoricidine was observed at doses that bordered upon toxicity, a characteristic also observed with narciclasine. Subcutaneous administration of the prodrug lycoricidine triacetate **3** in carboxymethylcellulose (0.4% CMC in PBS) at 300 and 150 mg/kg/day was neither toxic nor effective in the low-dose challenge model. In the same model, lycorine **15**, administered at 45, 30 and 15 mg/kg/day (i.p., in 0.4% CMC) and pseudolycorine **16**, administered at 32, 16 and 8 mg/kg/day (s.c., in 2% ethanol/saline), did not significantly increase survivor number or prolong the ADD compared to the diluent-treated infected control mice. None of these dose levels were toxic to the mice and no weight loss was observed.

Pancratistatin **4** and 7-deoxypancratistatin **5** were evaluated in the high- and low-dose challenge models, while mono-acetate **6** was evaluated in the latter model only. When a 2% ethanolic-saline solution of pancratistatin was administered to JE virus-infected mice (s.c., once daily for 7 days, day -1 to +5, low-dose challenge model), the drug was lethal to 60% of the mice at 6 mg/kg/day, showed little efficacy and no toxicity at 2 mg/kg/day, but, when administered at 4 mg/kg/day, produced 100% survival over 28 days. No drug toxicity was observed at this dose. When the treatment was repeated at 2, 4 and 6 mg/kg/day, no toxicity was observed at any dose level; the 2 mg/kg dose was ineffective; the 4 mg/kg dose increased survival to 30%; and the 6 mg/kg dose significantly increased survival to 90%. In all cases, there was no survival of diluent-treated control mice. Prophylactic administration of pancratistatin at 2-8 mg/kg/day to JE-infected mice (i.p., b.i.d. x 9, day -1 to +7, high-dose challenge model), resulted only in 50% survival at 6 mg/kg/day, (cf. 0% survival in diluent-treated mice). Little or no efficacy was observed at other dose levels. Similarly, prophylactic administration of 7-deoxypancratistatin **5** at 40, 20 and 10 mg/kg/day in HPC (s.c., once daily, 7 days, day -1 to +5) produced neither toxicity nor efficacy at 10 mg/kg, an increased survival rate to 60% and a prolonged ADD at 20 mg/kg, and an increased survival to 80% (cf. 0% survival in diluent-treated mice) at 40 mg/kg/day. While the highest dose was not lethally toxic, some weight loss occurred. The results with **5** in the low-dose challenge model were not observed when the drug was evaluated in the high-dose challenge model.[34] Full details of these experiments are described elsewhere.[1] The monoacetate **6** (precursor of **5**) was evaluated in the low-dose challenge model. When administered at 200, 100 and 50 mg/kg/day (s.c., in 0.4% CMC, once daily, 7 days, day -1 to +7), toxicity was observed at all dose levels; doses of 10 mg/kg/day and less were ineffective and non-toxic. However, doses of 30 and 15 mg/kg (administered s.c.) were well-tolerated and produced increased survival to 60% ($p = 0.085$) and 70% ($p = 0.035$) respectively, cf. 20% survival in the diluent-treated controls; 15 mg administered i.p. did not alter survival significantly, indicative of the effect of varying routes of administration.

Overall, the *in vivo* data from the murine JE models was indicative of efficacy; however, considerable data variability was observed arising from several possible sources. These include animal models with differing routes of administration of virus and drug as well as quantities thereof, differing ages of mice, poor solubility of drugs leading to non-uniform distribution, poor bioavailability and non-uniform pharmacokinetics. No efforts were made in these experiments to separately determine the effects of these variables.

Conclusions

New members of this class of compounds need to be synthesized or isolated which, it is hoped, would: retain their antiviral activity yet exhibit decreased host cell toxicity (increased selectivity); be more water-soluble hence avoiding the current formulation problems encountered in clinical trials of pancratistatin;[11] possess increased and more uniform bioavailability in general; and, be permeable to the blood brain barrier to be efficacious against encephalitic components of viral disease. The *in vitro*, bioassay-directed fractionation process of plant, marine and other extracts is lengthy and costly. Therefore, one would optimally need to ascertain the *in vivo* potential of unknown compounds present in these extracts at an early stage of the fractionation process. Thus, the proper use and timing of evaluation in animal models need to be addressed. Natural products are also isolated in relatively small amounts, requiring an extensive investment of time and effort. Here it would be advantageous to have the capability of *in vivo* evaluation in small animal model systems requiring small amounts of scarce compounds. This would be especially advantageous in the pursuit of chemotherapy against dengue viral infections.

Narciclasine, lycoricidine (7-deoxynarciclasine), pancratistatin, 7-deoxypancratistatin, isonarciclasine (and 7-deoxy analogue), *cis*- and *trans*-dihydronarciclasine (and 7-deoxy analogues), lycorine, pseudolycorine and pretazettine all exhibited *in vitro* activity against the three flaviviruses, JE, YF and dengue-4 (except for 7-deoxyisonarciclasine against JE). Most were also active *in vitro* against the bunyaviruses, PT and RVF. Only the 7-deoxy analogues of narciclasine, pancratistatin and *cis*-dihydronarciclasine, **2**, **5** and **10b** respectively were active against SF virus. Narciclasine, pancratistatin and 7-deoxypancratistatin exhibited significant activity (>80% survival) in mice infected with a low-dose challenge of JE virus. In the high-dose challenge model, only pancratistatin reproduced its activity. However, as observed, the effective and toxic concentrations *in vitro* versus *in vivo* were not dissimilar. There apparently is a narrow therapeutic margin or poor selectivity. In spite of this, to our knowledge, this provides the most significant example of effective, non-immunomodulatory, antiviral chemotherapy of Japanese encephalitis viral infections in mice.

References

1. An initial manuscript describing this subject has been submitted for publication by B. Gabrielsen, G.R. Pettit, G. Groszek, T.P. Monath, J.W. Huggins, M.J. Phelan, D.F. Kefauver, E.M. Schubert, J. DaRe, B. Ugarkar, M.A. Ussery, M. Hollingshead, J.J. Kirsi, and W.M. Shannon, Antiviral (RNA) activity of selected Amaryllidaceae isoquinoline constituents and synthesis of related substances, *J. Nat. Prod.*, in press.

2. a) J.W. Huggins, RNA viruses that cause hemorrhagic, encephalitic, and febrile disease, *in*: "Antiviral Agents and Viral Diseases of Man," 3rd ed., G.J. Galasso, R.J. Whitley, and T.C. Merigan, eds., Raven Press, Ltd., New York (1990); b) J.W. Huggins, Z.X. Zhang, and T.P. Monath, Inhibition of ebola virus replication *in vitro* and in a SCID mouse model by S-adenosylhomocysteine hydrolase inhibitors, *Antiviral. Res.* S(1):122 (1991).

3. a) R.W. Sidwell, G.R. Revankar, and R.K. Robins, Ribavirin: review of a broad-spectrum antiviral agent, *in*: "Viral Chemotherapy," vol. 2, D. Shugar, Pergamon Press, New York (1985); b) J.W. Huggins, Prospects for treatment of viral hemorrhagic fevers with ribavirin, a broad-spectrum antiviral drug, *Revs. Infect. Dis.* 11(S4):S750 (1989).

4. a) G.R. Pettit, C.L. Herald, and C.R. Smith. "Biosynthetic Products for Cancer Chemotherapy," vol. 6, Elsevier Scientific Pub. Co., Amsterdam (1989); b) G.R. Pettit, V. Gaddamidi, D.L. Herald, S.B. Singh, G.M. Cragg, J.M. Schmidt, F.E. Boettner, M. Williams, and Y. Sagawa, Antineoplastic agents, 120. Pancratium Littorale, *J. Nat. Prod.* 49(6):995 (1986); c) G.R. Pettit, V. Gaddamidi, and G.M. Cragg, Antineoplastic agents, 105. Zephyranthes grandiflora, *J. Nat. Prod.* 47(6):1018 (1984).

5. G. Ceriotti, Narciclasine: an antimitotic substance from Narcissus bulbs, *Nature* 213:595 (1967).

6. M.I. Suffness and G.A. Cordell, Antitumor alkaloids, *in*: "The Alkaloids," vol. XXV, pp. 198-212, A. Brossi, ed., Academic Press, Inc., New York (1985).

7. S.F. Martin, The Amaryllidaceae alkaloids, *in*: "The Alkaloids," vol. 30, pp. 251-376, A. Brossi, ed., Academic Press, Inc., New York (1987) and references therein.

8. A. Baez and D. Vasquez, Binding of [^{3}H]narciclasine to eukaryotic ribosomes. A study on a structure-activity relationship, *Biochim. et Biophys. Acta.* 518:95 (1978).

9. T.S. Papas, L. Sandhaus, M.A. Chirigos, and E. Furusawa, Inhibition of DNA polymerase of avian myeloblastosis virus by an alkaloid extract from Narcissus tazetta L., *Biochem. Biophys. Res. Commun.* 52(1):88 (1973).

10. A. Mondon and K. Krohn, Zur kenntnis des narciclasins, *Chem. Ber.* 108:445 (1975).

11. B.W. Fox, Medicinal plants in tropical medicine. 2. Natural products in cancer treatment from the bench to the clinic, *Trans. Roy. Soc. Trop. Med. Hyg.* 85:22 (1991).

12. a) S. Ohta and S. Kimoto, Synthetic studies on lycoricidine and related compounds. II. Total synthesis of racemic lycoricidine, *Chem. Pharmacol. Bull.* 24:2977 (1976); b) H. Paulsen and M. Stubbe, Chirale synthese von (+)-lycoricidin, *Tetrahedron Lett.* 23(31):3171 (1982); c) H. Paulsen and M. Stubbe, Synthese von enantiomerenreinem (+)-lycoricidin aus D-glucose, *Liebigs Ann. Chem.* 535 (1983).

13. S. Danishefsky and J.Y. Lee, Total synthesis of racemic pancratistatin, *J. Am. Chem. Soc.* 111:4829 (1989).

14. B.G. Ugarkar, J. DaRe, and E.M. Schubert, Improved synthesis of lycoricidine triacetate, *Synthesis* 715 (1987).

15. V. Van Rheenan, R.C. Kelly, and D.Y. Cha, An improved catalytic OsO$_4$ oxidation of olefins to cis-1,2-glycols using tertiary amine oxides as the oxidant, *Tetrahedron Lett.* 1973 (1976).

16. K. Krohn and A. Mondon, Synthese des isonarciclasins und verwandter verbindungen, *Chem. Ber.* 109:855 (1976).

17. K. Isobe, J. Taga, and Y. Tsuda, Synthesis of racemic dihydrolycoricidines from the common intermediate to ethanophenanthridine alkaloids, *Heterocycles* 9(5):625 (1978).

18. G.R. Pettit, G.M. Cragg, S.B. Singh, J.A. Duke, and D.L. Doubek, Antineoplastic agents, 162. Zephyranthes candida, *J. Nat. Prod.* 53(1):176 (1990).

19. S. Ghosal, K.S. Saini, and S. Razdan, Crinum alkaloids: their chemistry and biology, *Phytochemistry* 24:2141 (1985).

20. E. Furusawa, S. Furusawa, S. Tani, H. Irie, K. Kitamura, and W.C. Wildman, Isolation of Pretazettine from Narcissus tazetta L., *Chem. Pharm. Bull.* 24:336 (1976).

21. S. Ghosal, A. Shanthy, A. Kumar, and Y. Kumar, Palmilycorine and lycoriside: acyloxy and acylglucosyloxy alkaloids from crinum asiaticum, *Phytochemistry* 24:2703 (1985).

22. a) G. Stork and D.J. Morgans Jr., A simple, stereospecific synthesis of the skeleton of the lycorine alkaloids, *J. Am. Chem. Soc.* 101:7110 (1979); b) T. Sano, N. Kashiwaba, J. Toda, Y. Tsuda, and H. Irie, A modified total synthesis of racemic lycorine, *Heterocycles* 14:1097 (1980) and references therein; c) S.F. Martin and C. Tu, General strategies for alkaloid synthesis via intramolecular [4+2] cycloadditions of enamides. Application to the formal total synthesis of racemic lycorine, *J. Org. Chem.* 46:3763 (1981); d) S. Martin, C. Tu, M. Kimura, and S. Simonsen, Intramolecular [4+2] cycloadditions as a general strategy for alkaloid synthesis. A novel formal synthesis of lycorine, *J. Org. Chem.* 47:3634 (1982); e) Y. Tsuda, T. Sano, J. Taga, K. Isobe, J. Toda, S. Takagi, M. Yamaki, M. Murata, H. Irie, and H. Tanaka, Total synthesis of the Amaryllidaceae alkaloids, lycorine and zephyranthine, *J. Chem. Soc. Perkin Trans.* I:1358 (1979); f) O. Moller, E.M. Steinberg, and K. Torssell, Synthesis of lycorine-type alkaloids. II. Synthesis of D,L-lycorine, *Acta Chem. Scand.* Ser. B. 32:98 (1978).

23. A. Jimenez, L. Sanchez, and D. Vasquez, Yeast ribosomal sensitivity and resistance to the Amaryllidaceae alkaloids, *FEBS Lett.* 60(1):66 (1975).

24. A. Jimenez, A. Santos, G. Alonso, and D. Vasquez, Inhibitors of protein synthesis in eukaryotic cells. Comparative effects of some Amaryllidaceae alkaloids, *Biochim. et Biophys. Acta* 425:342 (1976).

25. I. Ieven, A.J. Vlietinck, D.A. Vanden Berghe, J. Totte, R. Dommisse, E. Esmans, and F. Alderweireldt, Plant antiviral agents. III. Isolation of alkaloids from *Clivia Miniata* regel (Amaryllidaceae), *J. Nat. Prod.* (Lloydia) 45(5):564 (1982).

26. R. Vrijsen, D. Vanden Berghe, A. Vlietinck, and A. Boeye, Lycorine: A eukaryotic termination inhibitor?, *J. Biol. Chem.* 261(2):505 (1986).

27. E. Furusawa, S. Furusawa, J.Y.B. Lee, and S. Patanavanich, Therapeutic activity of Pretazettine on Rauscher leukemia: Combination of antiviral activity and cellular protein inhibition, *Chemotherapy* 24:259 (1978).

28. E. Furusawa, H. Irie, D. Combs, and W.C. Wildman, Therapeutic activity of pretazettine on Rauscher leukemia: Comparison with the related Amaryllidaceae alkaloids, *Chemotherapy* 26:36 (1980).

29. a) E. Furusawa and W. Cutting, The higher plants with antiviral and anti-lethal activity on virus infections in mice, *Ann. of the N.Y. Acad. Sci.* 173:668 (1970); b) E. Furusawa, S. Furusawa, S. Morimoto, and W. Cutting, Therapeutic activity of Narcissus alkaloid on Rauscher leukemia and comparison with standard drugs, *Proc. Soc. Exp. Biol. Med.* 136:1168 (1971).

30. T.H. Kinstle, W.C. Wildman, and C.L. Brown, Mass spectra of Amaryllidaceae alkaloids. The structure of Narcissidine, *Tetrahedron Lett.* 4659 (1966).

31. F.J. Malinoski, S.E. Hasty, M.A. Ussery, and J.M. Dalrymple, Prophylactic ribavirin treatment of dengue type 1 infection in rhesus monkeys, *Antiviral Res.* 13:139 (1990).

32. a) T.P. Monath, Japanese encephalitis--a plague of the Orient, *N. Engl. J. Med.* 319(10):641 (1988); b) A.D.T. Barrett, Vaccines against flaviviruses, *Current Opinion In Infectious Diseases* 3:374 (1990); c) C.H. Hoke, A. Nisalak, N. Sangawhipa, S. Jatansen, T. Laorakapongse, B.L. Innis, S. Kotchasenee, J.B. Gingrich, J. Latendresse, K. Fukai, and D.S. Burke, Protection against Japanese encephalitis by inactivated vaccines, *N. Engl. J. Med.* 319:608 (1988).

33. a) D.G. Harrington, D.E. Hilmas, M.R. Elwell, R.E. Whitmire, and E.L. Stephen, Intranasal infection of monkeys with Japanese encephalitis virus: clinical response and treatment with a nuclease-resistant derivative of Poly(I).Poly(C), *Am. J. Trop. Med. Hyg.* 26(6):1191 (1977). b) J.L. Taylor, C. Schoenherr, and S.E. Grossberg, Protection against Japanese encephalitis virus in mice and hamsters by treatment with carboxymethylacridanone, a potent interferon inducer, *J. Infect. Dis.* 142(3):394 (1980).

34. T.P. Monath, unpublished results.

AMINOSUGAR ATTENUATION OF HIV INFECTION

Gary S. Jacob,* Peter Scudder,* Terry D. Butters,§ Ian Jones£ and David C. Tiemeier+

*Monsanto, 800 N. Lindbergh Blvd., St. Louis, MO 63167, § G. D. Searle Research Group, Glycobiology Institute, Dept. of Biochemistry, University of Oxford, South Parks Road, Oxford OX1 3QU, England, +G. D. Searle, 700 Chesterfield Village Parkway, St. Louis, MO 63198, £NERC Institute of Virology, Mansfield Road, Oxford OX1 3SR, England

SUMMARY

The envelope glycoproteins of HIV mediate cellular uptake of virus, and fusion of infected with non-infected CD4+ T cells. Inhibitors of oligosaccharide processing have been found to attenuate HIV infections in vitro and putatively work by altering the structure and function of these glycoproteins. One such compound, SC-48334 (N-butyl-1-deoxynojirimycin) inhibits α-glucosidase I and has been reported to reduce the infectivity of virus recovered from infected cell cultures in a dose-dependent manner. The envelope precursor, gp160, is heavily glycosylated and its endoproteolytic processing to gp120 and gp41, necessary for infection, is retarded in the presence of SC-48334. However, processing is not abolished and mature envelope glycoproteins are still present. Using baculovirus-insect-derived gp120 probed with antibodies, we find suggestive evidence that a local conformational change is induced in the V3 loop of gp120 which may affect cell-surface proteolytic cleavage thought to be involved in cell-virus fusion. Finally, we present preliminary information on a prodrug form of SC-48334 which does not inhibit intestinal disaccharidases.

INTRODUCTION

Over the last few years, a number of aminosugar derivatives capable of inhibiting glycoprotein processing have been reported to inhibit HIV replication and cellular cytopathicity in vitro[1-6]. These compounds include castanospermine[1-3] and its derivatives[6], 1-deoxynojirimycin[1-3] (DNJ), N-alkylated derivatives of DNJ[4,5] and other related aminosugars[4,5] which all have the common property that they are potent inhibitors of specific glycosidases involved in glycoprotein processing. These compounds are effective because HIV and other enveloped viruses contain glycoproteins in the envelope which are obligatory for successful replication. In the case of HIV, a single viral envelope glycoprotein is coded for by the ENV gene, and is expressed in host CD4+ T lymphocytes as a precursor translation product requiring further host cellular proteolysis to generate the mature forms observed in virion particles (see Fig. 1). For HIV, this processing occurs at an REKR*AVGIG site in the gp160 precursor molecule, producing a transmembrane component, gp41 and the surface glycoprotein gp120. The mature gp120, is responsible for binding to CD4$^+$ receptors on helper T cells and is noncovalently anchored to gp41 embedded within the virion envelope (see Fig. 2).

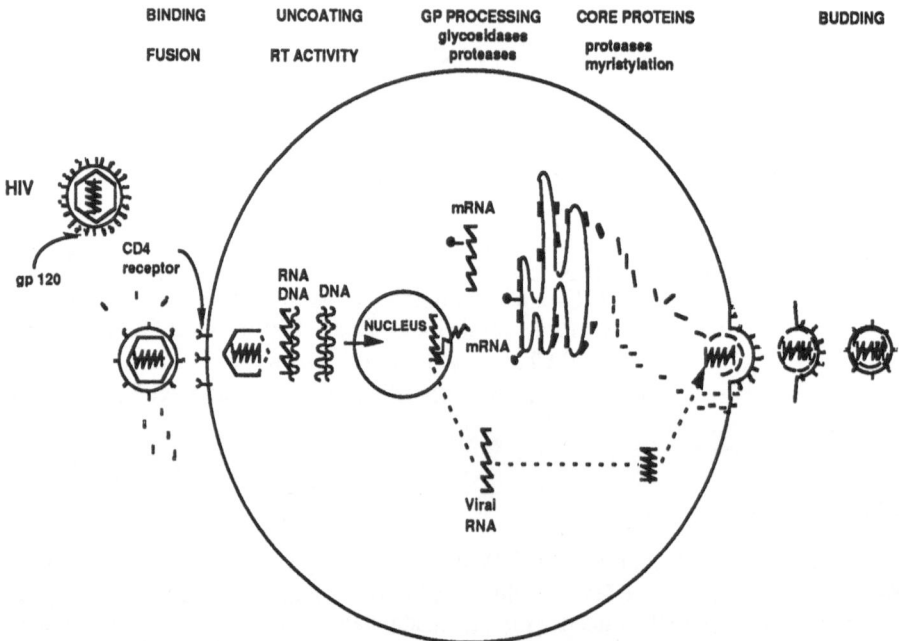

Fig. 1. Replication cycle of the human immunodeficiency virus.

All glycoproteins, whether constitutive or virally encoded, traffic the same cellular post-translational pathway (Fig. 3). This pathway, located in the endoplasmic reticulum and the Golgi complex, is responsible for processing N-linked oligosaccharides to the various mature forms observed in glycoproteins (Fig. 4). The process begins with cleavage of glucose residues from the terminus of Glc3Man9GlcNAc2-N-linked residues (see Fig. 3 for structure), mediated by two endoplasmic reticulum α-glucosidases. Further processing occurs in the Golgi by the action of two α-mannosidases, after which specific glycosyltransferases catalyse the synthesis of mature high-mannose, hybrid, and complex-type oligosaccharides.

GLYCOSYLATION CHARACTERISTICS OF HIV GLYCOPROTEINS

Glycosylation of Gp120

Understandably, the glycosylation pattern observed for gp120 can be expected to be dependent on the cell line used to express the molecule and depends on whether gp120 was obtained from supernatant virions or from cell extracts. The mature molecule, containing

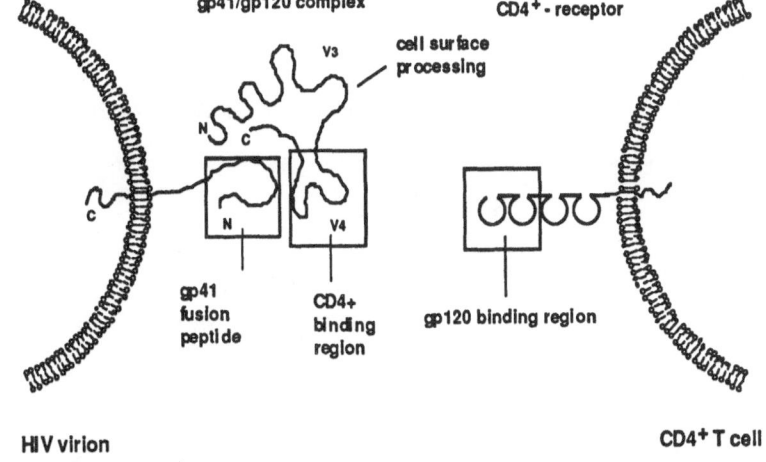

Fig. 2. Schematic representation of HIV-CD4$^+$ T cell interaction.

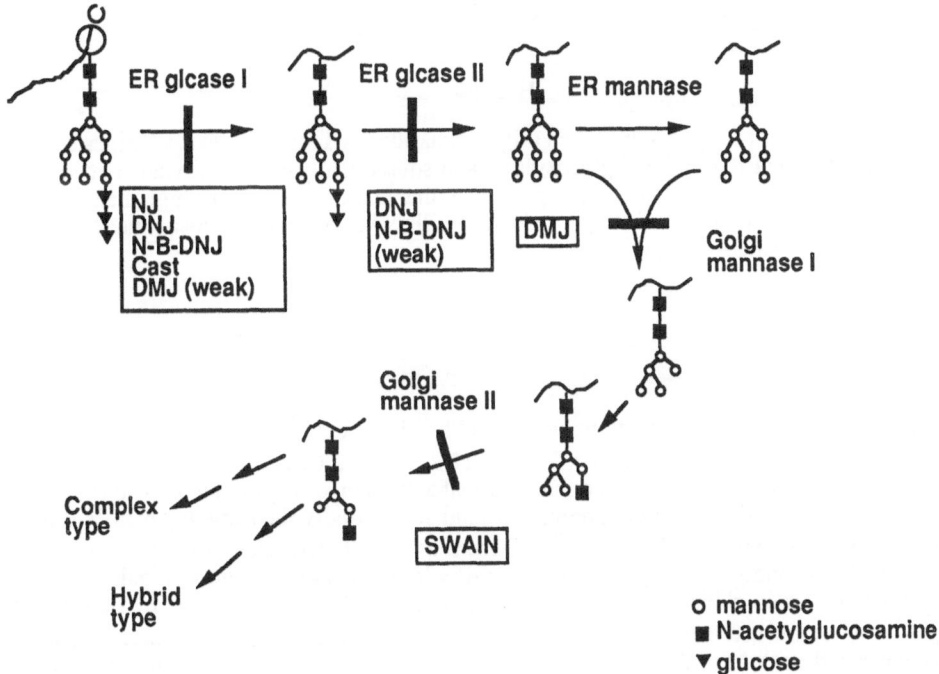

Fig. 3. Oligosaccharide processing of N-linked glycans by ER and Golgi glycosidases. Oligosaccharide processing is blocked selectively using inhibitors of α-glucosidases I and II and α-mannosidases I and II. NJ, nojirimycin; DNJ, deoxynojirimycin; Cast, castanospermine; DMJ, deoxymannojirimycin; N-B-DNJ, N-butyldeoxynojirimycin; Swain, swainsonine; glcase, α-glucosidase; mannase, α-mannosidase

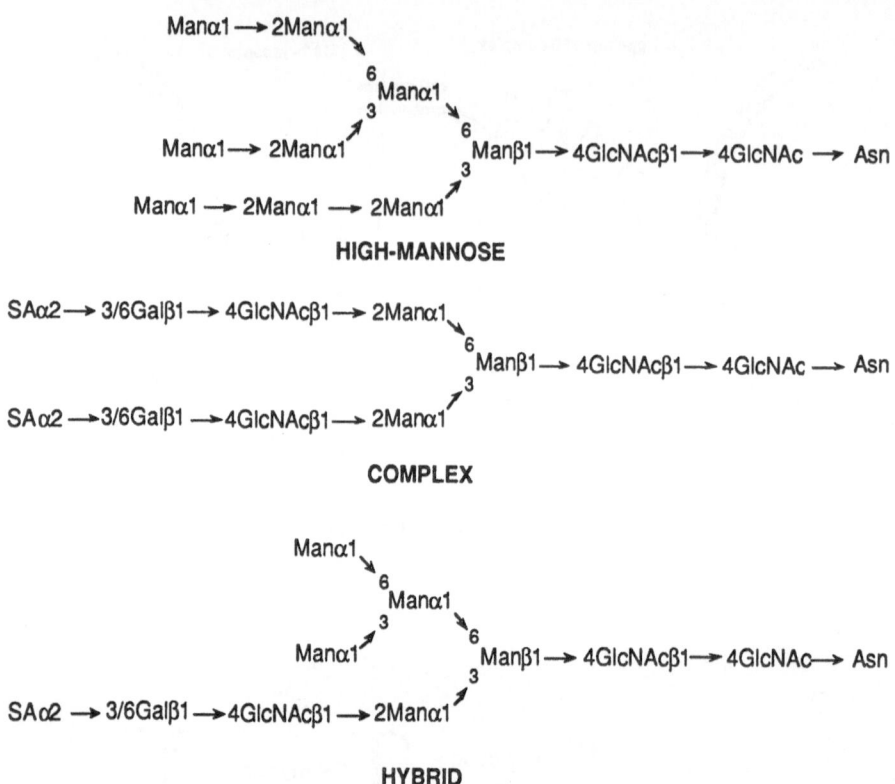

Fig. 4. General classes of N-linked oligosaccharide. N-Linked oligosaccharides are processed by ER and Golgi glycosidases and subsequently elongated via the action of specific glycosyltransferases to generate either complex, hybrid or high mannose structures. SA, sialic acid; Gal, galactose; GlcNAc, N-acetylglucosamine; Man, mannose; Asn, asparagine.

about 50% by weight carbohydrate, was found by Geyer et al.[7] to contain primarily high-mannose structures when obtained from an H-9 CD4+ T cell extract. Whereas, gp120 from supernatant-derived virions contained about a 2:1 ratio of high-mannose and complex-type glycans. Feizi and her collaborators found substantially the same types of N-linked sugars from transfected CHO cells[8] and from an H-9 T cell line[9], although the molar percentages of individual sugar types were different in the two cell lines. Finally, the site specificity of recombinant gp120 expressed in CHO cells was determined by Leonard et al.[10] who established that all 24 of the potential N-glycosylation sites on the molecule were glycosylated, with 13 of them containing complex oligosaccharides, and the rest containing high-mannose/hybrid structures.

We have explored the glycosylation of recombinant gp120 expressed in a baculoviral-insect system[11]. This glycoprotein obtained from insect cells was earlier demonstrated to have an affinity for soluble CD4+ similar to the mammalian-expressed glycoprotein[12], although it shows a somewhat lower molecular weight based on SDS-PAGE analysis. The lower molecular weight is consistent with the absence of complex sugars on glycoproteins expressed in insect cells, and was verified by a complete analysis of the oligosaccharides present on gp120 (Table 1).

Table 1. N-Linked Oligosaccharides Obtained from Gp120 Expressed in Lepidopteran Cells. Oligosaccharides were released from gp120 using hydrazine, reduced with NaB$_3$H$_4$ and isolated by Bio-Gel P-4 chromatography. Structures were deduced by glycosidase sequencing.

Structure	% of Total
Manα1 \ 6 Manβ1 — 4GlcNAcβ1 — 4GlcNAc.ol 3 | α1,6 Manα1 / Fuc	30
± Manα1 — 2Manα1 \ 6 Manα1 \ 3 6 Manα1 — 2Manα1 / Manβ1 — 4GlcNAcβ1 — 4GlcNAc.ol ± Manα1 — 2Manα1 — 2Manα1 / 3	[a]17
Manα1 \ 6 Manβ1 — 4GlcNAcβ1 — 4GlcNAc.ol 3 Manα1 /	13
Manα1 \ 6 Manα1 \ Manα1 / 3 6 Manβ1 — 4GlcNAcβ1 — 4GlcNAc.ol Manα1 — 2Manα1 — 2Manα1 / 3	12
Manα1 — 2Manα1 \ 6 Manα1 \ Manα1 — 2Manα1 / 3 6 Manβ1 — 4GlcNAcβ1 — 4GlcNAc.ol Manα1 — 2Manα1 — 2Manα1 / 3	11
Manα1 \ 6 Manα1 \ Manα1 / 3 6 Manβ1 — 4GlcNAcβ1 — 4GlcNAc.ol Manα1 — 2Manα1 / 3	9
Manα1 \ 6 Manα1 \ Manα1 / 3 6 Manβ1 — 4GlcNAcβ1 — 4GlcNAc.ol Manα1 / 3	8

[a]Obtained as a mixture of two Man8 isomers (indicated by ±).

141

Table 2. Plant Alkaloids Which Inhibit Oligosaccharide Processing of Glycoproteins

Compound	Structure	Sugar Mimic
Nojirimycin		glucose
Deoxynojirimycin		glucose
N-Alkylated-DNJ		glucose
Deoxymannojirimycin		mannose
Swainsonine		mannosyl cation
Castanospermine		glucose

[a]R = alkyl

Glycosylation of Gp41

Glycosylation of the transmembrane glycoprotein, gp41, has not been carefully analysed, to date. Fenouillet et al.[13],using endoglycosidase F and N-glycanase, two enzymes typically used to release oligosaccharides from glycoproteins, were able to shift the molecular weight of gp41 to 35 kD, consistent with the removal of four glycan chains, corresponding to the four potential glycosylation sites observed in the molecule. However, the specific role of these N-linked glycans in defining the structure and function of the overall transmembrane molecule has yet to be investigated, as well as the effect which glycosylation inhibitors such as the aminosugars may have on their oligosaccharide processing.

AMINOSUGAR INHIBITORS AS ANTIVIRAL AGENTS

Aminosugar Inhibitors of N-Glycan Processing

The pathway for post-translational processing of N-linked sugars (Fig. 3) was determined, in part, by the use of compounds capable of blocking various steps in the pathway, and over the years a number of aminosugar derivatives have been isolated and synthesized which are inhibitors of the processing α-glucosidases and α-mannosidases[14,15]. A number of these compounds and the monosaccharides they mimic are displayed in Table 2. Essentially, they are analogues of their corresponding sugars with the oxygen in the ring replaced by a nitrogen. The use of these inhibitors to modify the N-glycosylated component of glycoproteins has been explored by a great number of investigators over the last 10-15 years; however, the discovery that these inhibitors had potential antiviral and antimetastatic activity gave rise to a renewed interest in their therapeutic potential, and has fueled an increased interest in the particular way, or ways, in which these compounds exert their biological effect.

Antiviral Activity of N-Butyldeoxynojirimycin and Possible Mechanisms of Action

Effect of N-Butyldeoxynojirimycin (SC-48334) on Glycosylation of Gp120. The compound which we have explored at length, and which is the major subject of this paper, SC-48334, is a potent inhibitor of α-glucosidase I[16], the enzyme which is responsible for hydrolysis of the terminal Glcα1-2Glc linkage (see Fig. 3). It has been suggested that inhibition of α-glucosidase I and inhibition of HIV replication are directly related, and that derivatives of DNJ exert their antiviral effect by interfering with the proper post-translational processing of the viral envelope glycoprotein, gp160. Indeed, Gruters et al.[1] and Walker et al.[2] showed that gp120 produced in the presence of DNJ and castanospermine, a DNJ analogue, displayed an altered mobility on SDS-PAGE, consistent with approximately a 5-10 kD increase in molecular weight. This would be consistent with a gp120 molecule devoid of any processing beyond the $Glc_3Man_9GlcNAc_2$ stage; however this point has never been rigorously validated by analysis of oligosaccharides released from the molecule. This shift in molecular weight was also found for H9 T cells grown in the presence of SC-48334[17]. And an even more pronounced difference (about 17 kD) in molecular weight is seen in baculoviral-insect expressed gp120 from cells grown in the presence of this compound (T. D. Butters, I. Jones, G. S. Jacob unpublished results) (Fig. 5).

It has been tempting from the outset to ascribe the anti-viral activity of DNJ analogues to an altered glycosylation of gp120 giving rise to reduced affinity for the CD4$^+$ receptor on the target helper T cell. However, Walker et al.[2] found that although no difference in the level of virus-associated envelope glycoprotein was observed in cells grown in the presence of castanospermine, the virus produced in the presence of this inhibitor could still bind to the CD4 receptor, although the binding constant was not quantitatively measured.

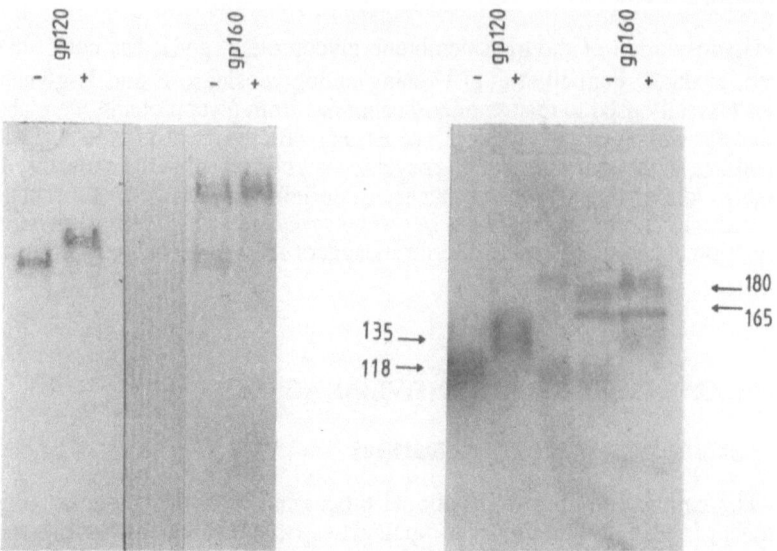

Fig. 5. SDS-PAGE analysis of recombinant, baculovirus-expressed HIV envelope proteins. Recombinant baculovirus infected Sf8 cells were cultured in the presence (+) or absence (-) of 0.5 mg/ml SC-48334 and the expressed envelope proteins separated by SDS-PAGE. The gel in the right hand panel was electrophoresed for a longer time to increase separation. Gp120 and gp160 were detected by Western blot.

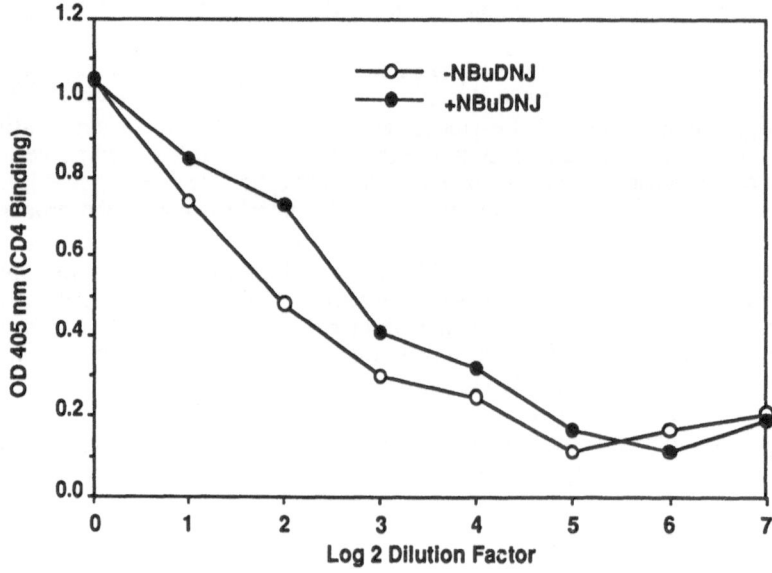

Fig. 6 Effect of SC-48334 on CD4 binding to baculovirus-expressed gp120. Soluble CD4 was captured on microtiter plates and supernatants from infected Sf8 cells cultured in the presence or absence of 0.5 mg/ml SC-48334 were added for 1 hour. Antigen binding was determined using an alkaline phosphatase-conjugated second antibody.

We have further explored this point using gp120 expressed in the baculoviral-insect system, in the presence and absence of SC-48334 (T. D. Butters, I. Jones, G.S. Jacob, unpublished results). Although the molecular mass of gp120 derived from insect cells treated with SC-48334 was significantly greater than the control gp120, both glycoproteins demonstrated similar affinities for soluble CD4+ (Fig. 6).

Effect of SC-48334 on Proteolytic Processing of Gp160. A second and more subtle effect of glycosylation inhibitors such as SC-48334 is the potential for altering other processes which might be affected by the different glycosylation pattern of the precursor gp160 molecule. This molecule undergoes a number of proteolytic steps throughout the virus' replicative life cycle (Fig. 7). After elimination of its signal peptide, the first proteolysis involves processing at an REKR*AVGIG site in the gp160 precursor molecule, producing a transmembrane component, gp41, and the surface glycoprotein gp120. The biological significance of this processing event relates to the fusogenic character of the free hydrophobic N-terminus of gp41, essential for viral infectivity. Hence, unsuccessful processing of gp160, as shown in previous mutagenesis experiments of the REKR cleavage site[18-21], leads to intact virions incapable of fusing with the host cell.

The effect of SC-48334 on processing of gp160 to gp120 and gp41 has been explored by Ratner and his colleagues[17], and although a large decrease was seen in the amount of gp120 and gp41 present on the surface of infected cells (with a concurrent increase in unprocessed gp160), the amounts of virus-associated envelope glycoproteins appeared normal. This effect suggests that cell-cell fusions involving infected and noninfected cells might be reduced by compounds such as SC-48334, an observation which was earlier made by Walker et al.[2], although, in this case, these researchers also reported low cell-surface expression of envelope glycoproteins when HIV-infected cells were grown in the presence of the aminosugar, castanospermine.

Cell-Surface Processing of Gp120. A third, and less well documented cleavage, which has gained increased attention recently, involves cell surface proteolysis of gp120 to gp70 and gp50[22-24]. Because the cleavage presumably occurs after docking of the gp120 to the CD4+ receptor, and is dependent on the presence of a cell surface protease, a partial explanation for the cell tropism displayed by HIV may be attributable to this processing event. There appear to be at least two closely-spaced sites within the V3 hypervariable domain of gp120 where cleavage can occur. This region encodes a GPGRAFVT stretch which can be acted upon by either a trypsin (GPGR*AFVT) or chymotrypsin-like (GPGRAF*VT) activity. The significance of this cell surface processing event is not yet known, but it has been suggested that processing of gp120 plays a pivotal role in the HIV-cell fusion reaction, perhaps by expediting access of the fusogenic N-terminal peptide of gp41 to the target T cell membrane[22-24]. Interestingly, point mutation of the proline residue in the GPGR sequence of an HIV-1 proviral clone altered the infectivity of this specific clone against certain T-cell lines but not against others[25].

Effect of SC-48334 on the V3 Loop. Using a recombinant gp120 expressed in a baculoviral-insect system in the presence or absence of SC-48334 (0.5 mg/ml), we screened a number of different antibodies to probe for structural or possible conformational changes in gp120 which might help to explain this drug's action[26]. Although the molecular mass of SC-48334-treated gp120 was approximately 17 kD greater than the untreated fraction, both fractions bound equally well to sCD4, and to a polyvalent antiserum (MAb221) against a 44 amino acid stretch at the C-terminus of gp120, indicating no gross conformational changes in the overall molecule from cells grown in the presence of this drug (Fig. 8). However, we found that the immunoreactivities of the two gp120s to a V3 loop-specific monoclonal antibody, ADP358, were considerably different. These results suggest that SC-48334 induces an alteration in glycosylation, possibly within the hypervariable V3 loop, which results in a masking of the antigenic determinant recognised by ADP358. A possible consequence of this alteration is that the constitutive cell surface protease (or proteases) may no longer be able to cleave the V3 loop, and that fusion of the viral and cellular membranes is thus prevented.

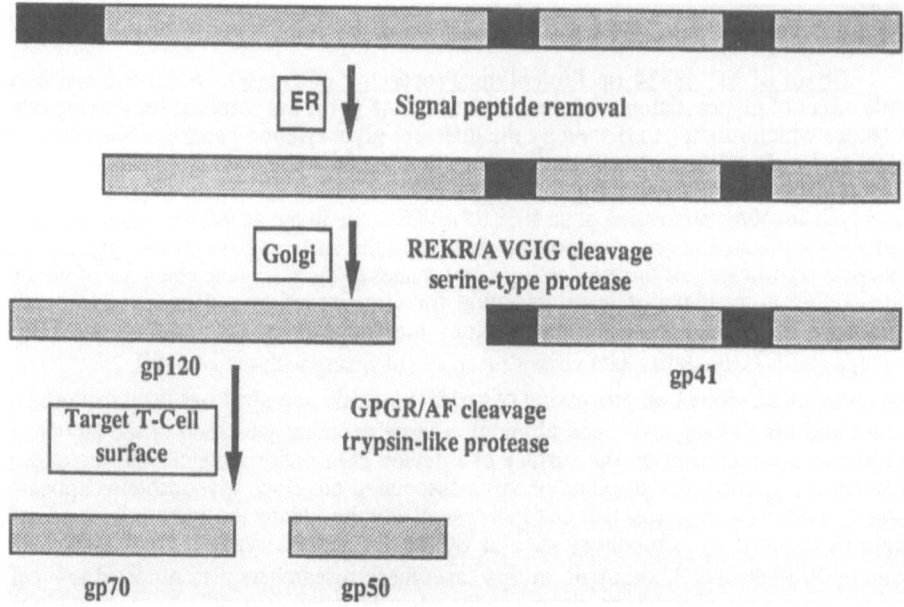

Fig. 7. Endoproteolytic processing of gp160

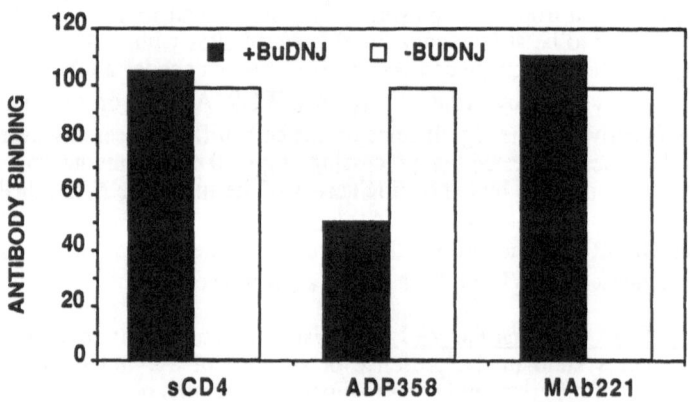

Fig. 8. Binding of monoclonal antibodies to gp120 expressed in the presence and absence of SC-48334. Gp120 was expressed using recombinant baculoviruses with or without the addition of 0.5 mg/ml SC-48334. Antigen was immobilized to a solid phase using anti-gp120 serum D7324 and incubated with soluble CD4, ADP358 or MAb 221. Antibody binding was expressed as the percent of binding observed with the SC-48334 treated compared with the control which lacked SC-48334.

A potential drawback to the use of α-glucosidase inhibitors such as SC-48334 is that these compounds can also inhibit intestinal α-glucosidases such as maltase and sucrase[27-29], enzymes involved in the final steps of carbohydrate digestion, the hydrolysis of disaccharides to monosaccharides in the gut. Although similar inhibitors of intestinal disaccharidases such as BAY m 1099 (a derivative of DNJ) have been explored clinically as a means of controlling glucose levels in diabetic patients[30-34] intestinal problems do occur (flatulence, diarrhea, meteorism) which scale with dose[33]. Consequently, a potential drawback to the therapeutic use of SC-48334 for controlling infectious diseases such as AIDS, wherein long-term treatment of patients who may already exhibit gastrointestinal problems, is the degree and severity of the gastrointestinal distress associated with the use of this drug.

One possible way to alleviate gastrointestinal side-effects would be to administer a prodrug which does not inhibit gut α-glucosidases but which is absorbed and metabolized to the active agent. We have synthesized one such prodrug candidate, SC-49955 (N-butyldeoxynojirimycin-6-phosphate, NBDNJ6P) (P. Scudder and G. S. Jacob, submitted for publication). A study of the inhibitory properties of this and related compounds against porcine sucrase and maltase and the processing α-glucosidase I suggests that SC-49955 may be devoid of intestinal side effects associated with administration of SC-48334.

Synthesis of SC-49955 (N-Butyldeoxynojirimycin-6-Phosphate)

A summary of the protocol for the synthesis of SC-49955 from DNJ is shown in Fig. 9, and was achieved using a two-step process involving enzymatic phosphorylation of DNJ followed by chemical butylation. It was necessary that phosphorylation precede alkylation since SC-48334 does not act as a substrate for yeast hexokinase, which catalyses the transfer of a phosphate group from ATP to the C-6 position of DNJ. Enzymatic conversion of DNJ to DNJ-6-phosphate was only 45% using ATP and hexokinase alone at pH 8.0 but, by including phosphocreatine and creatine phosphokinase in the reaction mixture to regenerate ATP, the efficiency was improved to 75%.

Effect of SC-48334, SC-49955 and Related Compounds on the Activity of α–Glucosidases

Porcine Microsomal α–Glucosidase I. Castanospermine, DNJ and SC-48334 were all strong inhibitors of porcine liver α–glucosidase I (Table 3). The K_i values for DNJ and SC-48334 were similar to those reported by Schweden et al.[16] who used the calf liver enzyme for their studies. Phosphorylation of DNJ and SC-48334 (to give DNJ-6-P and SC-49955, respectively) dramatically reduced the ability of these compounds to inhibit α-glucosidase I, but the effect of phosphorylation of castanospermine could not be determined since it was not a substrate for yeast hexokinase.

Porcine Intestinal Sucrase and Maltase. In agreement with studies on intestinal disaccharidases isolated from rabbit small intestine[35] and a human intestinal cell line[36], we found that DNJ and castanospermine are potent inhibitors of porcine sucrase with K_i values in the sub-micromolar range. In addition, we have established that SC-48334 is an equally effective inhibitor of porcine intestinal sucrase, and that all three compounds are potent inhibitors of porcine intestinal maltase, each displaying a K_i value of less than 1 µM. In the case of SC-48334, phosphorylation at the C-6 position to give SC-49955 dramatically increased the K_i value against both sucrase and maltase to 1.5 mM and 0.33 mM, respectively. Phosphorylation of DNJ at the C-6 position also significantly increased its K_i value against sucrase from 0.05 µM to 0.5 mM but, interestingly, only raised the value against maltase ten-fold from 0.09 µM to 1 µM. These results are summarised in Table 3.

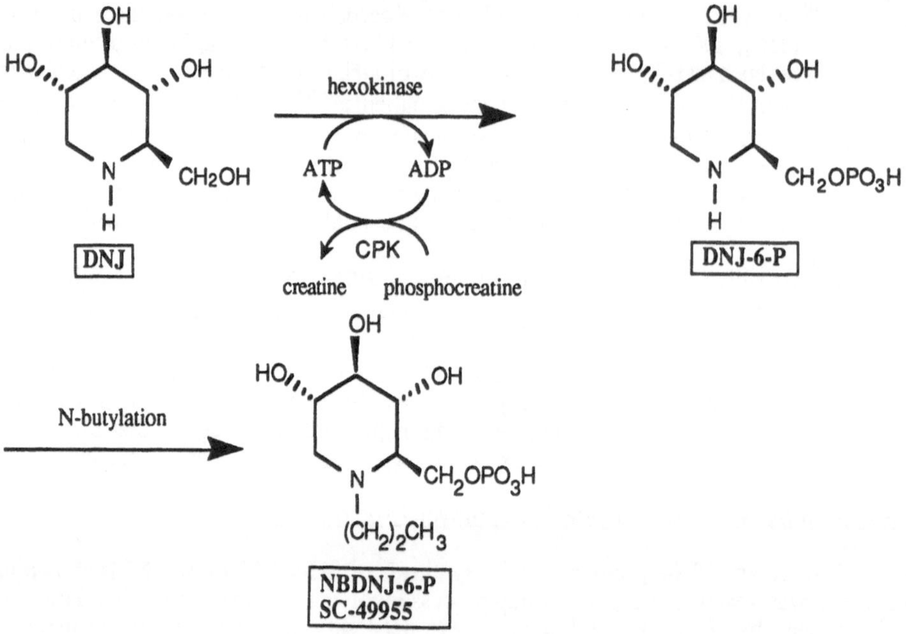

Fig. 9. Synthesis of N-butyldeoxynojirimycin-6-phosphate, SC-49955. Deoxynojirimycin was phosphorylated with yeast hexokinase using creatinine phosphokinase (CPK) to regenerate ATP. The product, deoxnojirimycin-6-phosphate, (DNJ-6-P) was purified by gel-filtration chromatography and converted into the N-butyldeoxynojirimycin-6-phosphate (SC-49955) by treatment with butyraldehyde in the presence of a palladium black catalyst. Purification of SC-49955 was done by ion-exchange chromatography

Table 3. Inhibition Constants of Aminosugar Derivatives Against α–Glucosidase I, Sucrase and Maltase

Compound	K_i (µM)		
	α–Glucosidase I	Sucrase	Maltase
Castanospermine	0.04	0.064*	0.34
DNJ	0.83	0.05	0.09
SC-48334	0.22	0.18	0.92
DNJ-6-P	6600	520	1.0
SC-49955	4000	1500	330

*Non-competitive inhibitor[35], no pre-incubation

Susceptibility of SC-49955 to Alkaline Phosphatase

To demonstrate that SC-49955 could potentially be metabolized to SC-48334, the phosphorylated derivative was incubated with bovine alkaline phosphatase, an enzyme capable of hydrolyzing a variety of monophosphoester linkages[37]. High performance anion exchange chromatography of the reaction products demonstrated complete dephosphorylation to give the parent compound, SC-48334. Other tissue and cellular phosphatases were not tested against SC-49955, however rabbit liver glucose-6-phosphatase was shown to efficiently hydrolyze a related compound, DNJ-6-P. Thus, we expect that the orally administered phosphorylated derivatives of DNJ, and presumably other amino-sugars, should be susceptible to the action of endogenous phosphatases.

Therapeutic Potential of a Prodrug Form of SC-48334

The above work describes our efforts to circumvent the problems of acute intestinal disruption associated with the administration of SC-48334, an anti-HIV agent presently undergoing phase 2 clinical trials in AIDS patients. Our strategy has been to synthesize, using a combined enzymatic and chemical approach, a phosphorylated prodrug form of SC-48334, which unlike the parent compound exhibits only very weak inhibition of two gut α-glucosidases namely sucrase and maltase. Thus, it is anticipated that oral administration of SC-49955 would result in minimal side effects, i.e. intestinal disruption, that are believed to be the result of inhibition of intestinal disaccharidases. This being the case, and assuming that SC-49955 crosses the intestinal barrier, levels of the prodrug equal to or higher than presently being used for SC-48334 are conceivable (SC-48334 may need to be given in doses as high as 160 mg/Kg/day since this compound has a short serum half-life and is rapidly cleared from the body via excretion in the urine). This minimal side-effect profile is essential since AIDS patients will need to be treated continuously with this or similar anti-HIV drugs.

To our knowledge only one other phosphorylated prodrug specifically designed to break down after ingestion to generate the active agent has been explored[38]. However, in this case the purpose of phosphorylation was to increase uptake and hence bioavailability. This anti-rhinoviral agent, a phosphorylated analogue of Chalcone Ro 09-4010, was administered to human volunteers and shown to be absorbed intact into their tissues, where it was ultimately converted by endogenous phosphatases to the bioactive material.

Since endogenous phosphatases are widely distributed in the tissues and body fluids[37], hydrolysis could occur at any number of sites including the intestine, serum or in the tissues following absorption. If conversion to SC-48334 does occur at the intestinal level, it will be important to determine its rate of conversion, its location relative to gut disaccharidases and, most importantly, whether the level of SC-48334 in the gut is still sufficiently low to abrogate symptoms of intestinal distress.

Since α-glucosidase inhibitors are potential therapeutic agents for the treatment of a diverse array of disease states including cancer[39], diabetes[30-34], AIDS[1-6] and other viral diseases[15], we anticipate that an increased emphasis will be made to develop these compounds in a form that is "physiologically acceptable". The work described represents a first step in this direction.

CLINICAL EVALUATION OF SC-48334

SC-48334 has been carried through a multi-center Phase I tolerability study and analysis of the results is underway. Based on early findings a multi-center Phase II study of the compound in combination with AZT has been initiated. The results of that study are expected in early 1992. In addition, the effect of SC-48334 on various markers in SIV-infected monkeys is under study and the results are expected by the end of 1991.

ACKNOWLEDGEMENTS

We thank D. Mueller and R. Partis for providing the SC-48334 used in these studies, and F. Platt and D. Neville for providing technical support for both the prodrug and baculoviral-derived gp120 projects. In addition, we would like to thank R. A. Dwek and G. W. J. Fleet for their continued support of this work and gratefully acknowledge J. Rotsaert for help with preparation of the manuscript.

REFERENCES

1. R.A. Gruters, J.J. Neefjes, M. Tersmette, R.E.Y. de Goede, A. Tulp, H.G. Huisman, F. Miedema and H.L. Ploegh, Interference with HIV-induced syncytium formation and viral infectivity by inhibitors of trimming glucosidase, Nature, 330:74 (1987).

2. B.D. Walker, M. Kowalski, W.C. Goh, K. Kozarsky, M. Krieger, C. Rosen, L Rohrschneider, W.A. Haseltine, and J. Sodroski, Inhibition of human immunodeficiency virus syncytium formation and virus replication by castanospermine, Proc. Natl. Acad. Sci. USA, 84:8120 (1987).

3. A.S. Tyms, E.M. Berrie, T.A. Ryder, R.J. Nash, M.P. Hegarty, T.L. Taylor, M.A. Mobberly, J.M. Davis, E.A. Bell, D.J. Jeffries, D. Taylor-Robinson, D. and L.E. Fellows, Castanospermine and other plant alkaloid inhibitors of glucosidase activity block the growth of HIV. Lancet, i:1025 (1987).

4. G.W.J. Fleet, A. Karpas, A., R.A. Dwek, L.E. Fellows, A.S. Tyms, S. Perturrsson, S.K. Namgoong, N.J. Ramsden, P.W. Smith, J.C. Son, F. Wilson, D.R. Witty, G.S. Jacob and T.W. Rademacher, Inhibition of HIV replication by aminosugar derivatives, FEBS Letts., 237:128 (1988).

5. A. Karpas, G.W.J. Fleet, R.A. Dwek, S. Pertusson, S.K. Namgoong, N.G. Ramsden., G.S. Jacob and T.W. Rademacher, Aminosugar derivatives as potential anti-human immunodeficiency virus agents. Proc. Natl. Acad Sci. USA, 85:9229 (1988).

6. P.S. Sunkara, D.L. Taylor, M.S. Kang, T.L. Bowlin, P.S. Liu, A.S. Tyms, and A. Sjoerdsma, Anti-HIV activity of castanospermine analogues., Lancet, i:1206 (1989).

7. H. Geyer, C. Holschbach, G. Hunsmann and J. Schneider, Carbohydrates of human immunodeficiency virus. Structures of oligosaccharides linked to the envelope glycoprotein 120, J. Biol. Chem., 263: 11760 (1988).

8. T. Mizuochi, M.W. Spellman, M. Larkin, J. Solomon, L.J. Basa and T. Feizi, Carbohydrate structures of the human-immunodeficiency-virus (HIV) recombinant envelope glycoprotein gp120 produced in chinese-hamster ovary cells, Biochem. J., 254:599 (1988).

9. T. Mizuochi, T.J. Matthews, M. Kato, J. Hamako, K. Titani, J. Solomon and T. Feizi, Diversity of oligosaccharide structures on the envelope glycoprotein gp120 of human immunodeficiency virus 1 from the lymphoblastoid cell line H9, J. Biol. Chem., 265:8519 (1990).

10. C.K. Leonard, M.W. Spellman, L. Riddle, R.J. Harris, J.N. Thomas and T.J. Gregory, Assignment of intrachain disulfide bonds and characterization of potential glycosylation sites of the type 1 recombinant human immunodeficiency virus envelope glycoprotein (gp120) expressed in Chinese hamster ovary cells, J. Biol. Chem., 265:10373 (1990).

11. T.D. Butters, I. Jones, I., V.A. Clarke and G.S. Jacob, The effect of glycosidase inhibition on the N-glycans of HIV gp 120 expressed in lepidopteran cells, Glycoconjugate J., 8:240 (1991).

12. Y. Morikawa, H.A. Overton, J.P. Moore, A.J. Wilkinson, R.L. Brady, S.J. Lewis and I.M. Jones, Expression of HIV-1 gp120 and human soluble CD4 by recombinant baculoviruses and their interaction in vitro, Aids Res. Hum. Retro, 6:765 (1990).

13. E. Fenouillet, J. C. Gluckman and E. Bahraoui, Role of N-linked glycans of envelope glycoproteins in infectivity of human immunodeficiency virus type 1, J. Virol., 64:2841 (1990).

14. A.D. Elbein, Glycosylation inhibitors for N-Linked glycoproteins, in "Methods in Enzymology", Academic Press, New York (1987).

15. R. Datema, S. Olofsson and P.A. Romero, Inhibitors of protein glycosylation and glycoprotein processing in viral systems, Pharmac. Ther., 33:221 (1987).

16. J. Schweden, C. Borgmann, G. Legler and E. Bause, Characterization of calf liver glucosidase I and its inhibition by basic sugar analogs, Arch. Biochem. Biophys., 248:335 (1986).

17. L. Ratner, N.V. Heyden and D. Dedera, Inhibition of HIV and SIV infectivity by blockade of α-glucosidase activity, Virology, 181:180 (1991).

18. J.M. McCune, L.B. Rabin, M.B. Feinberg, M. Lieberman, J.C. Kosek, G.R. Reyes and I.L. Weissman, Endoproteolytic cleavage of gp160 is required for the activation of human immunodeficiency virus, Cell, 53:55 (1988).

19. E.O. Freed, D.J. Myers and R. Risser, Mutational analysis of the cleavage sequence of the human immunodeficiency virus type 1 envelope glycoprotein precursor gp160, J. Virol., 63:4670 (1989).

20. V. Bosch and M. Pawlita, Mutational analysis of the human immunodeficiency virus type 1 env gene product proteolytic cleavage site, J. Virol., 64:2337 (1990).

21. H.-G. Guo, F.D. Veronese, E. Tschachler, R. Pal, V.S. Kalyanaraman, R.C. Gallo and M.S. Reitz, Characterization of an HIV-1 point mutant blocked in envelope glycoprotein cleavage, Virology, 174:217 (1990).

22. T. Hattori, A. Koito, K. Takatsuki, H. Kido and N. Katunuma, Involvement of tryptase-related cellular protease(s) in human immunodeficiency virus type 1 infection, FEBS Letts., 248:48 (1989).

23. P.E. Stephens, G. Clements, G.T. Yarranton and J.P. Moore, A chink in HIV's armour?, Nature, 343:219 (1990).

24. G.J. Clements, M.J. Price-Jones, P.E. Stephens, C. Sutton, T.F. Schulz, P.R. Clapham, J.A. McKeating, M.O. McClure, S. Thomson, M. Marsh, J. Kay, R.A. Weiss and J.P. Moore, The V3 loops of the HIV-1 and HIV-2 surface glycoproteins contain proteolytic cleavage site: a possible function in viral fusion?, AIDS Res. Hum. Retro., 7:3 (1991).

25. L.A. Ivanoff, D.J. Looney, C. McDanal, J.F. Morris, F. Wong-Staal, A.J. Langlois, S. R. Petteway, and T. J. Matthews, Alteration of HIV-1 infectivity and neutralization by a single amino acid replacement in the V3 loop domain, AIDS Res. Hum. Retro., 7:595 (1991).

26. I.M. Jones and G.S. Jacob, Anti-HIV drug mechanism, Nature, 352:198 (1991).

27. A.M. Scofield, L.E. Fellows, R.J. Nash,. and G.W.J. Fleet, Inhibition of mammalian digestive disaccharidases by polyhydroxy alkaloids, Life Sciences, 39:645 (1988).

28. Y, Yoshikuni, Y. Ezure, Y. Aoyagi, and H. Enomoto, Inhibition of intestinal α-glucosidase and postprandial hyperglycemia by N-subtitued moranoline derivatives, J. Pharmacobio-Dyn., 11:356 (1988).

29. Y. Yoshikuni, Inhibition of intestinal α-glucosidase and postprandial hyperglycemia by moranoline and its N-alkyl derivatives, Agric. Biol. Chem., 52:121-128 (1988).

30. P.H.Joubert., C.P.Venter., H.F.Joubert and I.Hillebrand, The effect of a 1-deoxynojirimycin derivative on postprandial blood glucose and insulin levels in healthy black and white volunteers, Eur. J. Pharmacol., 28:705 (1985).

31. P.H. Joubert, W.J. Bam and N. Manyane, Effect of an alpha-glucosidase inhibitor (Bay m 1099) on post-prandial blood glucose and insulin in type II diabetics, Eur. J. Clin. Pharmacol. 30:253 (1986).

32. Ch. Schnack, G. Röggla, A. Luger and G. Schernthaner, Effect of the α-glucosidase inhibitor 1-deoxynojirimycin (Bay m 1099) on postprandial blood glucose, serum insulin and C-peptide levels in type II diabetic patients, Eur. J. Clin. Pharmacol. 30:417 (1986).

33. I. Hillebrand, K. Boehme, K.H. Graefe, and K. Wehling, The effect of new α-glucosidase inhibitors (BAY m1099 and BAY o 1248) on meal-stimulated increases in glucose and insulin levels in man, Klin. Wochenschr., 6:393 (1986).

34. G. Dimitriadis, S. Raptis, A. Raptis, E. Hatziagelaki, A, Mitrakou, P. Halvatsiotis, S. Ladas and I. Hillebrand, Effects of two new α–glucosidase inhibitors on glycemic control in patients with insulin-dependent diabetes mellitus. Klin. Wochensch. 64:415 (1986).

35. G. Hanozet, H-P., Pircher, P. Vanni, B. Oesch, and G. Semenza, An example of enzyme hysteresis. The slow and tight interaction of some fully competitive inhibitors with small intestinal sucrase. J. Biol. Chem., 256:3703 (1981).

36. G. Trugnan, M. Rousset and A. Zweibaum, Castanospermine: a potent inhibitor of sucrase from the human enterocyte-like cell line Caco-2, FEBS Letts., 195:28 (1986).

37. H.N. Fernley, Mammalian alkaline phosphatases, in "The Enzymes", P.D. Boyer, ed., Academic Press, New York and London (1971).

38. R.J. Phillpotts, P.G. Higgins, J.S. Willman, D.A.J. Tyrrell and I. Lenox-Smith, Evaluation of the antirhinovirus Chalcone Ro 09-0415 given orally to volunteers, J. Antimicrobial Chemotherapy 14:403 (1984).

39. R.J. Bernacki, M.J. Niedbala and W. Korytnyk, Glycosidases in cancer and invasion, Cancer and Metast. Rev., 4:81 (1985).

PLANT PROTEINS WITH ANTIVIRAL ACTIVITY
AGAINST HUMAN IMMUNODEFICIENCY VIRUS

Sylvia Lee-Huang,[1*] Hao-Chia Chen,[2] Hsiang-fu Kung,[3] Philip L. Huang,[4]
Peter L. Nara,[5] Bao-Qun Li,[6] Peter Huang,[4] Henry I. Huang,[4] and
Paul L. Huang[4†]

[1] Department of Biochemistry, New York University School of Medicine,
New York, New York 10016

[2] Endocrinology and Reproduction Research Branch, National Institutes of
Child Health and Human Development, Bethesda, MD 20892

[3] Laboratory of Biochemical Physiology, National Cancer Institute-Frederick
Cancer Research and Development Center, Frederick, MD 21701

[4] American Biosciences, New York, New York 10021

[5] Laboratory of Tumor Cell Biology, National Cancer Institute-Frederick
Cancer Research and Development Center, Frederick, MD 21701

[6] Biological Carcinogenic Development Program Resources, Inc., National
Cancer Institute-Frederick Cancer Research and Development Center,
Frederick, MD 21701

[*] Correspondence Author

[†] Present Address: Harvard Medical School and Massachusetts General
Hospital, Boston, MA. 02114

INTRODUCTION

Human Immunodeficiency Virus (HIV), the etiological agent of acquired immunodeficiency syndrome (AIDS), is a member of the lentiviruses, a subfamily of retroviruses[1-6]. Unlike other retro-oncoviruses, HIV is not known to cause cancer in humans and other animals, but it does present a formidable challenge to the host. HIV integrates its genetic information into the genome of the host. The viral genome contains

Natural Products as Antiviral Agents, Edited by C.K. Chu
and H.G. Cutler, Plenum Press, New York, 1992

many regulatory elements that allow the virus to control its rate of replication in both resting and dividing cells. Most importantly, HIV infects and invades cells of the immune system and renders the patient susceptible to opportunistic infections and neoplasms.

It is estimated that 2 million people in the United States are currently infected with HIV, and between 5 to 10 million people are infected worldwide. Recent projections indicate that the majority of those now infected will develop AIDS within a 7 year follow-up period. In 1991 alone, over 150,000 cases of AIDS have been reported domestically and more than half of these patients have died. It is apparent that AIDS is an unprecedented threat to national as well as global health. The search for effective therapies to treat AIDS is of paramount importance.

HIV is tropic and cytopathic to T4 lymphocytes, cells of the immune system that express the cell surface differentiation antigen CD4 (T4, Leu3). The interaction between CD4 and the viral envelope glycoprotein gp120 seems to be the basis for the viral tropism[7]; it mediates the infection of susceptible cells by HIV, and is also responsible for viral-induced fusion of infected and uninfected cells[7,8]. Cell fusion results in the formation of multinucleated syncytia and progressive depletion of CD4 cells, leading to irreversible immuno-incompetence with a high mortality rate.

In addition to CD4-positive T cells, HIV is also capable of infecting other types of cells including mononuclear phagocytic cells[7,8] such as blood monocytes, tissue macrophages, lymph node dendritic reticulum cells, and Langerhans cells of the skin. HIV is also neurotropic, capable of invading monocytes and macrophages in the central nervous system, causing severe neurologic damage. Monocytes/macrophages are major reservoirs of HIV. These cells may interact and fuse with CD4 bearing T cells, causing syncytia formation and T cell depletion, thus contributing to the development and pathogenesis of AIDS.

Many stages of the HIV life cycle are potential targets for the development of drug interventions. Significant efforts have been focused on the use of nucleoside analogues such as AZT (3'-azidothymidine) and dideoxynucleoside derivatives (ddA, ddT, ddC). Although these drugs have significantly improved the survival of AIDS patients, several limitations complicate their clinical applications. The reported toxic side effects have been so severe that many AIDS patients are unable to tolerate full-dose treatment. Even those who initially respond to AZT often stop responding after the first year, suggesting that the virus may become resistant to the drug[9]. The emergence of AZT-resistant HIV strains also limits the long term effectiveness of this drug[10]. Furthermore, these drugs inhibit the viral enzyme reverse transcriptase (RT)[11,12] which is required for the early stage of viral infection; thus, they specifically inhibit new viral infection. Once viral infection has been established within a cell, viral replication is then carried out using host cell enzymes, and RT inhibitors would be expected to have limited effect. In addition, fusion of an infected macrophage with a CD4-positive target cell may result in passage of the extrachromosomal proviral DNA into uninfected cells, bypassing reverse transcription of viral RNA by viral RT. Indeed, *in vitro* studies have demonstrated HIV replication even in the continued presence of nucleoside analogues in prolonged culture.

Drugs that target other viral processes are also being developed, including soluble CD4 and dextran sulfate to inhibit viral binding, alpha interferon and ampligen to inhibit viral budding, castanospermine to inhibit the processing of the viral glycoprotein, and anti-sense RNA to hybrid-arrest viral RNA[12]. These drugs are at various stages of testing. While some preliminary studies are encouraging, some of these drugs show restricted efficacy and nonspecific toxicity. These drugs act against a single site in the viral replicative cycle, and are ineffective in the elimination of latent virus in already infected cells. Thus, the search for drugs that are capable of attacking multiple sites in the HIV life cycle is urgently needed.

In this chapter, we report the isolation, purification and characterization of a new class of anti-HIV agents from plants. These compounds exhibit dose-dependent inhibition of both HIV-1 infection and replication, as measured by syncytium formation, viral core protein p24 expression and HIV-associated reverse transcriptase (RT) activity. Furthermore, these agents are not toxic to normal, uninfected cells, nor to intact animals. The therapeutic indices of these compounds are in the order of 10^4, suggesting that they may be a group of potentially useful agents in the treatment of AIDS. The N-terminal amino acid sequences of these proteins have been determined and their structure-function features have been studied.

PLANTS AS SOURCES OF MEDICINES

About three quarters of the world population rely mainly on plants and plant extracts for health care. Currently, more than 120 clinically useful prescription drugs are derived from plants. Most of these were developed because of their use in traditional medicine[13]. Some common examples are shown in Table 1. They include the antitumor drugs vinblastine and vincristine from a species of rose, rosy periwinkle known in China as Chang Chun Hwa, meaning "forever spring flower," the cardiac medications digitalis from the foxglove plant and quinidine from the bark of the cinchona tree, and the narcotic sedative codeine and morphine derivatives from the opium poppy.

Table 1. Examples of useful drugs derived from plants

Drug	Medical Use	Plant Source
Vinblastine	Antitumor	*Vinca rosea*
Vincristine	Antitumor	*Vinca rosea*
Digitalis	Cardiac	*Digitalis purpurea*
Quinidine	Cardiac, antimalaria	Cinchona tree bark
Codeine	Narcotic, sedative	Opium poppy

A NEW CLASS OF ANTI-HIV PROTEINS FROM PLANTS

In the past several years, we have searched for antitumor and antiviral activities in a variety of plant extracts. From about 200 plants examined, we have isolated and purified to homogeneity several compounds with potent anti-HIV activity and low cytotoxicity. Some of these compounds are proteins, whereas others are heterocyclic compounds. In this chapter, we will discuss our studies on the anti-HIV proteins. These proteins are isolated from distinct and unrelated medicinal plant species. Their names are derived from the genus of their plant source and their molecular weight in kilodaltons, as listed in Table 2.

Table 2. Anti-HIV plant proteins and their sources

Plant Source	Anti-HIV Protein
Momordica charantia	MAP 30 (Momordica Anti-HIV Protein, 30 kD)
Trichosanthes kirilowii	TAP 29 (Trichosanthes Anti-HIV Protein, 29 kD)
Gelonium multiflorum	GAP 31 (Gelonium Anti-HIV Protein, 31 kD)
Dianthus caryophyllus	DAP 32, 30 (Dianthus Anti-HIV Proteins, 32, 30 kD)

MAP 30 (Momordica Anti-HIV Protein, MW 30 kD) is isolated from the matured fruits and seeds of the bitter melon *M. charantia*, a medicinal plant indigenous to China. Extracts of this plant have been used for centuries as antiviral, antitumor and immunopotentiating agents[14]. In recent years, several proteins isolated from this plant were found to inhibit the formation of prostate adenocarcinoma in rats[15] and lymphoma in mice[16].

TAP 29 (Trichosanthes Anti-HIV Protein, MW 29 kD) is isolated from the root tubers of the Chinese traditional medicinal plant *Trichosanthes kirilowii*. The medical uses of this plant have been recorded since 1596[14]. They include the induction of labor in childbirth[17] and the treatment of trophoblastic tumors[18]. TAP 29 is different from trichosanthin in size, cytotoxicity and known N-terminal amino acid sequence, although both compounds are isolated from the same plant.

GAP 31 (Gelonium Anti-HIV Protein, MW 31 kD) is isolated from the seeds of *Gelonium multiflorum*, an ancient medicinal plant from the Himalaya mountains. This plant has also been widely used for medicinal purposes in Germany and Switzerland.

DAP 32 and DAP 30 (Dianthus Anti-HIV Proteins, MW 32 and 30 kD) are isolated from the leaves of *Dianthus caryophyllus*, commonly known as carnation. DAP 32 and DAP 30 are found most abundantly in the original variety of pink carnation. Extracts from *G. multiflorum* and *D. caryophyllus* have been reported to inhibit the multiplication of Herpes simplex virus-1 and poliovirus I in Hep-2 cells[19].

ISOLATION AND PURIFICATION OF THE ANTI-HIV PROTEINS

The anti-HIV agents we have isolated are basic proteins (pI 8-10) with similar sizes (MW 26-32 kD). Their isolation and purification principles are very similar. The general steps involved in their isolation and purification are shown in Table 3.

Table 3. Steps in the isolation and purification of anti-HIV proteins from plants

Initial extraction in 10mM sodium phosphate buffer, 0.15 M NaCl, pH 6-7.
Ammonium sulfate fractionation, 25-85% saturation
Cationic column chromatography on carboxymethyl cellulose
Affinity chromatography on Conconavalin A-Sepharose
Gel Filtration on Sephadex G75

Initial Extraction

Shelled seeds of *Momordica charantia* and *Gelonium multiflorum*, cleaned leaves of *Dianthus caryophyllus*, and root tubers of *Trichosanthes kirilowii*, were used as the starting material for the preparation of MAP 30, GAP 31, DAP 32, DAP 30 and TAP 29 respectively. Initial extraction was carried out in 10-20 mM sodium phosphate buffer, pH 6.4 to 7.4 containing 0.15 M NaCl (buffer A). All procedures were conducted at 4°C unless otherwise specified. The starting material was homogenized into fine powder in a blender for 5 to 10 minutes, it was then extracted with buffer A.

The amount of buffer A used depended on the type of plant tissue. For extraction from seeds, as in the preparation of MAP 30 and GAP 31, a ratio of about 5 ml buffer per gram of seed was used. For extraction from leaves, as in the preparation of DAP 30 and DAP 32, a ratio of 2.5 ml buffer per gram of leaf was used. For extraction from root tubers, as in the preparation of TAP 29, a ratio of 4 ml buffer per gram of root tuber was used.

Ammonium Sulfate Fractionation

The extract was stirred gently overnight followed by centrifugation at 16,000g for 30 minutes to remove cell debris. The supernatant was fractionated by ammonium sulfate precipitation. The fraction between 25-85% saturation of ammonium sulfate contained the bulk of the anti-HIV activity. The precipitate was dissolved in a minimum volume of 5 mM sodium phosphate, pH 6.4 (buffer B) and dialyzed exhaustively against the same buffer.

Cationic Column Chromatography on Carboxymethyl Cellulose

The dialyzed solution was centrifuged at 12,000g for 30 minutes to remove any precipitate formed during dialysis. The solution was concentrated using a Centriprep 10 concentrator (Amicon). The sample was then loaded onto a column of carboxymethyl cellulose (CM 52, Whatman) equilibrated with buffer B. For standard preparation.s of 60-100 g of starting material, a column of 1.6 x 36 cm was used. The column was washed with the same buffer until the baseline of A_{280} was reached. The majority of the impurities was excluded from the column while the anti-HIV protein was retained. The latter was eluted with a linear gradient composed of 0 to 300 mM NaCl in buffer B (15 x column volume). Under these conditions, MAP 30 eluted at 70 to 100 mM NaCl and GAP 31 eluted at 200 to 260 mM NaCl. DAP 30 and DAP 32 eluted at about 140 mM and 180 mM NaCl, respectively. TAP 29 eluted at 100 to 130 mM NaCl.

The elution profiles of these anti-HIV proteins vary with the size of the column, sample, gradient, as well as flow rate and other chromatographic conditions. The active fractions were pooled and concentrated by ammonium sulfate precipitation to 85% saturation. The sample was then dialyzed against buffer B for further purification. In this and subsequent steps, the chromatographic procedures were conducted at room temperature, whereas fractionation and dialysis were carried out at 4°C.

Affinity Chromatography on Con A-Sepharose CL 4B

Some of the anti-HIV agents are glycoproteins, such as GAP 31, DAP 32 and DAP 30. These compounds were effectively purified by affinity chromatography on Con A-Sepharose CL4B (Pharmacia-LKB). Impurities were excluded from the column, whereas the anti-HIV protein was selectively bound to Con A and retained on the column. GAP 31 was eluted with 0.1 M α-methylmannoside in buffer B. DAPs 30 and 32 were eluted with 0.25 M α-methylmannoside in buffer B or by 60 mM sodium borate in 10 mM Tris-HCl buffer, pH 7.4.

Gel Filtration on Sephadex G 75

The pooled sample of active fractions was dialyzed against 20mM sodium phosphate buffer, pH 7.2 (buffer C). The precipitate formed during dialysis was removed by centrifugation at 12,000g for 30 minutes. The sample was then concentrated using a Centricon B15 concentrator (Amicon) to about 2ml, and subjected to gel filtration on Sephadex G75 superfine. A column of 1.5 x 86 cm in buffer C was used, with a flow rate of 2 ml/hour. 1 ml fractions were collected. Homogeneous preparations of DAP 32, DAP 30, GAP31, MAP 30 and TAP 29 eluted between 0.42 and 0.48 column volumes, depending on the size, shape and hydration of the anti-HIV proteins.

PHYSICAL CHARACTERIZATION: HOMOGENEITY, SIZE AND SUBUNIT STRUCTURE

The size, homogeneity and subunit structure of these anti-HIV proteins were determined by SDS-PAGE in the presence and absence of the reducing agent β–mercaptoethanol. As seen in Figure 1, a single band with molecular weights about 30, 29, 31, 30 and 32 kD was obtained for MAP 30, TAP 29, GAP 31, DAP 30 and DAP 32 respectively, in the presence of β-mercaptoethanol. Identical results were obtained under non-reducing conditions, indicating that these molecules are homogeneous single chain polypeptides. A single band corresponding to 26 kD was obtained for trichosanthin.

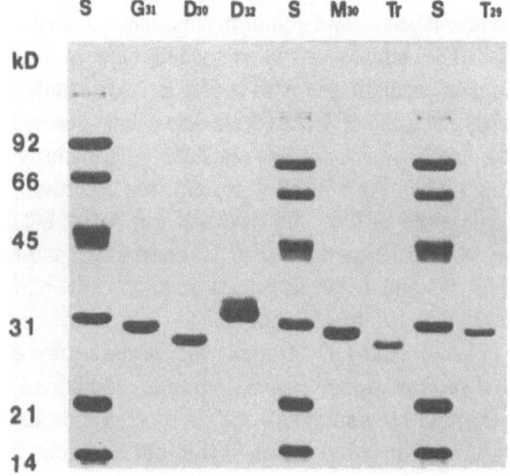

Figure 1. SDS-PAGE of GAP 31, DAP 30, DAP 32, MAP 30, and TAP 29 Electrophoresis was carried out in the presence of β-mercaptoethanol in 10% acrylamide at a constant voltage of 90 V for 5 hour. The gel was stained with silver stain. Lanes: S, molecular weight standards, 2 µg each; G31, GAP 31, 2µg; D30, DAP 30, 2 µg; D32, DAP 32, 3 µg; M30, MAP 30, 3 µg; T29, TAP 29, 2 µg; and Tr, trichosanthin, 2 µg.

ANTI-HIV ACTIVITY, CYTOTOXICITY AND TOXICITY

Cell, Virus and Media

The anti-HIV activity of these plant proteins was measured in terms of their effects on viral infection and replication. Infection was measured by microtiter syncytium formation in the infectious cell center assays[20]. Replication was measured by viral core protein p24 expression[21] and viral-associated RT activity[22].

The CEM-ss (syncytium sensitive, Leu-3 positive) cell line was used as the indicator cells for the microtiter syncytial-forming assay. The H9 cell line was used for p24 expression and viral-associated RT activity assays. HIV-1 virus was prepared and stocked as described previously[21]. Cell lines were cultured in RPMI-1640 medium containing 100 U/ml of penicillin-streptomycin and 10% heat-inactivated fetal calf serum. The cytotoxicity of these compounds was measured by their effect on cellular DNA and protein syntheses in uninfected cells.

The Effect of Anti-HIV Proteins on Syncytium Formation

The indicator cells were pre-treated with 50 μl of the anti-HIV protein at various concentrations for 90 minutes. At the end of this time, 50 μl of a frozen pre-titered HIV stock from HxB3/H9 cells, corresponding to about one hundred syncytial forming units, was added to each well. After 60 minutes, the supernatant containing the virus and antiviral agents was then removed, and the cells were washed with complete medium to remove residual free virus. The wells were then filled with 200 μl medium containing the antiviral agent at the same original concentration. The plates were incubated at 37°C in a humidified incubator at 5% CO_2. Focal syncytium formation, representing single infectious virion units, was scored at day 5 by examination under an inverted microscope. Control wells of uninfected indicator cells were also treated with each of these compounds at various concentrations for the determination of the cytotoxicity of these anti-HIV agents.

This assay quantitates acute cell free HIV-1 infection, and is based on the interaction between fusigenic virus-infected cells expressing the HIV envelope gene products and uninfected adjacent cells bearing CD4 molecules. The results were expressed as percent infectious cell center (% ICC). The % ICC is defined as the ratio of average number of syncytia in anti-HIV agent treated samples to the average number of syncytia in untreated control samples. The results are shown in Figure 2. All of the anti-HIV agents exhibited dose-dependent inhibition of syncytium formation. ID_{50}s of 0.31, 0.34, 0.28, 0.83, and 0.76 nM were obtained for MAP 30, TAP 29, GAP 31, DAP 30 and DAP 32 respectively. No cytotoxic or cytostatic effect was observed under the assay conditions. These results suggest that these anti-HIV proteins affect initial HIV infection as well as the transmission of viral gene products by cell contact or release of free virions.

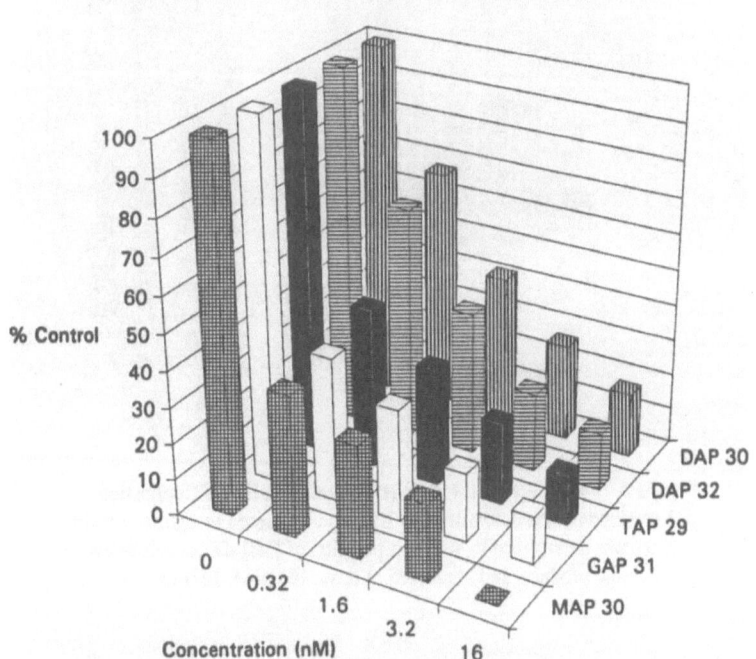

Figure 2. The effects of the Anti-HIV Proteins on HIV-1 Infection
Infectious cell center (% ICC) are expressed in terms of Vn/Vo, where Vn and Vo are average numbers of syncytia in the drug-treated and untreated samples respectively. Results given in this figure represent averages of triplicates from two independent experiments.

The Effects of the Anti-HIV Proteins on p24 Expression and HIV-RT Activity

The effect of the anti-HIV agents on HIV-1 replication and transmission *in vitro* was tested by assaying viral core protein p24 expression and viral-RT activity in suspension cultures of HIV-infected H9 cells. A log phase culture of the target cells was inoculated with a viral stock at a multiplicity of infection of 5×10^{-3}. Cells at 5×10^7/ml were incubated with the inoculum at 37°C for 60 minutes to allow viral absorption, and unbound virus was removed by washing with medium. The cells were then resuspended in culture medium and plated at 1×10^5/ml with or without drug for the duration of the experiment. Under the assay conditions, viral production peaks on day 4. Thus, the cultures were harvested on day 4 and cell-free supernatants of the cultures were collected. The amount of p24 was measured by radioimmunoassay and expressed in terms of ng/ml. HIV-RT activity was determined by the incorporation of [3H]-labeled thymidine into trichloroacetic acid (TCA) precipitable material, using poly(rA).p(dT)12-18 as primer-template, expressed in terms of cpm/ml. The effect of these anti-HIV proteins on viral core protein p24 expression and HIV-RT activity are shown in Figure 3.

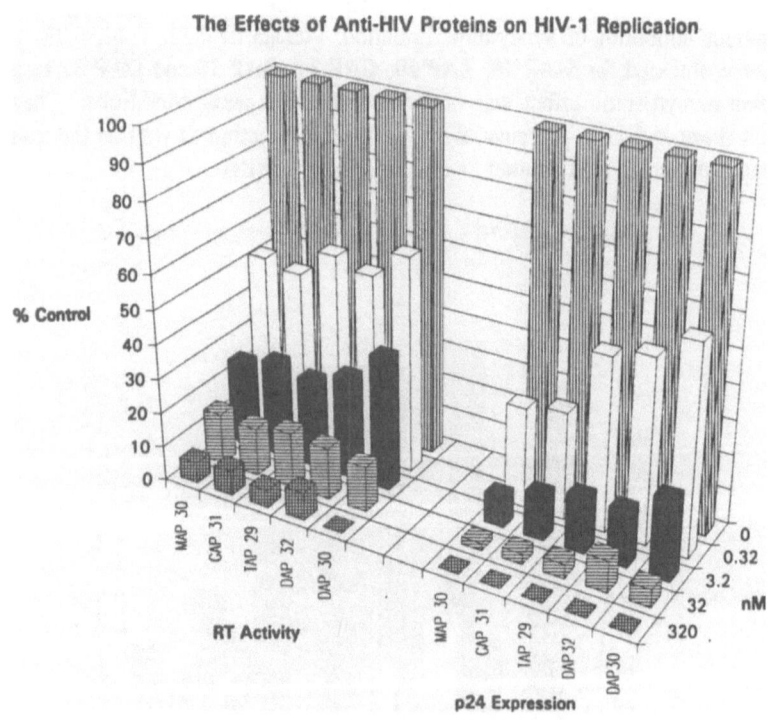

Figure 3. The Effects of the Anti-HIV Agents on HIV-1 Replication
p24 production was determined by RIA and expressed in ng/ml. Viral-RT
activity was determined by the incorporation of [3H]-dTTP into TCA-
precipitable products and expressed in terms of cpm x 10^3/ml.

Each of these agents demonstrated a dose-dependent inhibition on viral core protein p24 expression and HIV-RT activity. The ID$_{50}$s for MAP 30, TAP 29, GAP 31, DAP 30 and DAP 32 were 0.22, 0.37, 0.23, 0.86 and 0.71 nM for p24 expression, and 0.33, 0.46, 0.32, 0.88 and 0.76 nM for HIV-RT activity. The reduction in p24 expression was not due to cytotoxic or cytostatic effect. No inhibition of DNA or protein synthesis was detected at the dose level of the assay. The decrease in HIV-RT activity is likely to be due to a reduction in virion production, which is also evidenced by reduced p24 expression.

Cytotoxicity: Effects on Cellular DNA and Protein Syntheses

To ascertain that the anti-HIV activities of these proteins are virus-specific, the cytotoxicities of these compounds were studied. Cytotoxicity was measured by the effects of these agents on cellular DNA and protein syntheses in uninfected H9 cells. Target cells were grown in the absence and presence of various amounts of anti-HIV agents. The cultures were pulse-labeled with 1 μCi of [3H]-labeled thymidine or leucine for 8 hours prior to harvesting on day 4. Cellular incorporation of these labeled precursors was determined by scintillation counting. Cell viability was determined by trypan blue dye exclusion. These results are shown in Figure 4.

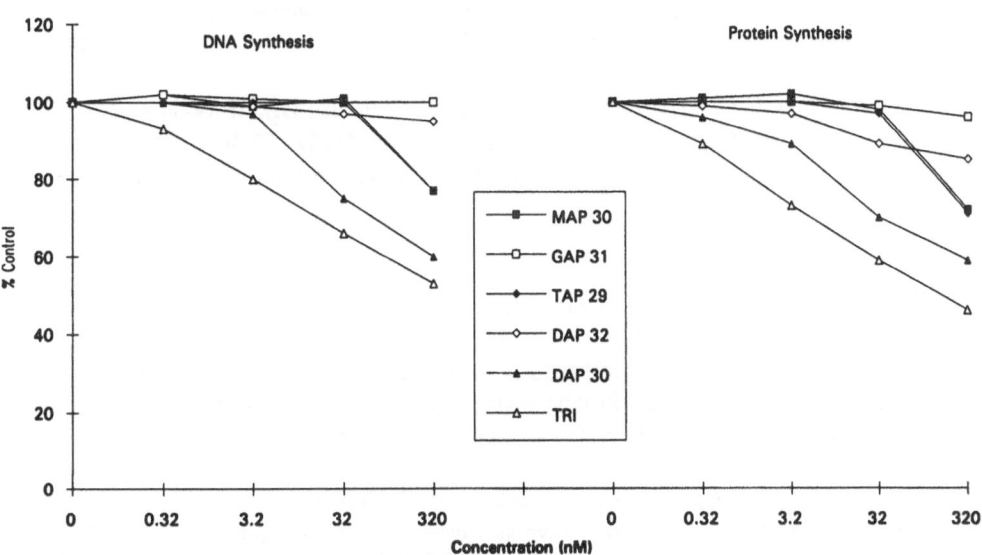

The Effects of Anti-HIV Proteins on Cellular DNA and Protein Syntheses

Figure 4. Cytotoxicities of the Anti-HIV Agents
H9 cells were seeded into 96-well plates at 2×10^4/well, pulse labelled with 1μCi of [3H]-thymidine or leucine for 8 hours, and harvested at day 4. Cellular synthesis of DNA and protein were determined by scintillation counting. Results are normalized to values obtained for control cultures without drug.

Incorporation of [3H]-thymidine or [3H]-leucine into TCA-precipitable DNA or protein was measured and expressed as cpm/ml. Experiments were carried out over a concentration range of three orders of magnitude. From 0.32 to 32 nM, most of these anti-HIV agents caused no detectable effect on cellular incorporation of labeled thymidine or leucine, while they inhibited p24 production and HIV-RT activity by over 90%. At 320 nM, or

1,000 x ID$_{50}$, still no inhibition of cellular DNA or protein synthesis was observed for GAP 31. At this concentration, DAP 32 demonstrated 5 and 15% reduction on the incorporation of [3H]-labeled thymidine and leucine respectively; MAP 30 and TAP 29 showed about 20 to 25% inhibition of these reactions. DAP 30 and trichosanthin exhibited dose-dependent inhibition of cellular synthesis of these macromolecules, at 320 nM, about 40% and 50% inhibition were observed respectively.

Cytotoxicity to uninfected cells in culture may be expressed as toxic dose 50 (TD$_{50}$), the dose at which cellular protein and DNA synthesis is inhibited by 50%. The therapeutic index of these anti-HIV agents, defined as the TD50 divided by the ID50, exceeds 1,000 using any of the three assays of antiviral activity.

Toxicity to Intact Animals

The toxicity of these anti-HIV agents to intact animals was studied on 6-8 week old CF1 mice. Filter-sterilized anti-HIV agents in buffer A were injected intraperitoneally at doses of 0.1, 1, 10 and 100 mg per 100 gm body weight every three days. Control animals received similar injections of sterile buffer A. These experiments were carried out using four mice in each group. Animals were weighed and examined for gross pathological alterations. The animals appeared alert and fed normally. Upon completion of the experiment, the animals were sacrificed and examined. No lesions were observed in any organ in MAP 30, GAP 31, DAP 32 and TAP 29 treated animals. Mild and moderate hepatic hyperplasia was found in trichosanthin treated animals. These results are summarized in Table 5.

GAP 31 showed the least toxicity to intact mice, with an LD$_{50}$ of 59-64 mg/kg. The LD$_{50}$s for MAP 30, TAP 29, DAP 32 and DAP 30 are 56-62, 37-44, 42-46 and 12-16 mg/kg respectively. In comparison, the LD$_{50}$s for trichosanthin and ricin are 5-7 mg/kg and 2-3 ug/kg.

N-TERMINAL AMINO ACID SEQUENCE

The N-terminal 44-66 amino acid sequences of MAP 30, TAP 29, GAP 31, DAP 32 and DAP 30 have been determined. Protein sequencing was carried out by automated Edman degradation using an Applied Biosystems model 470A protein sequencer with on-line PTH analyzer. The results are shown in Figure 5. Comparison of the N-terminal amino acid sequences of these proteins to the N-terminal sequences of trichosanthin (TRI) and ricin A chain (RIC A) residues 7 to 51 are shown as well[23-25]. Boxed regions represent identical or conserved amino acid residues. About 10 to 15 % homology was found among these anti-HIV proteins.

Most homology was found in aromatic (tyrosine and phenylalanine), hydrophobic, and hydroxyl-containing amino acids. GAP 31 shows little homology with DAP 30 and DAP 32 (about 7% identical and 12% conserved). Although the sequences of the two DAPs are highly homologous with 54.5% identity in the 55 amino acid overlap, some unique differences were also found.

For MAP 30, about 34% homology to ricin A chain and 57% homology to trichosanthin were found when both identical and conserved residues are considered. When only identical residues are considered, the homology is reduced to 25% and 43% respectively. Like ricin and trichosanthin, MAP 30 also inhibits *in vitro* translation of eukaryotic cells. Distinct from these compounds, MAP 30 is not toxic to intact normal cells[26].

The N-terminal amino acid sequence of TAP 29 is different from the sequence of trichosanthin[23, 27-29]. In addition to three conservative substitutions, namely, Arg-29 to Lys, Ile-37 to Val, and Pro-42 to Ser, residues 12 to 16 were different. At this position, TAP 29 has -Lys-Lys-Lys-Val-Tyr-, whereas trichosanthin has -Ser-Ser-Tyr-Gly-Val-. Although the two proteins exhibit similar anti-HIV activity, they differ significantly in cytotoxicity as measured by effects on cellular DNA and protein syntheses[27,30].

N-terminal Amino Acid Sequence of the Anti-HIV Plant Proteins
Their Comparison with Other Ribosome-Inactivating Proteins

```
            1                    10                   20
DAP 30    A T A Y T L N L A M P S A S Q Y S X F L D Q I R N N
DAP 32  A V K T I T L N L V S P S A N R Y A T F L T E I R D N
GAP 31  G L D T V S F S T K G A T Y I T Y V N F L N E L R V K
MAP 30      D V N F D L S T A T A K T Y T K F I E D F R A T
TAP 29      D V S F R L S G A T S K K K V Y F I S N L R K A
TRI         D V S F R L S G A T S S S Y G V F I S N L R K A
RIC A       I I N F T T A G A T V Q S Y T N F I R A V R G R

            31                   40                   50
DAP 30    T S L I Y G G T D V A V I G A P S T T D K F L R L N F
DAP 32    R S L D Y S H S G I D V I G A P S S R D S X L N I N F
GAP 31    E G N S H G I P S L R K S S D D P G S S F V V A G
MAP 30    S H K V Y D I P L I R S S I S A P
TAP 29    E K K L Y D I P L V R S S X S G S
TRI       E R K L Y D L P L I R S S L P G S
RIC A     G A D V R H E I P V R L P L P I N
```

Figure 5. N-terminal Amino Acid Sequences of the Anti-HIV Proteins
The sequences of the anti-HIV proteins, trichosanthin (TRI), and ricin A chain (RIC A) are aligned to maximize homology. The numbers correspond to amino acid positions in DAP 32 and GAP 31. Boxed regions are identical or conserved residues.

Multiple forms of ribosome-inactivating proteins have been isolated from *Momordica charantia*[31], *Gelonium multiflorum*[32] and *Dianthus caryophyllus*[33]. However, no amino acid sequence data has been reported. Thus, no comparison can be made with the sequence of our anti-HIV compounds reported here.

Comparison of the amino acid sequences of these anti-HIV proteins to the EMBL data bank revealed that DAP 30 and DAP 32 show significant homology to saporin-6, a ribosome-inactivating protein isolated from *Saponaria officinalis*[34]. As shown in Fgure 6, in the N-terminal 60 amino acid overlap, 65% identity was found between DAP 30 and saporin-6 and 48.3% identity was found between DAP 32 and saporin-6. The finding of this homology led us to assay saporin-6 for antiviral activity and cytotoxicity; indeed, it exhibits the same level of anti-HIV activity and cytotoxicity as DAP 30.

STRUCTURAL AND FUNCTIONAL FEATURES

While the anti-HIV activity of these compounds is similar, their cytotoxicities differ considerably. Little is known about the basis of such differences. In our studies on the structure and function of these proteins, we have observed some interesting features about two pairs of the anti-HIV proteins. One pair consists of DAP 32 and DAP 30, and the other, TAP 29 and trichosanthin.

The members of each pair were isolated from the same source, and have similar anti-HIV activity, but very different cytotoxicity. The size and amino acid sequences of each member are also different. The nontoxic member of the pair is larger in size than the toxic one. The amino acid sequence at positions 10-16 also have unique features. One or more basic amino acid residues are found in the nontoxic anti-HIV proteins in this region, whereas these residues are absent in the toxic anti-HIV proteins. For example, DAP 32 has

Comparison of N-terminal Amino Acid Sequence of DAP 30 and DAP 32 with

Saporin-6 isolated from *Saponaria offocinalis*

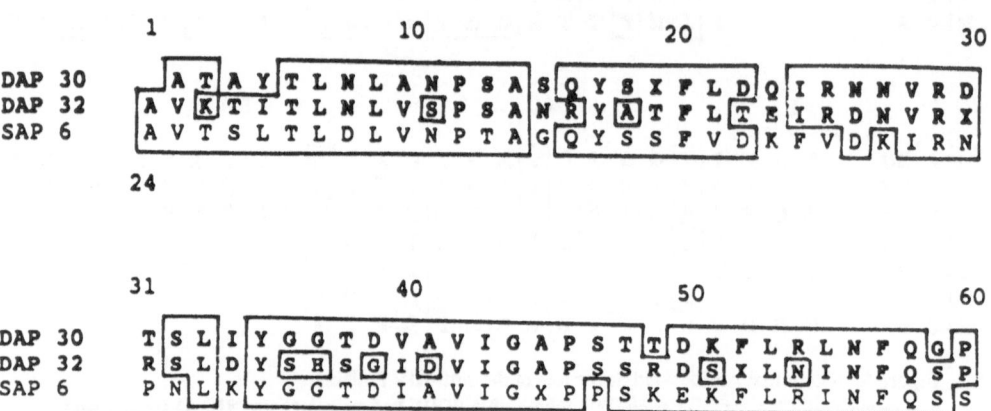

Figure 6. Amino acid sequence homology between DAP 30, DAP 32, and saporin-6 (SAP 6) Sequences are aligned to show maximum homology. Boxed regions are identical or conserved residues.

an Arg at position 16, whereas the corresponding residue in DAP 30 is Tyr. TAP 29 has -Lys-Lys-Lys- at positions 12-14, whereas the corresponding residues in trichosanthin are -Ser-Ser-Tyr-. In view of these results, we further examined the sequence of other nontoxic anti-HIV proteins such as MAP 30 and GAP 31 as well as other relatively toxic proteins such as saporin-6 and ricin A chain. These comparisons are shown in Table 4.

Interestingly, all the nontoxic compounds have one or more lysine or arginine residues in this region. GAP 31 has a Lys at position 10, and MAP 30 has two Lys residues at positions 12 and 16. These residues are absent in the corresponding regions of the toxic proteins saporin-6, trichosanthin, and ricin A chain. Consequently, these results raise the possibility that the absence of basic residues in this unique region may play a role in the cytotoxicity of these anti-HIV proteins. Whether this is because the basic residues contribute potential tryptic sites, positive charge density, or other properties to the proteins is currently being studied by site-specific modification and mutagenesis.

Table 4. Comparison of the N-Terminal Sequence at Positions 10-16 of Anti-HIV Proteins with Other Proteins

Protein	Amino Acid Sequence 10-16	Cytotoxicity
TAP 29	-Thr-Ser-**Lys**-**Lys**-**Lys**-Val-Tyr-	-
MAP 30	-Thr-Ile-**Lys**-Thr-Tyr-Thr-**Lys**-	-
GAP 31	-**Lys**-Gly-Ala-Thr-Tyr-IIe-Thr-	-
DAP 32	-Val-Ser-Pro-Ser-Ala-Asn-**Arg**-	-
DAP 30	-Asn-Pro-Ser-Ala-Ser-Gln-Tyr-	+
SAP-6	-Asn-Pro-Thr-Ala-Gly-Gln-Tyr-	+
TRI	-Thr-Ser-Ser-Ser-Tyr-Gly-Val-	++
RIC A	-Thr-Val-Gln-Ser-Tyr-Thr-Asn-	++++

Basic amino acids, lysine (Lys) and arginine (Arg), are bold faced. SAP-6, saporin-6; TRI, trichosanthin; RIC A, ricin A chain.

OTHER BIOLOGICAL ACTIVITIES

Ribosome-inactivating Activity

N-terminal amino acid sequence analysis indicates that these anti-HIV proteins share homology with ricin A chain, saporin-6, and other ribosome-inactivating proteins (RIPs). These anti-HIV proteins also posess ribosome-inactivating activity, as measured by assaying cell-free protein biosynthesis in a rabbit reticulocyte lysate system with globin message.

The reaction was carried out in a total volume of 25 µl containing 2mM $Mg(OAc)_2$, 80 mM KOAc, 2.5 mM spermidine, 34.5 mg/ml creatine phosphate, 26 mg/ml GTP, 250 mM HEPES buffer, 1 µCi of [3H]-leucine and 1 µg of globin mRNA. The incubation was conducted at 37°C for 30 minutes. The translation products were resolved by thin-layer chromatography. The incorporation of [3H]-labeled leucine into TCA precipitable product was measured by scintillation counting. The results are shown in Figure 7.

Dose-dependent inhibition of [3H] leucine incorporation was observed in all cases. ID_{50}s of 3.3, 3.7, 4.1, 3.2 and 2.3 nM were obtained for MAP 30, TAP 29, GAP 31, DAP 30 and DAP 32 respectively. These values are about 6-10 fold higher than the ID_{50} for the anti-HIV activity of these proteins.

DNA Topological Enzyme Activity

Comparison of the amino acid sequences of these anti-HIV proteins to the EMBL data bank further revealed extensive homology between GAP 31 and *Drosophila* DNA topoisomerase II[35]. As shown in Figure 8, a homology of 48%, with 40% identity, was found between residues 1-40 of GAP 31 and residues 660-699 of *Drosophila* DNA topoisomerase II. These amino acids lie within the DNA breakage and rejoining domain, and are conserved between topoisomerase II and other topoisomerases[35], suggesting that they are functionally important. We thus examined whether these anti-HIV compounds have topological activity on DNA substrates.

We found that GAP 31, DAP 30, and DAP 32 irreversibly relax and decatenate supercoiled DNA and catalyze double-stranded breakage to form linear DNA. The relaxed products are topologically inactive and no longer serve as substrates for DNA gyrase to form supercoils (Figure 9). The details of these studies will be reported elsewhere.[36] The topological effect of these anti-HIV proteins is similar to that of cellular topoisomerase in the presence of topoisomerase poisons. The ability of these anti-HIV agents to interrupt essential topological interconversions of DNA may provide a novel mechanism for their

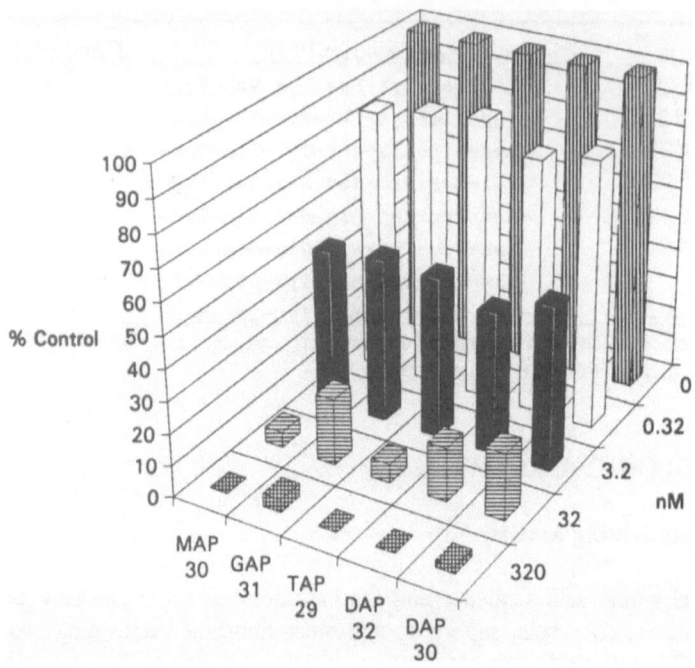

Figure 7. Ribosome-inactivating Activity of the Anti-HIV Proteins
The effect of the anti-HIV proteins on *in vitro* translation of globin message in a rabbit reticulocyte lysate system was measured. Inhibition of protein biosynthesis was determined by the incorporation of [3H]-leucine into TCA-insoluble material as a function of anti-HIV protein concentration.

antiviral and antitumor actions. The presence of this new DNA topological enzyme activity in these plant proteins also suggests that their anti-HIV activity may not be merely a consequence of ribosome inactivation.

Dual Activity Against DNA and RNA

RIPs inhibit ribosomal function via an N-glycosidase activity by hydrolytic cleavage of the glycosidic linkage between the ribose and the adenine or guanine at A4324 or G4323 on the 28S rRNA[37,38]. This action requires that the proteins bind to RNA, recognize a specific nucleotide sequence in rRNA, and cleave the glycoside linkage. The new topological activity of the anti-HIV proteins that we have discovered is specific for DNA. It requires that the anti-HIV proteins bind to DNA, recognize a specific site in DNA, and topologically act on it.

The dual capability of these proteins to act on both RNA and DNA substrates may be related to their potent antiviral activity at multiple stages of the viral life cycle. The DNA topological activity that we have found does not appear to be a universal property of all ribosome-inactivating proteins. Further studies are in progress to define the roles of this unique DNA topological activity in the anti-HIV action of these anti-HIV proteins.

Comparison of N-terminal Amino Acid Sequence of GAP 31 with

DNA Topoisomerase II

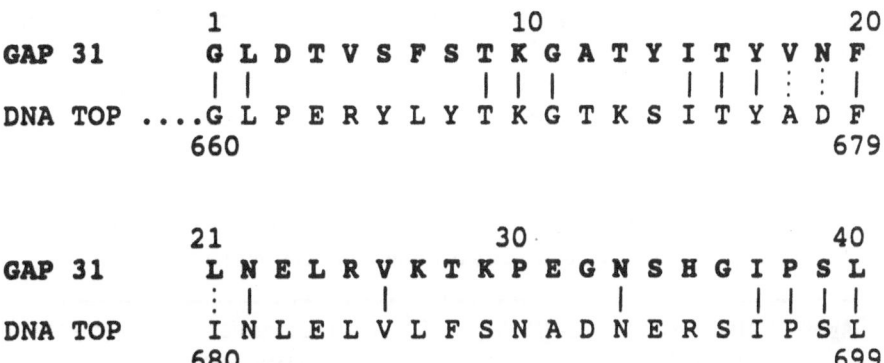

Figure 8. Amino acid sequence homology between GAP 31 and *Drosophila* DNA topoisomerase II Sequences are aligned to show maximum homology. Solid lines indicate identical amino acids. Dotted lines indicate conserved amino acids

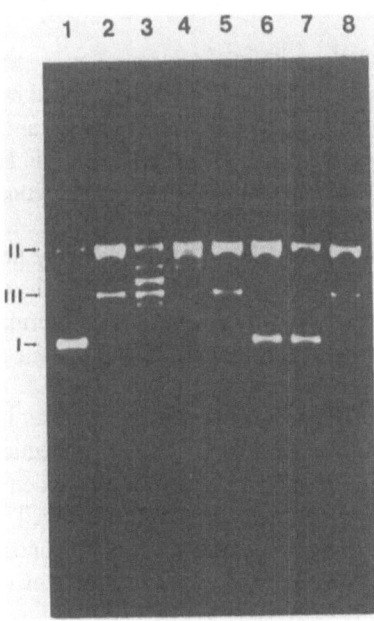

Figure 9. DNA topological enzyme activity of the anti-HIV proteins
Lane 1, pUC19 supercoiled DNA (substrate). Lanes 2, 3 and 4 substrate treated with GAP 31, DNA topoisomerases I and II. Lanes 5, 6 and 7, samples from lanes 2, 3 and 4 treated with DNA gyrase for 2 hours. Lane 8, same as in lane 5, except it was treated with DNA gyrase for 12 hours. These results show that GAP 31 irreversibly converted supercoiled DNA (form I) to topologically inactive relaxed circular (form II) and linear (form III) DNA and that cannot be converted back to supercoils by DNA gyrase. In contrast, topoisomerase I or II relaxed DNA was readily supercoiled by the gyrase.

SUMMARY AND DISCUSSION

In summary, we have isolated and purified to homogeneity a new class of anti-HIV agents from different and unrelated plant species. These agents, MAP 30, TAP 29, GAP 31, DAP 32, and DAP 30, inhibit HIV-1 infection as well as replication, as measured by syncytium formation and viral core protein p24 expression and HIV-associated reverse transcriptase. In the dose range of the assays, these compounds show little or no cytotoxicity in tissue culture or toxicity to intact animals. Although these compounds exhibit ribosome inactivating activity in cell-free translation systems, they have no detectable effect on intact normal cells. The values of ID_{50}, TD_{50} and LD_{50} of these compounds are summarized in Table 5.

Table 5. A Comparison of Ribosome-inactivating Activity, Anti-HIV Activity, Cytotoxicity, and Toxicity of the Anti-HIV Proteins

Agent	RIP Activity ID_{50} (nM)	Anti-HIV Activity ID_{50} (nM)			Cytotoxicity TD_{50} (nM)	Toxicity LD_{50} (mg/kg)
		Syncytia	p24	RT		
MAP 30	3.3	0.22	0.22	0.33	>3000	56-62
TAP 29	3.7	0.34	0.37	0.46	>1000	37-44
GAP 31	4.1	0.28	0.23	0.32	>3200	59-64
DAP 30	3.4	0.76	0.85	0.88	910	12-16
DAP 32	2.3	0.76	0.71	0.76	>3200	42-46
TRI	3.7	0.34	0.37	0.46	340	5-7
RIC						0.003

TRI, trichosanthin, and RIC, ricin

The mechanism of anti-HIV action of these compounds remains to be elucidated. The dual capability of these proteins to act on RNA, as evidenced by their ribosome-inactivating activity, as well as on DNA, as evidenced by their topological activity on supercoiled DNA raises the possibility that they act on specific sequences within HIV DNA or RNA. Clearly, they act at multiple stages of the viral life cycle, affecting both viral infection and replication.

The basis of their specificity for HIV-infected cells may lie in the selective binding or uptake of these agents by viral infected cells; alternatively, they may act on the biochemical differences in infected-cell metabolism as opposed to normal cell metabolism. These mechanisms of action are not mutually exclusive, and some combination may account for the anti-HIV action of these compounds.

This new class of plant-derived anti-HIV agents forms an important group of potential therapeutic drugs in the treatment of AIDS. Their therapeutic indices are in the range of 1,000 to 10,000. The anti-HIV activity and low cytotoxicity of these agents has been confirmed in human peripheral blood monocytes as well as in chronically infected macrophages. The question of how effective these compounds will be, with each other or in combination with other anti-HIV drugs, in the treatment of AIDS, can only be answered by clinical trials.

ACKNOWLEDGMENT

SLH acknowledges the partial support of her work by NIH grant R01 AI 31343.

REFERENCES

1. R.C. Gallo, The first human retrovirus, *Scientific American* 225: 88-89 (1989).
2. R.C. Gallo and L. Montagnier, AIDS in 1988, *Scientific American* 259: 41-48 (1988).
3. A.T. Haase, Pathogenesis of lentivirus infection, *Nature* 22: 130-136 (1986).
4. M. Gonda , The natural history of AIDS, *Natural History* 95: 4 (1986).
5. R.C. Gallo, S.Z. Salahuddin, M. Popovic, G.M. Shearer, M. Kaplan, B.F. Haynes, T.J. Palker, R. Redfield, J. Oleske, B. Safai, G. White, P. Foster, and P.D. Markham, Frequent detection and isolation of cytopathic retroviruses (HTLV-III) from patients with AIDS and at risk for AIDS, *Science* 224: 500 (1984).
6. J.M. Mann, J. Chin, P. Poit, and T. Quinn, The international epidemiology of AIDS, *Scientific American* 259: 4, 82-89 (1988).
7. J.N. Weber and R.A. Weiss, HIV infection: The cellular picture, *Scientific American* 259: 4, 101-109 (1988).
8. A.G. Dalgleish, P.C.L. Beverley, P.R. Clapham, D.H. Crawford, M.F. Greaves, and R.A. Weiss, The CD4 (T4) antigen is an essential component of the receptor for the AIDS retrovirus, *Nature* 312: 763-767 (1984).
9. B.A. Larder, G. Darby, and D.D. Richman, HIV with reduced sensitivity to zidovudine isolated during prolonged therapy, *Science* 243: 1731-1734 (1989).
10. R. Yarchoan, H. Mitsuya, C.E. Myers, and S. Broder, Clinical pharmacology of 3'- azido-2',3'-dideoxythymidine (zidovudine) and related dideoxynucleosides, *N. Engl. J. Med.* 321: 726-738 (1989).
11. H. Mitsuya and S. Broder, Strategies for antiviral therapy in AIDS, *Nature* 325: 773-778 (1987).
12. R. Yarchoan, H. Mitsuya, and S. Broder, AIDS therapies, *Scientific American* 259: 4, 110-119 (1988).
13. P.H. Abelson, Medicine from plants, *Science* 247: 513, (1990).
14. S.C. Li , "Pen Ts'ao Kang Mu (Chinese Pharmaceutical Compendium)" originally published in 1596, reprinted by People's Medical Publishing House, Beijing (1977).
15. M.A. Fletcher, K. Caldwell, A. Clafflin, and T. Malinin, Further characterization of a tumor cell growth inhibitor from the balsam pear, *Proc. Fed. Am. Soc. Exp. Biol.* 39: 414 (1980).
16. C. Jilka, B. Strifler, G.W. Fortner, E.F. Hayes, and D.J. Takemoto, In vivo antitumor activity of the bitter melon (*Momordica charantia*), *Cancer Research* 43: 5151-5155 (1983).
17. K.F. Cheng, Midtrimester abortion induced by Radix trichosanthis: Morphologic observations in placenta and fetus, *Obstet. Gynecol.* 59, 494-498 (1982).
18. Y. Wang, International Symposium on Organic Chemistry of Medicinal and Natural Products, Shanghai, China (1985).
19. L. Foa-Tomasi, G. Campadelli-Fiume, L. Barbieri, and F. Stirpe, Effect of ribosome-inactivating proteins on virus-infected cells. Inhibition of virus multiplication and of protein synthesis, *Arch. Virology* 71: 322-332 (1982).
20. P.L. Nara and P.J. Fischinger, Quantitative infectivity assay for HIV-1 and-2, *Nature* 332: 469-470 (1988).
21. P.L. Nara, W.C. Hatch, N.M. Dunlop, W.G. Robey, L.O. Arthur, M.A. Gonda, and P.J. Fischinger, A simple, rapid, quantitative syncytium-forming microassay for the detection of human immunodeficiency virus neutralizing antibody, *AIDS Res. Human Retroviruses* 3: 283-302 (1987).
22. A.D. Hoffman, B. Banapour, and J.A. Levy, Characterization of the AIDS-associated retrovirus reverse transcriptase and optimal conditions for its detection in virions, *Virology* 147: 326-335 (1985).
23. X. Zhang and J. Wang, Homology of trichosanthin and ricin A chain, *Nature* 321, 477-478 (1986).
24. R.Q. Qian, Z.W. Gu, and X.L. Zhang, Determination of N-terminal partial amino acid sequence of trichosanthin, *Acta Chemica Sinica* 39: 927-931 (1981).

25. Z.W. Gu, R.Q. Qian, S.W. Jin, W.W. Qian, S.Z. Xu, L.Q. Zhang, X.L. Zhang, Y.Z. Yao, Y.F. Liu, S. Q. Zhu, B.S. Cao, S. F. Wang, Q.H. Wang, W.J. Zhang, Y.Z. Liu, Y.H. Fu, and Y. Wang, Chemistry of trichosanthin IV. The amino acid sequence of trichosanthin, *Acta Chemica Sinica* 43: 943-945 (1984).

26. S. Lee-Huang, P.L. Huang, P.L. Nara, H-C. Chen, H-F. Kung, P. Huang, H.I. Huang, and P.L. Huang, MAP 30: A new inhibitor of HIV-1 infection and replication, *FEBS Letts.* 272: 12-18 (1990).

27. S. Lee-Huang, P.L. Huang, H-F. Kung, B-Q. Li, P.L. Huang, P. Huang, H.I. Huang, and H-C. Chen, TAP 29: An anti-human immunodeficiency virus protein from *Trichosanthes kirilowii* that is nontoxic to intact cells, *Proc. Natl. Acad. Sci. USA.* 88: 6570-6574 (1991).

28. E.J. Collins, J.D. Robertus, M. LoPresti, K.L. Stone, K.R. Williams, P. Wu, K. Hwang, and M. Piatak, Primary amino sequence of α-trichosanthin and molecular model for abrin A-chain and α-trichosanthin, *J. Biol. Chem.* 265:8665-8669 (1990).

29. T.P. Chow, R.A. Feldman, M. Lovett, and M. Piatak, Isolation and DNA sequence of a gene encoding α-trichosanthin, a type I ribosome-inactivating protein, *J. Biol. Chem.* 265: 8670-8674 (1990).

30. J. Palaca, Trials and tribulations of AIDS drug testing, *Science* 247:1406 (1990).

31. L. Barbieri, M. Zamboni, E. Lorenzoni, L. Montanaro, S. Sperti, and F. Stirpe, Inhibition of protein synthesis *in vitro* by proteins from the seeds of *Momordica charantia* (bitter pear melon), *Biochem. J.* 186: 443-452 (1980).

32. F. Stirpe, S. Olsnes, and A. Piphl, Gelonin, a new inhibitor of protein synthesis, nontoxic to intact cells, *J. Biol. Chem.* 255: 6947-6953 (1980).

33. F. Stirpe, D.G. Willams, L.J. Onyon, and R.F. Legg, Dianthins, ribosome-damaging proteins with anti-viral properties from *Dianthus caryophyllus* (carnation) *Biochem. J.* 195: 399-405 (1981).

34. L. Benatti, M.B. Saccardo, M. Dani, G. Nitti, M. Sassano, R. Lorenzetti, D.A. Lapppi, and M. Soria, Nucleotide sequence of cDNA coding for saporin-6, a type-1 ribosome-inactivating protein from *Saponaria officinalis, Eur. J. Biochem.* 183: 465- 470 (1989).

35. E. Wyckoff, D. Natalie, J.M. Nolan, M. Lee, and T.S. Hsieh, Stucture of the Drosophila DNA Topoisomerase II Gene. Nucleotide Sequence and Homology Among Topoisomerases II, *J. Mol. Biol.* 205: 1-13 (1989).

36. P.L. Huang, H-C. Chen, H-F. Kung, P.L. Huang, P. Huang, H.I. Huang, and S. Lee-Huang, Anti-HIV plant proteins catalyze topological changes of DNA into inactive forms, *Biofactors* 4: (1992). In Press.

37. Y. Endo, K. Mitsui, M. Motizuki, and K. Tsurugi, The mechanism of action of ricin and related toxic lectins on eukaryotic ribosomes, *J. Biol. Chem.* 262: 5908-5912 (1987).

38. Y. Endo,and K. Tsurugi, RNA N-Glycosidase activity of ricin A-chain, *J. Biol. Chem.* 262: 8128-8132 (1987).

ANTIVIRAL STUDIES WITH TRICHOSANTHIN, A PLANT DERIVED SINGLE CHAIN RIBOSOME INACTIVATING PROTEIN

M.S. McGrath*, K.C. Luk**, H.D. Abrams**, I. Gaston**, S. Santulli*,
S.E. Caldwell**, M. Piatak** and J. D. Lifson**
*San Francisco General Hospital , San Francisco, CA
**Genelabs Incorporated, Redwood City, CA

INTRODUCTION

Ribosome inactivating proteins (RIPs) are a family of plant derived proteins that have the capability to functionally inactivate ribosomes through an enzymatic mechanism (reviewed in 1,2). RIPs can be divided into two classes. Single chain RIPs (SCRIPs) such as trichosanthin consist of a single protein chain with enzymatic activity capable of mediating ribosome inactivation (1,2). Two chain RIPs, or toxins, as typified by ricin and abrin, consist of a B chain that mediates attachment to cell surfaces (through lectin-like binding to specific carbohydrate moieties) and perhaps facilitates translocation through cell membranes. The B chain is covalently linked to an A chain essentially comparable to a SCRIP, that contains ribosome modifying enzyme activity (1,2). Plant derived RIPs are N-glycosidases and depurinate a particular adenine residue in 28S rRNA in a site-specific fashion, resulting in structural modification and functional inactivation of the affected ribosomes, perhaps through impairment of ability to bind elongation factors (3-6).

Two chain RIPs inhibit translation in cell free systems, and are also extremely toxic to intact cells. In constrast, although they inhibit cell free translation with a molar potency comparable to two chain RIPs, SCRIPs are relatively non-toxic to most intact cells, presumably because in the absence of the cell binding and membrane translocation functions conferred on two chain RIPs by their B chains, SCRIPs are not taken up appreciably by intact cells (1,2,6,7).

Although RIPs are ubiquitously distributed within plants, with any single given plant species typically containing multiple related but distinct RIPs, the physiological function served by these proteins remains obscure (1,2,7). It has been speculated that the proteins may play a role in host defense against intracellular pathogens, perhaps being released from sequestered stores upon the occurance of cellular damage, leading to cell death and thereby limiting the replication of invading pathogens and their spread to other cells. Indeed, the first RIP to be characterized was pokeweed antiviral protein (PAP),

Natural Products as Antiviral Agents, Edited by C.K. Chu
and H.G. Cutler, Plenum Press, New York, 1992

derived from *Phytolacca americana*. The designation of the protein as PAP was based on its observed ability to confer protection to plants against challenge by tobacco mosaic virus (TMV) when rubbed on inoculated leaves (7,8).

Following the initial demonstration of anti-TMV activity of PAP, selective antiviral activity of several RIPs was subsequently demonstrated against a variety of different viruses, including both RNA and DNA viruses, and both enveloped and non-enveloped agents, in a variety of different assay systems (7-13). It has been speculated that preferential uptake of the compounds by virally infected cells, compared to healthy cells, leading to preferential cytotoxicity against the infected cells, may underly the observed selective antiviral activity, although this has not been conclusively demonstrated experimentally (14-17). Preferential inhibition of HIV replication in vitro has been demonstrated for trichosanthin and several other RIPs by our laboratory (18,19), and subsequently by others (20-22). In addition, we have demonstrated selective inhibition of viral replication mediated by trichosanthin and other SCRIPs in non-HIV systems, including both in vitro studies and animal model experiments employing human herpes viruses (M. Piatak, et. al., in preparation).

Despite the well documented in vitro antiviral activity of SCRIPS, concerns about potential toxicity have limited efforts at their clinical evaluation as antiviral agents, particularly for non-life-threatening indications. However, extensive prior human experience with at least one SCRIP suggested that this compound might be safely evaluated in vivo as a potential anti-HIV agent. Trichosanthin is a 27 kD basic (calculated pI 9.6) protein isolated from the root tubers of *Trichosanthes kirilowii,* Maximowicz. Trichosanthin containing preparations have been used for hundreds of years in Chinese traditional medicine, for a variety of clinical indications, including induction of abortion and treatment of trophoblastic tumors (23-28); the compound is selectively cytotoxic for syncytiotrophoblast cells in vitro and in vivo (24,28), presumably accounting for its abortifacient activity and utility in treatment of trophoblastic malignancies. More recently and importantly, highly purified trichosanthin preparations have been utilized extensively as a second trimester abortifacient, with excellent results with respect to safety and efficacy (25).

While numerous SCRIPs among those we evaluated showed selective anti-HIV activity in vitro, the extensive experience with prior human use of trichosanthin, along with additional considerations, led us to select it as a lead compound for clinical testing (29). A determined amino acid sequence had been published for the protein (23,30), although our own complete determination of amino acid sequence contained several discrepancies relative to the published sequence (31,32). The protein could be prepared in good yields and purity from a reasonably abundant plant source by relatively straightforward purification approaches. In addition, trichosanthin is not glycosylated (23,31,32), which represents a favorable characteristic as glycosylation had been implicated as an important feature leading to preferential hepatic uptake and toxicity in testing of other RIPs (33). The lack of glycosylation, along with the lack of cysteine residues in the protein (23,32,32), also suggested that prokaryotic recombinant expression (34) would be feasible.

The present communication reviews some of our observations relevant to the anti-HIV activity of trichosanthin, in various in vitro assay systems. Additional information relating to the potential mechanism(s) of antiviral activity of the compound, and a brief review of some aspects of the clinical experience to date in evaluation of the compound in HIV infected patients will also be presented.

ANTI-HIV ACTIVITY OF TRICHOSANTHIN, IN VITRO

Trichosanthin preparations

Unless otherwise indicated, all studies were conducted with various lots of GLQ223, a highly purified form of trichosanthin, formulated for injection. All GLQ223 supplies were prepared and provided by the Pharmaceutical Development Group of Genelabs Incorporated, Redwood City, CA.

Anti-HIV activity of trichosanthin in acute HIV-1 infection of T lymphoblastoid cells

Inhibition of HIV replication was evaluated in acutely infected T lymphoblastoid cells, utilizing the VB cell line and the HIV-1$_{DV}$ isolate at calculated multiplicities of infection of 0.01-0.001, essentially as described (18). Supernatant HIV p24 content was evaluated as a readout parameter reflecting viral replication. [^{3}H]-leucine incorporation was evaluated as an index of overall protein synthesis to assess selectivity of any observed inhibition of HIV replication reflected in decreased culture supernatant p24 content. In a further approach to assessment of the selectivity of any observed inhibition of HIV p24 production, parallel experiments were conducted in which supernatant HIV p24 production and [^{3}H]-leucine incorporation were evaluated for cultures treated with varying concentrations of the non-selective protein synthesis inhibitor cycloheximide.

As shown in Figure 1A, trichosanthin treatment resulted in concentration dependent decreases in the amount of HIV p24 antigen found in culture supernatants, with substantial inhibition of p24 production observed at concentrations that did not meaningfully inhibit overall cellular protein synthesis by either infected or uninfected cultures, as reflected by [^{3}H]-leucine incorporation. In contrast, for the non-selective protein synthesis inhibitor cycloheximide, the observed inhibition of [^{3}H]-leucine incorporation was approximately equal to the observed inhibition of HIV p24 production (Figure 1B). In combination, these observations indicate that trichosanthin mediates a selective inhibition of HIV replication relative to total cellular protein synthesis in acutely infected T lymphoblastoid cells, and suggest that the observed selectivity does not simply reflect a trivial effect such as preferential inhibition of translation of viral or other relatively low abundance transcripts in the face of a low level of global inhibition of protein synthesis.

To assess whether the observed selective inhibition of HIV p24 production was accompanied by inhibition in the production of other viral proteins, we conducted studies with metabolically labelled acutely HIV-1 infected VB cells (M.O.I.= 0.01), and parallel

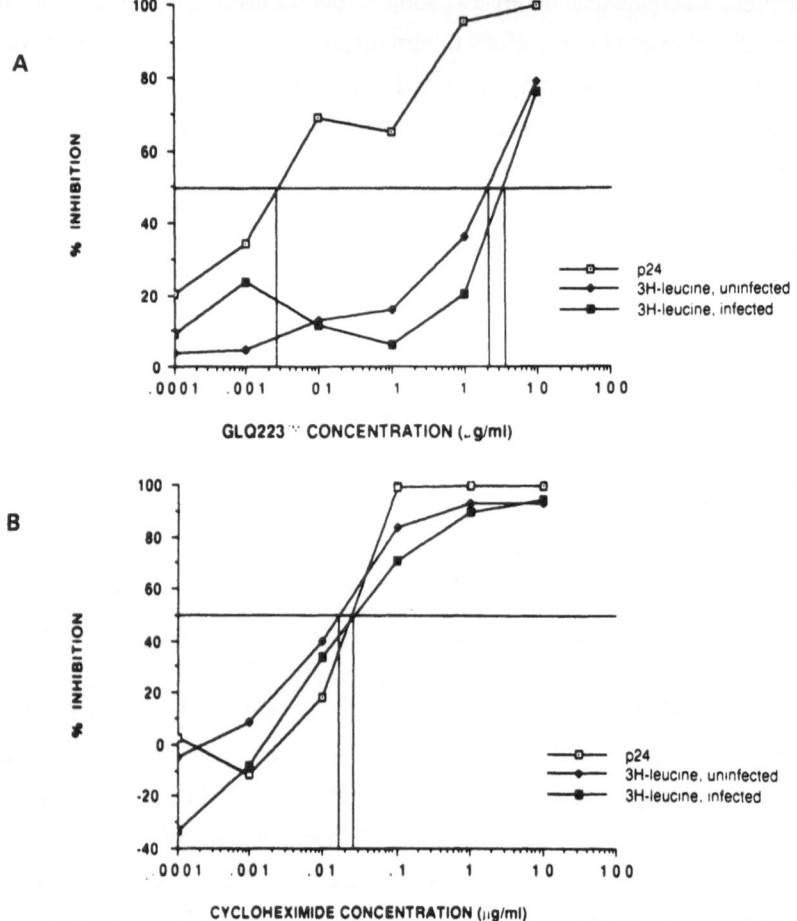

Figure 1. A. GLQ223 treatment mediated inhibition of HIV p24 production, and effects on [³H]-leucine incorporation by infected and uninfected cells. Comparison of concentrations required for 50% inhibition of HIV p24 production and 50% inhibition of [³H]-leucine incorporation yields an in vitro selective index of approximately 1000, indicating antiviral selectivity. B. Cycloheximide treatment mediated inhibition of HIV p24 production and [³H]-leucine incorporation by infected and uninfected cells. Comparison of concentrations required for 50% inhibition of HIV p24 production and 50% inhibition of [³H]-leucine incorporation yields an in vitro selective index of approximately 1, indicating essentially no antiviral selectivity.

cultures of uninfected cells. Metabolic labelling was performed during the last four hours of a 48 hour culture period following inoculation. For metabolic labelling, cells were washed in phosphate buffered saline, then resuspended for 15 minutes in medium lacking methionine and cysteine, supplemented with 10% (v/v) heat inactivated, dialyzed fetal calf serum. Cultures were then resuspended in medium lacking methionine and cysteine, supplemented with 10% (v/v) heat inactivated, dialyzed fetal calf serum and [³⁵S]-methionine and [³⁵S]-cysteine (100 μCi each). On completion of the four hour labelling incubation, cells were washed twice with cold PBS and cell pellets frozen at -70° C for subsequent analysis.

Frozen cell pellets were lysed by thawing in cold lysis buffer (50 mM NaCl, 20 mM Tris pH 7.4, 0.5% NP-40, 0.5% deoxycholate, 0.5% aprotinin and 0.2 mM PMSF), with vigorous vortexing. After pelleting of nuclei and debris, soluble cytoplasmic lysates were used for analysis. Aliquots of cytoplasmic lysates of metabolically labelled cells were subjected to SDS-PAGE and fluorographic analysis both before and after immunoprecipitation of HIV proteins, using standard techniques and a reference human serum sample containing antibody reactive with HIV proteins (Centers for Disease Control, Atlanta, GA). Attempted repeat immunoprecipitation from the lysates demonstrated that the initial immunoprecipitation was quantitative under the conditions employed. Aliquots of either total labelled cytoplasmic proteins or immunoprecipitated HIV proteins were electrophoresed on reducing 10% SDS gels, using equal volumes of each sample, derived from a comparable number of cells, loaded in each lane. Following electrophoresis gels were fixed, treated with intensifier (Enhance, DuPont) and fluorographs produced by exposing vacuum dried gels to preflashed Kodak X-Omat X-ray film at -70°C (35).

As shown in Figure 2A, trichosanthin treatment had no detectable effect on the overall profile of total radiolabelled cytoplasmic proteins synthesized in either treated uninfected cells or in HIV infected cells, as assessed by fluorography following SDS-PAGE of total cytoplasmic lysates. This visual impression was confirmed by scintillation counting of trichloroacetic acid precipitated aliquots of total cytoplasmic lysates, with approximately comparable counts observed irrespective of drug treatment. In marked contrast, the results of the radioimmunoprecipitation component of these metabolic labelling studies indicate that under the experimental conditions tested, trichosanthin treatment resulted in a selective, concentration dependent inhibition of production of all HIV proteins detectable by radioimmunoprecipitation, under conditions not detectably affecting total cellular protein synthesis by infected or uninfected cells (Figure 2B).

To assess effects of trichosanthin treatment on HIV replication at the RNA level in acutely infected T lymphoblastoid cell cultures, VB cells were inoculated with HIV-1$_{DV}$ as described above, at a calculated M.O.I. of 0.01-0.5. Cells were cultured in the continuous presence of the desired concentration of trichosanthin for 48 hours, at which time total cellular RNA was extracted, essentially as described, with lysis of harvested, washed cells in guanidinium thiocyanate followed by extraction in phenol/chloroform and ethanol precipitation (36). When cultured for greater than 48 hours, untreated infected cultures showed extensive virus induced cytopathic effects and cell death; RNA extracted from such cultures showed extensive degradation and was not suitable for Northern blot analysis. Pelleted ethanol precipitated RNA was washed, dissolved in distilled water and stored at -70°C until use. Equal amounts of total RNA extracted from each culture were denatured and electrophoresed on 1.5% agarose gels, then transferred to nitrocellulose filters for Northern blot hybridization (37). Northern blots were probed with a [^{32}P]-labelled full length HIV-1 probe derived from HXB2 (38). As shown in Figure 3, trichosanthin treatment resulted in a concentration dependent decrease in the amount of HIV RNA detectable in identically inoculated cultures. This effect on synthesis and/or accumulation of HIV RNA was observed under treatment conditions that gave a commensurate decrease in levels of HIV

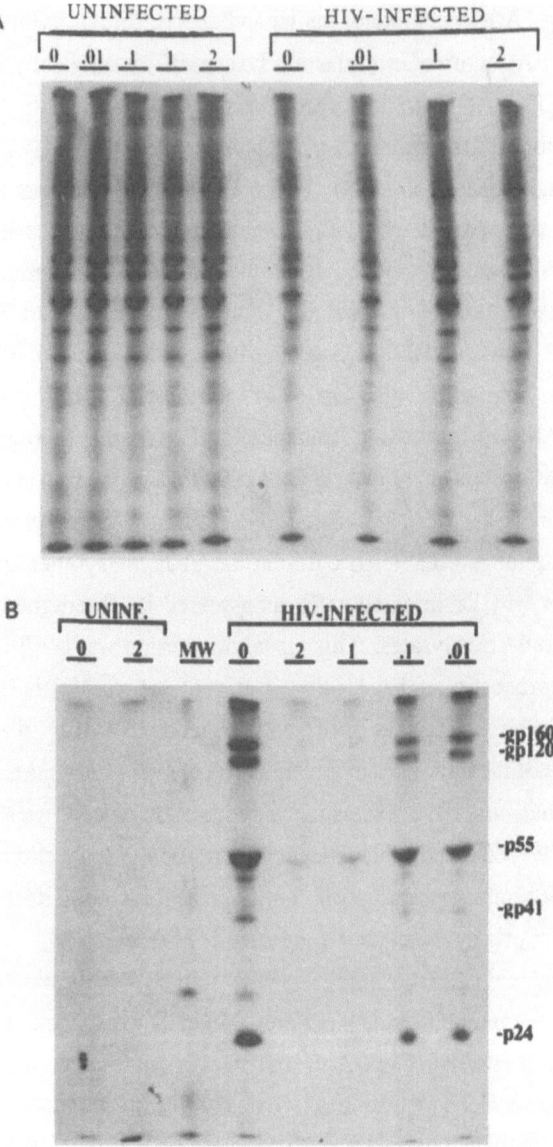

Figure 2. A. Fluorograph of total [³⁵S]-metabolically labelled proteins from uninfected and HIV infected cells treated with the indicated concentrations of GLQ223 (µg/ml), after SDS-PAGE separation. GLQ223 treatment did not result in appreciable differences in the overall profile of proteins synthesized, or in the intensity of bands representing particular proteins. B. Fluorograph of [³⁵S]-metabolically labelled proteins from uninfected and HIV infected cells treated with the indicated concentrations of GLQ223, followed by immunoprecipitation of HIV proteins and SDS-PAGE separation. No HIV proteins were immunoprecipitated from uninfected cells. In the infected cells, GLQ223 treatment resulted in a concentration dependent decrease in the amounts of all HIV proteins detected by quantitative immunoprecipitation.

p24 antigen observed in supernatants of treated cultures, but did not appreciably impair total cellular protein synthesis as measured by [³H]-leucine incorporation or decrease levels of γ-actin mRNA detectable by Northern blot hybridization (not shown), suggesting relative specificity of the effect.

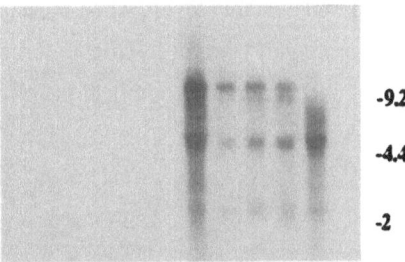

Figure 3. Northern blot analysis of HIV RNA in uninfected and acutely infected VB cells, with and without trichosanthin treatment. No HIV RNA is detected in mock infected control cells treated with 0 μg/ml (Lane 1), 3.14 μg/ml (Lane 2), 1.57 μg/ml, (Lane 3), 0.157 μg/ml (Lane 4), or 0.016 μg/ml (Lane 5). Untreated infected cells show the three characteristic size classes of HIV RNA (Lane 6) while cells treated with 3.14 μg/ml (Lane 7), 1.57 μg/ml, (Lane 8), 0.157 μg/ml (Lane 9), or 0.016 μg/ml (Lane 10) show a concentration dependent decrease in detectable HIV RNA species.

Anti-HIV activity in chronically infected monocyte derived macrophages

Cells of mononuclear phagocyte lineage are believed to represent an important reservoir of HIV replication in vivo, and to play an important role in the pathogenesis of AIDS (18,19,39,40). As the pharmacology of antiviral agents, including RIPs, may vary considerably in different target cells (41,42), it was of interest to evaluate trichosanthin for anti-HIV activity in chronically infected macrophges. In contrast to acutely infected T cells, that typically undergo extensive cytopathology and are killed during the course of infection, macrophages infected with HIV are relatively resistant to viral cytopathology, allowing establishment of a state of chronic productive infection, a feature that facilitiates their ability to serve as a reservoir of productively infected cells in vivo (18,19,39,40).

To evaluate the effects of trichosanthin on HIV replication in macrophages, cultures of monocyte derived macrophages, chronically infected with HIV-1$_{DV}$, were prepared and cultured in suspension, essentially as described (18,19,39,40). Briefly, peripheral blood mononuclear cells were isolated by density centrifugation. Monocytes were isolated by glass adherence, then recovered by scraping on ice; recovered adherent cells were placed into culture medium supplemented with human serum, and cultured in suspension in PTFE culture vessels (18,19,39,40). Cells were inoculated with HIV-1$_{DV}$, then cultured for 10-14 days, by which time typically 25-35% of cells expressed HIV p24 antigen as measured by permeabilized cell immunocytofluorometric analysis (18,40). At this point, levels of viral expression in untreated cultures remain reasonably steady for several weeks with regular medium changes (40); such cultures were considered to be chronically infected.

Chronically infected cultures were exposed to the desired concentrations of trichosanthin for three hours, then washed and returned to culture. Effects on viral replication at the protein level were evaluated five days later, as assessed by permeabilized cell immunocytofluorometric analysis for cytoplasmic expression of HIV p24 antigen, as described (18,40). As shown in Figure 4, a single three hour exposure of chronically HIV-1 infected monocyte derived macrophages to 500 ng/ml trichosanthin was sufficient to reduce HIV p24 expression to undetectable levels when measured five days later. These treatment conditions did not appreciably affect [^{3}H]-leucine incorporation by uninfected macrophage cultures, indicating the antiviral selectivity of the effect (19).

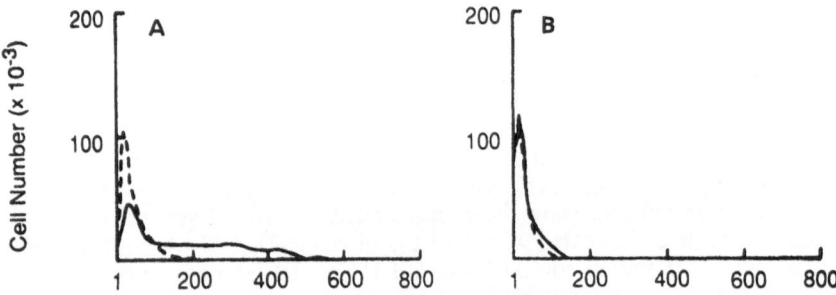

Figure 4.   Decreased HIV p24 expression by chronically HIV infected monocyte derived macrophages treated with GLQ223, measured by immunocytofluorometric analysis. A. Expression of cytoplasmic HIV p24 as detected by anti-p24 monoclonal antibody (solid line) in sham treated cells, compared to background fluorescence using isotype matched monoclonal antibody of irrelevant specificity (dotted line). Linear fluorescence scale measured in arbitrary fluorescence units. B. GLQ223 treated cells show no detectable HIV p24 expression (solid line), as compared to sham treated control cells (solid line, Panel A); HIV p24 attributable fluorescence in treated infected cells (Panel B, solid line) is comparable to background fluorescence from irrelevant specificity control antibody (Panel B, dotted line). Modified from McGrath et al, 1989, with permission.

Effects on viral RNA levels in chronically HIV infected macrophages were evaluated by RNA slot blot as described (19,39). Briefly, chronically HIV-1 infected cultures of monocyte derived macrophages were prepared, then treated with trichosanthin for three hours, followed by washing, with no further drug exposure, as described above. Cells were cultured for a subsequent five days, then total cellular RNA was extracted from 2 X 10^6 cells for each treatment condition, as described (19,39). RNA slot blot analysis was performed as described using a 3.5 kb [^{32}P]-labelled hybridization probe derived from the 3'-end of the HXB2 HIV-1 clone (19,39). As shown in Figure 5, five days following treatment of chronically HIV-1 infected macrophage cultures with 500 ng/ml of trichosanthin for 3 hours, no HIV RNA was detectable by slot blot hybridization of RNA extracted from treated cells.

Trichosanthin is a SCRIP with empirically observed activity to inhibit HIV replication as measured at both protein and RNA levels. The mechanisms that account for the observed antiviral activity of SCRIPs in general have not been unequivocally elucidated. The ribosome inactivating properties of RIPs, and the mechanistic basis of this activity are relatively well understood, however, there is no a priori reason why ribosome inactivation and consequent inhibition of translation should result in a *selective* inhibition of viral replication. Nevertheless, the simplest mechanistic explanation for the observed antiviral activity of RIPs, including the anti-HIV activity of trichosanthin, would involve preferential

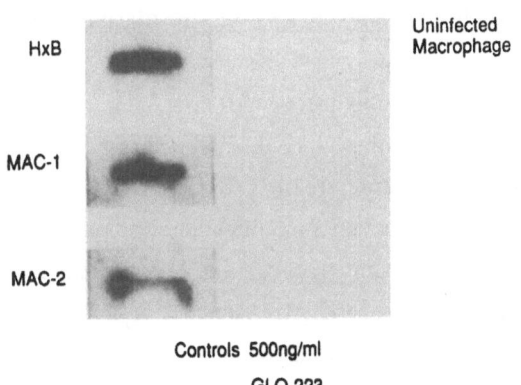

Figure 5. Slot blot analysis of HIV RNA expression by HIV infected monocyte derived macrophages. Positive control ("HXB") shows strong hybridization of HIV probe to RNA from chronically infected H9 cells. Negative control ("uninfected macrophage") shows no hybridization. RNA extracted from sham treated HIV infected control macrophage preparations from two donors ("MAC-1, MAC-2") show hybridization, while RNA extracted from parallel cultures from the same donors following GLQ223 treatment in vitro shows no detectable hybridization. Reproduced from McGrath, et al, 1990, with permission.

inactivation of ribosomes, leading to impairment of translation and eventual cytotoxicity in infected, compared to uninfected cells. In principle, antiviral selectivity could be conferred at either the level of preferential uptake of RIPs by infected, compared to uninfected cells, or through preferential intracellular mechanisms favoring ribosome inactivation in infected, but not uninfected cells. Indeed, it has been suggested that in some in vitro systems, viral infection may alter cell membrane permeability, facilitating enhanced, preferential uptake of RIPs by virally infected cells, compared to uninfected cells (14-17), although in initial studies with radiolabelled trichosanthin we have been unable to demonstrate preferential uptake by HIV infected cells (J.D. Lifson, et al, unpublished observations).

Alternatively, it is theoretically possible that the observed antiviral properties of trichosanthin may be mediated through mechanisms unrelated to ribosome inactivation per se. Such mechanisms might involve the catalytic activity responsible for ribosome modification, but directed against a target substrate other than 28S rRNA, leading to a selective inhibition of viral replication. A final formal possibility is that the antiviral properties of trichosanthin may be due to a mechanism completely unrelated to ribosome inactivation, in which the catalytic activity and site involved in ribosome modification do not play a role.

In preliminary studies, we have begun to explore the relation of ribosome inactivation to the observed ability of trichosanthin to inhibit HIV replication, in vitro, and to investigate the potential involvement of other mechanisms in the compound's antiviral activity.

Mechanism of action studies: Relation of ribosome modification to antiviral activity

A molecular mechanism to account for how RIPs inactivate ribosomes has been proposed. The proteins cleave the N-glycosidic bond of a specific adenine residue in a highly conserved stem-loop structure in 28S rRNA (3,4,43,44). This depurination presumably results in a conformational modification, leading to functional inactivation of ribosomes, perhaps through impairment of capacity to bind elongation factors. Site specific depurination also renders the phosphodiester backbone of 28S rRNA selectively susceptible to aniline cleavage at the site of depurination, leading to release of a diagnostic fragment consisting of the 3'-end of 28S rRNA, easily detectable by Northern blot hybridization (Figure 6). Assessment of the relative extent of ribosome modification by this method affords a reasonably convenient, semi-quantitative means of monitoring cellular uptake of trichosanthin, and intracellular bioactivity of the protein. This in turn allows comparative evaluation of the extent of antiviral activity and ribosome inactivation under different treatment conditions, and provides an approach to the question of whether the antiviral activity of trichosanthin is directly related to its ribosome inactivating properties.

VB cells were inoculated with HIV-1$_{DV}$ at a calculated M.O.I. of 0.5, and treated with varying concentrations of trichosanthin, essentially as described (18). Cells were harvested two days later for analysis, at a time when early characteristic HIV induced cytopathology was evident in untreated control cultures. When cultured for longer periods, untreated infected cultures showed extensive virus induced cytopathic effects and cell death; RNA extracted from such cultures showed extensive degradation and was not suitable for analysis. RNA was extracted essentially as described, with lysis of harvested, washed cells in guanidinium thiocyanate followed by extraction in phenol/chloroform and ethanol precipitation (36). Pelleted ethanol precipitated RNA was washed, dissolved in distilled water and stored at -70°C until use. Aliquots of isolated RNA samples were analyzed with and without prior treatment with aniline to achieve site specific hydrolysis of 28 S rRNA

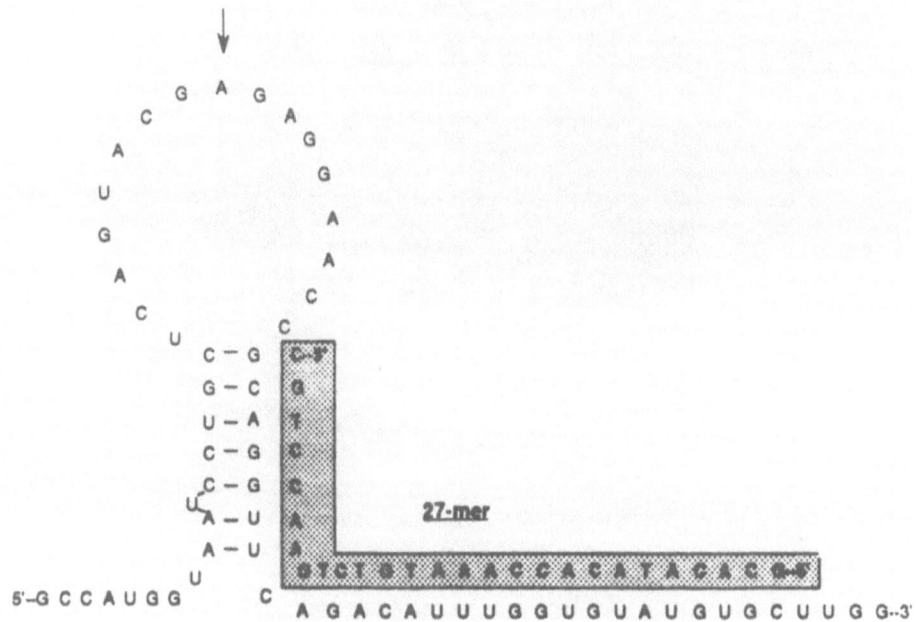

Figure 6. Figure shows the sequence of a highly conserved region of 28S rRNA, capable of forming a stem loop structure that includes the target site adenine depurinated by RIPs (arrow) (3,4). Shaded sequence demonstrates 27-mer oligonucleotide probe (Luk-e-3) complementary to the 3'- portion of the stem loop structure, used as a specific diagnostic probe to demonstrate site specific depurination of 28S rRNA by trichosanthin, after aniline treatment of isolated RNA.

depurinated by trichosanthin (3,4,5,43). Equal amounts of extracted total cellular RNA were treated with either distilled water or double distilled aniline (1 M, buffered to pH 4.5 with glacial acetic acid), essentially as described (45). After 15 minutes incubation at 37° C, mixtures were placed on ice. Treated RNA samples were reprecipitated with ethanol, washed, then pelleted and resuspended in distilled water for analysis by Northern blot hybridization. Samples were denatured and electrophoresed on 1.5% agarose gels, followed by transfer to nitrocellulose filters. Filters were probed with a [^{32}P]-labelled 27-mer probe (Luk-e-3) designed to hybridize to the diagnostic 410 base fragment liberated from the 3'-end of 28 S rRNA following site specific depurination by trichosanthin and subsequent aniline hydrolysis (Figure 6).

As shown in Figures 7A and 7B for trichosanthin treated HIV infected and uninfected cultures respectively, trichosanthin treatment produced a concentration dependent increase in the amount of ribosome modification observed in both infected and uninfected cells, as measured by accumulation of the diagnostic 410 base fragment derived from the 3'-end of 28 S rRNA. Interestingly, there was no appreciable difference in the extent of ribosome modification observed for infected and uninfected cultures, and minimal degrees of ribosome modification were observed with trichosanthin treatment at concentrations associated with substantial inhibition of HIV replication.

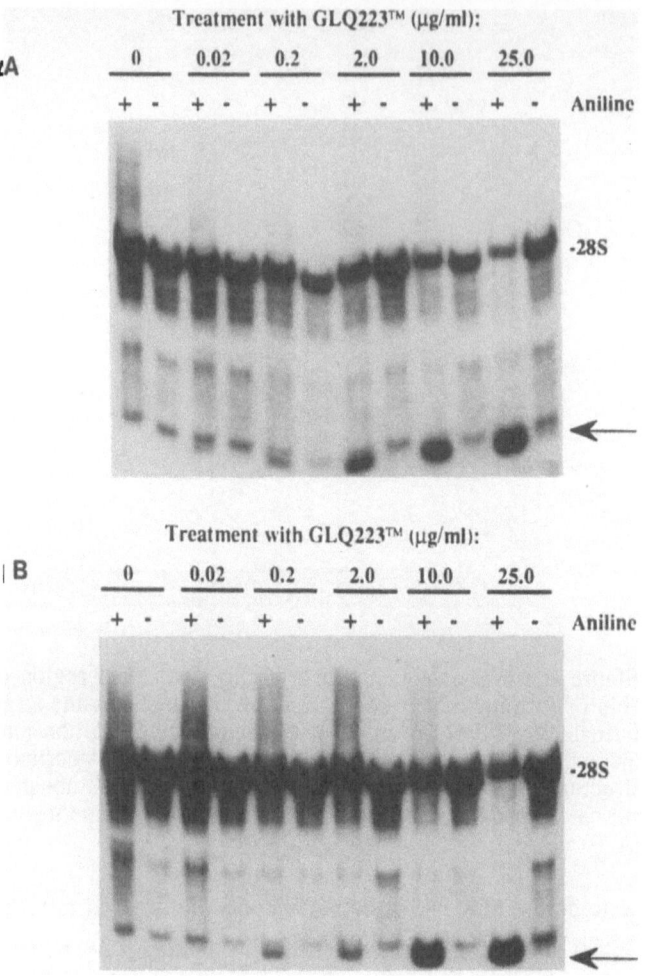

Figure 7. Extent of ribosome modification in infected and uninfected cells treated with trichosanthin. A. Infected cells. RNA was extracted from infected cells treated with the indicated concentrations of GLQ223. Isolated RNA was treated with aniline ("+"), or water ("-") as described above, then analyzed by Northern blot hybridization using the Luk-e-3 probe. The probe hybrizes to intact 28S rRNA (but not to 28S rRNA following trichosanthin mediated site specific depurination and cleavage by aniline treatment), and to the diagnostic 410 base fragment liberated from the 3'-end of 28S rRNA following trichosanthin mediated site specific depurination and cleavage by aniline treatment (arrow). Band just above the 410 base fragment band represents hybridization to an unidentified RNA species and is non-specific, as shown by hybridization in lanes with RNA from cells not treated with GLQ223, and a uniform signal, regardless of concentration of GLQ223 used to treat cells. B. Uninfected cells. Northern blot analysis of RNA from uninfected cells treated with GLQ223 shows results similar to those obtained from infected cell RNA preparations.

Preferential degradation of isolated HIV RNA by trichosanthin

The observation that treatment of HIV infected cultures with trichosanthin led to decreased levels of detectable HIV RNA species, as well as selective inhibition in the synthesis of HIV proteins, in infected T cells and macrophages raised the question of whether the compound exerted its effects at the level of protein synthesis (as might be expected if ribosome modification was involved in the compound's antiviral activity), or perhaps at other levels as well. The experiments described above do not permit discrimination between these possibilities, in part because the culture periods involved allowed time for multiple rounds of viral infection and replication. Thus, the levels of viral proteins and RNA measured when the cells were harvested at the conclusion of the culture period reflect macromolecules synthesized and accumulated through multiple rounds of viral replication. In spite of efforts to use relatively high calculated multiplicities of infection, in the experiments performed the proportion of cells productively infected at the time of analysis was less than 50%, in both acute and chronic infection models. In the acute infection model particularly, ribosome modification leading to decreased viral protein synthesis and decreased production of progeny virions during the culture period could lead to decreased levels of detectable HIV RNA species at the conclusion of the culture period, through diminished secondary infection in the cultures, even in the absence of any direct drug effect at the RNA level. Serial Southern blot analysis of acutely infected cultures for HIV DNA, revealing a greater inhibitory effect on the accumulation of detectable HIV DNA at later time points than was seen early in infection, was consistent with such a mechanism involving decreased production of progeny virus and subsequent spread within the culture (H. Abrams, et al, unpublished observations). Alternatively, mechanisms affecting production of HIV regulatory proteins, or mechanisms directly operative at the RNA level might also explain the observed effects on HIV RNA levels in trichosanthin treated cultures, although activity of SCRIPs against a wide variety of different viral pathogens makes potential specific mechanisms operative at the level of viral RNA sequences, or individual HIV regulatory proteins less likely.

Given our observation of decreased levels of HIV RNA in trichosanthin treated cultures, and in view of prior observations suggesting the existence of partially target specific nuclease-like activity mediated by RIPs (46), we evaluated the direct effects of trichosanthin on HIV and non-HIV RNA species, in experiments involving the direct treatment of isolated RNA with the compound. Acutely infected VB cells were prepared as described above, without trichosanthin treatment of cells during the culture period. At 48 hours post inoculation, RNA was extracted, essentially as described (36). Total cellular RNA (10 µg per experimental condition) from VB cells infected with HIV was incubated with trichosanthin at the desired concentrations in a total volume of 100 µl of 25 mM Tris-HCl, pH 7.6, containing 25 mM KCl and 5 mM MgCL2 for one hour at 37°C in sterile polypropylene tubes under RNase free conditions. After incubation, the RNA was extracted with phenol and ethanol precipitated. After pelleting, the precipitated RNA was resuspended in double distilled water, denatured and loaded onto a 1% formaldehyde-agarose gel for electrophoretic separation of RNA species and analysis by Northern blot hybridization, using hybridization probes for HIV and non-HIV sequences (CAD and γ-actin) (47,48).

As shown in Figure 8A, trichosanthin treatment of isolated total cellular RNA from HIV infected cells resulted in an apparent concentration dependent degradation of HIV RNA species, without a similar effect observed for CAD (Figure 8B) or γ-actin mRNA (not shown). A non-specific RNase activity might appear to produce a "preferential" degradation of HIV RNA species as compared to γ-actin, based solely on the relative abundances of the RNA species. However in studies with an endogenous RNA species (CAD (48)) or exogenous RNA species (lambda phage RNA markers) spiked into the RNA samples, and approximately matched with the HIV RNA species with respect to length and estimated abundance, the appearance of preferential degradation of HIV RNA species was confirmed (Figure 8B (CAD), and data not shown (lambda phage RNA marker experiments).

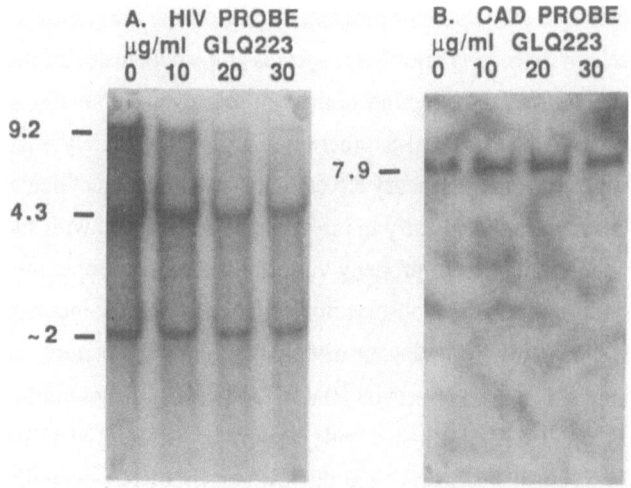

Figure 8. In vitro treatment of isolated RNA with GLQ223. A. Treatment of RNA isolated from HIV infected cells (no GLQ223 treatment of cells). Northern blot hybridization with full length HIV probe (38) shows progressive decrease in detectable HIV RNA species following in vitro treatment with increasing concentrations of GLQ223 (μg/ml). B. Reprobing of the same filter with a probe for CAD mRNA (48) shows essentially comparable signal, irrespective of concentration of GLQ223 used for in vitro treatment of RNA.

In separate experiments, the susceptibility of the apparent trichosanthin associated RNase activity to heat denaturation was evaluated. The experimental design was as above, with addition of trichosanthin to comparable aliquots of total cellular RNA extracted from HIV infected VB cells, except that the trichosanthin was subjected to mild (65°C for 20 minutes, followed by slow cooling) or vigorous (100°C for 10 minutes, followed by slow cooling) heat denaturation prior to being added to RNA samples. After incubation with treated trichosanthin preparations, Northern blot analysis was performed as above. As shown in Figure 9, in vitro treatment of isolated RNA from HIV infected VB cells with untreated trichosanthin resulted in a decrease in the amount of HIV RNA detectable by Northern blot; mild heat denaturation of the trichosanthin did not affect this activity, while vigorous heat denaturation completely eliminated the apparent degradation of HIV RNA mediated by trichosanthin. In contrast, as expected, neither mild or vigorous heat denaturation had any effect on the complete degradation of HIV RNA mediated by pancreatic RNase.

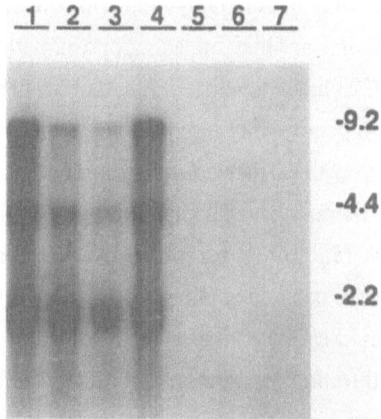

Figure 9. Apparent selective RNase like activity of GLQ223 is heat denaturable. Northern blot analysis with full length HIV probe of RNA isolated from HIV infected cells (no GLQ223 treatment of cells). Lane 4, no GLQ223 treatment; Lane 3, RNA exposure to untreated GLQ223 results in decreased detectable HIV signal; Lane 2, mild heat denaturation of GLQ223 prior to treatment of RNA did not abrogate degradation of HIV RNA; Lane 1, vigorous heat denaturation of GLQ223 prior to exposure of RNA inactivated the compound's ability to degrade HIV RNA. In contrast, the ability of pancreatic RNase to degrade HIV (and other) RNA species (Lane 5) was not affected by mild (Lane 6) or vigorous (Lane 7) heat denaturation of the enzyme prior to its use to treat isolated RNA.

DISCUSSION

SCRIPs are ubiquitous plant derived proteins whose antiviral activity in various in vitro systems has been well documented (1,2,6-15). We (18,19), and subsequently others (20-22), have demonstrated that the SCRIP trichosanthin, as well as other RIPs, inhibit HIV replication in vitro. Trichosanthin inhibits HIV replication as measured at both the protein and RNA levels, in both acutely infected T lymphoblastoid cells, and in chronically HIV infected monocyte derived macrophages (18,19). The anti-HIV activity of trichosanthin in chronically infected cells of mononuclear phagocyte lineage is particularly noteworthy in view of the important role that these cells are believed to play in the pathogenesis of HIV infection and AIDS (18,19,39,40), and given the fact that reverse transcriptase inhibitors show little or no antiviral activity in already infected cells (39,42). The striking effects of trichosanthin on HIV replication in macrophages may be related to enhanced uptake of the compound by these phagocytic cells. While there is an extensive history of prior medical use of both trichosanthin containing preparations and purified trichosanthin in China where the compound is used as abortifacient and has been used for the treatment of trophoblastic tumors (23-28), investigation of the potential antiviral therapeutic applications of this protein is still in the early stages.

Initial clinical studies in HIV infected patients with GLQ223, a highly purified form of trichosanthin formulated for injection, have shown an acceptable safety and tolerance profile for the compound (29). The predominant side effects include a constellation of constitutional complaints best characterized as a "flu like syndrome" and pruritus (29, and R. L. Williams, et al, in preparation). While symptoms can be severe, they appear to be reversible, self-limited, and may be relatively well controlled with the use of non-steroidal anti-inflammatory drugs and/or antihistamines. Approximately 12% of patients receiving repeated injection of GLQ223 have developed apparent immediate type hypersensitivity to this exogenous protein therapeutic. Although anaphylactic or anaphylactoid reactions are considered a contraindication to further administration of trichosanthin, attention to this possible complication of therapy and prompt intervention have prevented any serious adverse hypersensitivity reactions; no mortality or long term sequelae due to hypersensitivity reactions have occured in clincal evaluation of GLQ223 to date.

Apparent drug related serious adverse neurological reactions observed during the very early stages of clinical testing of trichosanthin in HIV infected patients appear to have been related to the advanced disease stage of the patients treated, as reflected by severely depleted CD4+ T cell levels, rather than reflecting intrinsic neurotoxicity of the compound itself (29,49,50). Serious drug attributed adverse neurological reactions have not been observed in extensive subsequent clinical testing in which administration of GLQ223 has been limited to patients with less advanced disease and ≥100 CD4+ T lymphocytes/mm^3 of blood (R.L. Williams, et. al., in preparation).

Ongoing in vitro studies with trichosanthin have included efforts to elucidate the mechanistic basis for the compound's antiviral activity. Perhaps the simplest potential mechanism of action would involve preferential uptake by HIV infected cells, with trichosanthin's known ribosome inactivating activity then leading to inhibition of translation and eventual cytotoxicity in infected, but not uninfected cells, thereby accounting for the observed antiviral selectivity of the compound. This is the simplest hypothesis in that it is consistent with the known molecular biochemistry of the protein and does not require invocation of novel and as yet undocumented molecular activities. While this mechanism would require an explanation for the postulated preferential uptake of trichosanthin by HIV infected cells, altered membrane permeability related to viral infection has been proposed as a mechanism to explain enhanced uptake of SCRIPs by infected cells compared to matched uninfected cells (9,14-17). In our preliminary studies with radiolabelled trichosanthin, we have not yet been able to document convincing evidence for preferential uptake by HIV infected cells, although the relatively low specific activity of the material tested limited the sensitivity of the analysis possible (J.D. Lifson, et. al., unpublished observations).

Assessment of the extent of RIP associated ribosome modification (rRNA depurination) by Northern blot analysis to monitor the accumulation of the diagnostic 410 base fragment derived from the 3'-end of 28S rRNA, following aniline hydrolysis of RNA isolated from trichosanthin treated cells, provides a semi-quantitative means to measure the uptake and intracellular bioactivity of trichosanthin. In studies with acutely HIV infected VB cells, the extent of ribosome modification appeared to be approximately comparable in infected and uninfected cultures treated with equal concentrations of trichosanthin. Minimal levels of ribosome modification were apparent in RNA from infected treated cultures that showed substantial inhibiton of viral replication, as measured by culture supernatant HIV p24 content and cytoplasmic HIV RNA content.

Superficially, these findings suggest that the observed antiviral activity of trichosanthin in these experiments was not directly related to preferential ribosome modification in the infected cells, compared to the uninfected cultures. However, despite the use of the highest titer viral stocks available, to achieve a relatively high calculated multiplicity of infection, the actual level of infection achieved, although not known with precision, was almost certainly less than desired. If only a small proportion of cells was productively infected during the initial cycle of viral replication, preferential uptake of trichosanthin by the minority population of infected cells leading to ribosome modification, inhibition of translation and eventual cytotoxicity would result in the effective deletion of this population from the culture. Apparent overall effects on viral replication measured later would be amplified in that even partial deletion of the population of cells infected during the first cycle of viral replication would preclude production of virus during secondary cycles of replication from infections attributable to progeny virions produced during the first cycle. In this scenario, with continued division of uninfected cells over the course of the experiment, the majority of the cells exposed to trichosanthin treatment in both the inoculated ("infected") and uninfected cultures would in fact be uninfected. Under these circumstances, failure to discern a distinct difference in the extent of ribosome modification between the "infected" and uninfected cultures would not be surprising. Similarly, evidence of minimal ribosome modification in cultures showing substantial inhibition of viral replication would no longer necessarily imply that the antiviral effects were separable from ribosome inactivation. Clearly, further studies, preferably in uniformly infected culture systems, are required to address these questions in experiments of this general design.

In an additional approach to addressing the mechanistic question of the relation of the ribosome modifying activity of trichosanthin to its antiviral properties, we have employed site directed mutagenesis of the gene encoding trichosanthin (M. Piatak et. al. in preparation). Molecular modelling of the structure of trichosanthin based on the determined crystal structure of the ricin A chain, which shares substantial sequence similarity with trichsosanthin, allowed identification of several putative active site residues considered likely to be crucial in the catalytic activity of the protein involved in ribosome modification (31,51,52). Semi-conservative mutations that were predicted to maintain overall protein confirmation, but expected to result in varying degrees of impairment of the protein's ribosome inactivating potency, were introduced at these sites. Expressed recombinant wild type protein and mutants were then tested in parallel for their ability to inhibit in vitro translation in a rabbit reticulocyte lysate system, and for anti-HIV activity. Ability to clearly dissociate ribosome modification activity from antiviral activity would be considered strong evidence that the antiviral properties of trichosanthin are not intrinsically tied to the compound's ribosome inactivating activity. In constrast, failure to dissociate the ribosome modifying and antiviral activities with mutant forms of trichosanthin would constitute less direct, implicit evidence that the activities are tightly associated. This would be the case if the antiviral properties of the compound involved ribosome modification per se, in infected cells, or if the same enzyme active site involved in ribosome modification mediated antiviral activity through action on a different (non-ribosomal) substrate.

The mutant forms of trichosanthin had approximately the predicted level of impairment in their ability to inactivate ribosomes as measured in the reticulocyte lysate translation inhibition assay. The mutant proteins also showed decreased antiviral activity, with the same rank order potency of the mutated proteins observed for both ribosome modification and antiviral activity (M. Piatak, et al, in preparation). These findings are thus most consistent with mechanistic models in which the antiviral properties of trichosanthin are directly or indirectly, but somehow inextricably linked to its ribosome inactivating activity.

The observations suggesting the possible existence of a heat denaturable RNase-like activity associated with trichosanthin, with an apparent target preference for HIV RNA species, are provocative, but of uncertain relevance to the observed antiviral activity of the compound. While the relatively high concentrations of trichosanthin required for this in vitro effect might seem to be one reason to question its relevance to antiviral phenomena, in fact this may not be the primary cause for believing that the observations may represent an epiphenomenon. Potency of RIPs to modify their 28S rRNA-derived target sequence varies depending on the actual substrate used. Thus, the ribonucleoprotein complex of an intact ribosome appears to present the 28S rRNA target in an optimal configuration for efficient modification by RIPs. Although site-specific RIP mediated depurination has been demonstrated with oligonucleotide substrates of appropriate sequence, several orders of magnitude higher concentrations of RIP were required than are needed for modification of 28S rRNA when present in intact ribosomes (44). Along these lines, it may be that HIV RNA complexed with protein in the intracellular environment may be a more efficient target substrate for the putative trichosanthin-associated RNase like activity we have observed than is deproteinated isolated HIV RNA treated in vitro. Nevertheless, the fact that SCRIPs demonstrate antiviral activity against a relatively broad spectrum of viruses argues against sequence dependent preferential degradation of viral nucleic acids as a central mechanism accounting for the antiviral activity of the compounds.

It is of interest that the observed antiviral selectivity of trichosanthin and other RIPs in both HIV and non-HIV systems, as defined by the ratio of the concentration required to mediate a given level of inhibition of a parameter of viral replication (inhibitory concentration 50%, IC_{50}) to the concentration needed to exert a comparable level of inhibiton of a parameter of cellular metabolism (toxic concentration 50%, TC_{50}; IC_{50}: TC_{50} = in vitro selective index), may be heavily influenced by the assay system and target cells used for testing. Thus, while we have observed between two and three logs of selectivity in the inhibition of viral replication relative to inhibition of parameters of cellular metabolism by trichosanthin, others utilizing different assay systems, different target cells, and longer durations of drug exposure have noted lower degrees of selectivity (21,22). Other target cell parameters such as metabolic rate, culture density, and membrane biochemisty and biophysics may also affect the observed antiviral selectivity of SCRIPs, if preferential uptake by infected cells is the basis for the antiviral properties of the compounds. In studies to evaluate a panel of several SCRIPs for in vitro activity activity against human herpes viruses, the rank order for potency and selectivity of the compounds varied when they were tested against the same viruses, used to infect different target cells (M. Piatak, et al, unpublished observations).

If preferential uptake of SCRIPs by infected compared to uninfected cells is indeed the mechanism that accounts for the antiviral selectivity of SCRIPs, it is perhaps not surprising that different degrees of selectivity are observed for different target cells, which may be reasonably expected to vary in their uptake of SCRIPs. An extreme example of this phenomenon is provided by studies of anti-HIV activity of trichosanthin in MT-4 cells, a cell line chronically infected with HTLV-1 and widely used as an indicator cell line for evaluation of anti-HIV agents (21). Ferrari et al noted that although trichosanthin treatment resulted in selective inhibition of HIV replication in H9 and CEM-SS cells, no selective inhibition of HIV replication was observed in HIV infected MT-4 cells, where cellular toxicity was observed at comparable trichosanthin treatment concentrations for HIV infected and uninfected cells (21). This observation may have a ready explanation. It seems likely that if increased uptake of SCRIPs by virally infected cells is indeed the basis of their antiviral activity, then chronic infection of the MT-4 cell line with HTLV-1 may enhance trichosanthin uptake above what would be expected in cells not infected with HTLV-1; co-infection with HIV-1 may add little or no incremental enhancement in the uptake of trichosanthin.

In summary, we have characterized the anti-viral properties of the SCRIP trichosanthin in HIV related assay systems. The protein inhibits HIV replication in both acute and chronic infection models, as measured at both protein and nucleic acid levels, and shows activity in target cells of both lymphoid and mononuclear phagocyte lineage. While the exact mechanism of action that accounts for the antiviral activity of trichosathin remains to be elucidated, the available data are most consistent with models in which the antiviral activity of the compound appears to be inextricably linked to its ability to inactivate ribosomes. Our observations have implications that may be generalized to an improved understanding of the antiviral activity of SCRIPs in general. The outcome of the ongoing clinical evaluation of trichosanthin as a potential treatment for HIV infection and AIDS will determine the broader significance of these studies.

Acknowledgements: The authors thank the Pharmaceutical Development Group of Genelabs Incorporated for providing GLQ223, the Quality Control Group of Genelabs Incorporated from performing in vitro translation assays, the Process and Analytical Biochemistry Group of Genelabs Incorporated for purification of recombinant trichosanthins, A. Shiba for technical assistance, Dr. J. Irvin for providing PAP samples and for helpful discussions, Dr. G. Stark for providing the CAD plasmid probe, Dr. J. Robertus for thoughtful discussions regarding structural correlates of RIP activity, Dr. P. Dukor for unflagging support, and J. Wohltmann for her insightful comments on the mechanism of antiviral action of RIPs.

BIBLIOGRAPHY

1. F. Stirpe and L. Barbieri, Ribosome inactivating proteins up to date, FEBS, 195:1, (1986).

2. W.K. Roberts and C.P. Selitrennikoff, Plant proteins that inactivate foreign ribosomes, Biosci. Rep., 6:19, (1986).

3. Y. Endo, K, Mitsui, M. Motizuki, and K. Tsurugi, The mechanism of action of ricin and related toxic lectins on eukaryotic ribosomes. The site and the characteristics of the modification in 28S ribosomal RNA caused by the toxins, J. Biol. Chem., 262:5908, (1987).

4. Y. Endo, and K. Tsurugi, RNA N-glycosidase activity of ricin A chain. Mechanism of action of the toxic lectin ricin on eukaryotic ribosomes. J. Biol. Chem. 262:8128, (1987).

5. M.J. May, M.R. Hartley, L.M. Roberts, P.A. Krieg, R.W. Osborn, and J. M. Lord, Ribosome inactivation by ricin A chain: a sensitive method to assess the activity of wild-type and mutant polypeptides, EMBO J., 8:301, (1989).

6. T. G. Obrig, J. D. Irvin, and B. Hardesty, The effect of an antiviral peptide on the ribosomal reactions of the peptide elongation enzymes, EF-I and EF-II, Arch. Biochem. Biophys., 155:278, (1973).

7. J.D. Irvin, Pokeweed antiviral protein, Pharmac. Ther., 21:371, (1983).

8. B.M. Duggar and J.K. Armstrong, The effect of treating the virus of tobacco mosaic with the juice of various plants, Ann. Mo. Bot. Gard. 12:359, (1925).

9. M.A. Ussery, J.D. Irvin, and B. Hardesty, Inhibition of poliovirus replication by a plant antiviral peptide, Ann. N.Y. Acad. Sci. 284:431, (1977).

10. G. M. Aron, and J.D. Irvin, Inhibition of herpes simples virus multiplication by the pokeweed antiviral protein, Antimicrob. Agents Chemother., 17:1032, (1980).

11. F. Stirpe, D.G. Williams, L. J. Onyon, R.F. Legg, and W. A. Stevens, Dianthins, ribosome damaging proteins with anti-viral properties from dianthus caryophyllus L. (carnation), Biochem. J. 195:399, (1981).

12. Takemoto, C. Jilka, S. Rockenbach, and J.V. Hughes, Purification and characterization of a cytostatic factor with anti-viral activity from the bitter melon, Prep. Biochem. 13:371, (1983).

13. J.A. Tomlinson, V.M. Walker, T.H. Flewett, and G.R. Barclay, The inhibiton of infection by cucumber mosaic virus and influenza virus by extracts from Phytolacca americana, J. Gen. Virol., 22:225, (1974).

14. L. Carrasco, Membrane leakiness after viral infection and a new approach to the development of antiviral agents, Nature, 272:694, (1978).

15. C. Fernandez-Puentes and L. Carrasco, Viral infection permeabilizes mammalian cells to protein toxins, Cell, 20:769, (1980).

16. C. Fernandez-Puentes, Permeability to alpha sarcin in virus infected cells, Mol. Cell. Biochem., 50:185, (1983).

17. M. Yamaizumi, T. Uchida, and Y. Okada, Marcomolecules can penetrate the host cell membrane during the early period of incubation with HVJ (Sendai virus), Virology, 95:218, (1979).

18. M.S. McGrath, K.M. Hwang, S.E. Caldwell, I. Gaston, I, K.-C. Luk, P. Wu, V.L. Ng, S. Crowe, J. Daniels, J. Marsh, T. Deinhart, P. Lekas, J. C. Vennari, H.-W. Yeung and J.D. Lifson, GLQ223: An inhibitor of human immunodeficiency virus replication in acutely and chronically infected cells of lymphocyte and mononuclear phagocyte lineage, Proc. Natl. Acad. Sci (USA), 86:2844, (1989).

19. McGrath, S. Santulli, and I. Gaston, Effects of GLQ223 on HIV replication in human monocyte/macrophages chronically infected in vitro with HIV, AIDS Res. Hum. Retrovir., 6:1039, (1990).

20. J.M. Zarling, P.A. Moran, O. Haffar, J. Sias, D.D. Richman, C.A. Spina, D.E. Myers, P.V. Kuebelbeck, J.A. Ledbetter, and F.M. Uckun, Inhibition of HIV replication by pokeweed antiviral protein targeted to CD4+ cells by monoclonal antibodies, Nature, 347:92, (1990).

21. P. Ferrari, M.-A. Trabaoud, M. Rommain, E. Mandine, R. Zalisz, C. Desgranges, and P. Smets, Toxicity and activity of purified trichosanthin, AIDS, 5:865, (1991).

22. S. Lee-Huang, P.L. Huang, H.-F. Kung, B.-Q. Li, P.L. Huang, P. Huang, H.L. Huang, and H.-C. Chen, TAP 29: An anti-human immunodeficiency virus protein from Trichosanthes kirilowii that is nontoxic to intact cells, Proc. Natl. Acad. Sci. (USA), 88:6570, (1991).

23. Y. Wang, R.-Q. Qi, Z.-W. Gu, S.-W. Jin, L.-Q. Zhang, Z.-X. Xia, G.-Y. Tian, and C.-Z. Ni, Scientific evaluation of Tian Hua Fen (THF)--history, chemistry and application, Pure and Appl. Chem., 58:789, (1986).

24. The Second Laboratory, Shanghai Institute of Experiment Biology, Studies on the mechanisms of abortion induction by trichosanthin, Scientia Sinica, XIX:811, (1976).

25. Y.-C. Jin, Clinical study of trichosanthin, In: Advances in chinese medicinal materials research, H.M. Chang, H.W. Yeung, W.-W. Tso and A. Koo, eds., World Scientific Publishing Co., Singapore, 1985.

26. X.X. Zhong, D.Z. Wang, and J.H. Han, Follow-up study of mid term abortion induced by five different methods, Chinese J. Obstet. Gynecol. 15:216, (1980).

27. S.Z. Gao, Metastatic malignant mole treated with Radix trichosanthis, Chinese J. Obstet. Gynecol., 12:28, (1978).

28. K.-F. Cheng, Midtrimester abortion induced by radix trichosanthis: Morphologic observations in placenta and fetus, Obstetrics and Gynecology, 59:494, (1982).

29. J.O. Kahn, L.D. Kaplan, J. G.Gambertoglio, C.J. Arri, L. Turin, T. Kibort, R.L. Williams, J.D. Lifson, and P.A. Volberding, Safety and pharmacokinetics of GLQ223™ in subjects with acquired immunodeficiency syndrome (AIDS) and AIDS-related complex: A Phase I study, AIDS, 4:1197, (1990).

30. X. Zhang, and J. Wang, Homology of trichosanthin and ricin A chain, (letter), Nature, 321:477, (1986).

31. E.J. Collins, J.D. Robertus, M. LoPresti, K.L. Stone, K.R. Williams, P. Wu, K. Hwang, and M. Piatak, Primary amino acid sequence of trichosanthin and molecular models for abrin A-chain and trichosanthin, J. Biol. Chem., 265:8665, (1990).

32. T. P. Chow, R.A. Feldman, M. Lovett, and M. Piatak, Isolation and DNA sequence of a gene encoding trichosanthin, a type I ribosme inactivating protein, J. Biol. Chem., 265:8670, (1990).

33. D.C. Blakey, and P.E. Thorpe, Effect of chemical deglycosylation on the in vivo fate of ricin A chain, Cancer Drug Delivery, 3:189, (1986).

34. M. Piatak and N. Habuka, Expression of plant-derived ribosome-inactivating proteins in heterologous systems, in: Genetically engineered toxins, A.E. Frankel, ed. Marcel Dekker Publishers, New York, N Y, In press, 1992.

35. R.A. Laskey and A.D. Mills, Quantitative film detection of ^{3}H and ^{14}C in polyacrylamide gels by fluorography, Eur. J. Biochem., 56:335, (1975).

36. P. Chomczynski, and N. Sacchi, Single-step method of RNA isolation by acid guanidinim thiocyanate phenol chloroform extraction, Anal. Biochem., 162:156, (1987).

37. T. Maniatis, E.F. Fritsch, and J. Sambrook, Molecular cloning: A laboratory manual, Cold Spring Harbor Laboratory, Cold Spring Harbor, NY, (1982).

38. A.G. Fisher, L. Collalti, L. Ratner, R.C. Gallo, and F. Wong-Staal, A molecular clone of HTLV-III with biological activity, Nature, 316:262, (1985).

39. S.M. Crowe, M.S. McGrath, T. Elbeik, J. Kirihara, and J. Mills, Comparative assessment of antiretrovirals in human monocyte/macrophages and lymphoid cell lines infected with human immunodeficiency virus, J. Med. Virol., 29:176, (1989).

40. S. Crowe, J. Mills, and M.S. McGrath, Quantitative immunocytofluorographic analysis of CD4 antigen expression and HIV infection of human peripheral blood monocyte/macrophages, AIDS Res. Hum. Retrovir., 3:135, (1987).

41. D.D. Richman, R.S. Kornbluth and D. A. Carson, Failure of dideoxynucleosides to inhibit human immunodeficinecy virus replication in cultured human macrophages, J. Exp. Med., 166:1144, (1987).

42. C.-F. Perno, R. Yarchoan, D.A. Cooney, N.R. Hartman, S. Gartner, M. Popovic, Z. Hao, T.L. Gerrard, Y.A. Wilson, D.G. Johns, and S. Broder, Inhibition of human immunodeficiency virus (HIV-1/HTLV-III$_{Ba-L}$) replication in fresh and cultured human peripheral blood monocytes/macrophages by azidothymidine and related 2',3'-dideoxynucleosides, J. Exp. Med., 168:1111, (1988).

43. F. Stirpe, S. Bailey, S.P. Miller, and J.W. Bodley, Modification of ribosomal RNA by ribosome inactivating proteins from plants, Nucleic Acids Res., 16:1349, (1988).

44. Y. Endo,Y.-L. Chan, A. Lin, K. Tsurugi, and I.G. Wool, The cytotoxins α-sarcin and ricin retain their specificity when tested on a synthetic oligoribonucleotide (35-mer) that mimics a region of 28S ribosomal ribonucleic acid, J. Biol. Chem., 263:7917, (1988).

45. J.M. D'Alessio, RNA Sequencing, In: Gel electrophoresis of nucleic acids: a practical approach, D. Rickwood and B.D. Hame, eds., IRL Press Limited, Oxford, (1982).

46. T.G. Obrig, T.P. Moran, and R.J. Colinas, Ribonuclease activity associated with the 60S ribosome inactivating proteins ricin A, phytolaccin and shiga toxin, Biochem. Biophys. Res. Comm., 130, 879, (1985).

47. P. Gunning, P. Ponte, H. Okayama, J. Engel, H. Blau, and L. Kedes, Isolation and characterization of full length cDNA clones for human α-, β-, and γ-actin

mRNAs: Slekletal but not cytoplasmic actins have an animno-terminal cysteine that is subsequently removed, Mol. Cell. Biol., 3:787, (1983).

48   K. Shigesada, G.R. Stark, J.A. Maley, L.A. Niswander, and J.N. Davidson, Construction of a cDNA to the hamster CAD gene and its application toward defining the domain for aspartate transcarbamylase, Mol. Cell. Biol., 5:1735, (1985).

49.   V.S. Byers, A.S. Levin, L.A. Waites, B.A. Starrett, R. A. Mayer, J.A. Clegg, M.R. Price, R. A. Robins, M. Delaney, and R.W. Baldwin, A Phase I/II study of trichosanthin treatment of HIV disease, AIDS, 4:1189, (1990).

50.   L. Pulliam, B. Herndier, and M.S. McGrath, BLQ223 exacerbation of indirect HIV associated neurotoxicity in vitro, AIDS, In press, 1991.

51.   W. Montford, J.E. Villafranca, A.F. Monzingo, S.R. Ernst, B. Katzin, E. Reutenber, N.H. Xuong, R. Hamlin, and J.D. Robertus, The three-dimensional structure of ricin at 2.8 Å J. Biol. Chem., 262:5398, (1987).

52.   D. Schlossman, D. Withers, P. Welsh, A. Alexander, J. Robertus, and A. Frankel, Role of glutamic acid 177 of the ricin toxin A chain in enzymatic inactivation of ribosomes, Mol. Cell. Biol., 9:5012, (1989).

SCREENING OF NATURAL PRODUCTS AS HIV-1 AND HIV-2 REVERSE

TRANSCRIPTASE (RT) INHIBITORS

Ghee T. Tan, John M. Pezzuto and A. Douglas Kinghorn

Program for Collaborative Research in the Pharmaceutical Sciences, Department of Medicinal Chemistry and Pharmacognosy, College of Pharmacy, University of Illinois at Chicago, and the Specialized Cancer Center, University of Illinois College of Medicine at Chicago, Chicago, Illinois 60612

INTRODUCTION

The human immunodeficiency virus type-1 (HIV-1) is the etiologic agent for acquired immunodeficiency syndrome (AIDS)[1], which has reached pandemic proportions in recent years. A second immunodeficiency virus (designated HIV-2), which is serologically distinct from HIV-1, was later isolated from West African patients with AIDS[2,3]. Research efforts have intensified in an attempt to discover new therapeutic modalities for the treatment of this fatal disease. The complex replication process of the HIV in its host cell offers many potential target points for therapeutic intervention[4,5]. Examples include viral attachment and uncoating, reverse transcription of the viral genome, proviral DNA integration into the host cell DNA and subsequent transcription and translation of this genetic material into viral proteins. As these viral proteins in turn undergo processing, e.g., protein glycosylation, which then leads to the assembly and release of progeny virus particles, additional target points are obvious.

Emphasis will be placed on the HIV reverse transcriptase (RT) enzyme since it is specific and essential for the replication and infectivity of retroviruses including the HIV. The viral RT has no counterpart in the normal cell, and is hence unique to the virus. Since viral replication is essential for the progression of AIDS, HIV RT represents one of the most important targets for the development of selective anti-HIV compounds. Chemotherapeutic agents designed to inhibit the function of retroviral RT currently comprise the only entities used for the treatment of AIDS[5].

Pitfalls of current antiviral chemotherapy include the toxicity of agents like azidothymidine (AZT) and other nucleoside analogs, in addition to the emergence of viral resistance to AZT[5]. Most reverse transcriptase inhibitors also inhibit cellular DNA polymerases and the non-discriminatory nature of such agents accounts for their toxicity[6,7]. Hence, the discovery and characterization of agents capable of specifically inhibiting the HIV RT without mediating a toxic response remain a high priority. In addition, novel inhibitors of this enzyme may function by unique mechanisms of action.

Natural products represent a rich and largely untapped source of structurally novel chemicals which are worth investigating as specific inhibitors of HIV RT. As yet, very few natural products have been shown to inhibit HIV RT specifically. This chapter summarizes

Natural Products as Antiviral Agents, Edited by C.K. Chu
and H.G. Cutler, Plenum Press, New York, 1992

progress in the development of a bioassay and the screening of natural products for potent inhibitors of HIV-1 and HIV-2 RT. This includes the establishment of a protocol and definitions of concentration limits for activity, solvent selection, application of the bioassay to a bioactivity-guided fractionation procedure, and the development of strategies involved in the general screening of plant extracts. The sensitivity and selectivity of this bioassay procedure will also be described. In addition, various structural classes of HIV-1 and HIV-2 RT inhibitors will be presented, with a brief discussion on their possible mechanisms of action.

HIV-1 AND HIV-2: A COMPARISON

HIV-2 is similar to HIV-1 in morphology, ultrastructure and genomic organization, with both possessing homologous structural and replicative proteins including the RT. Biological properties are also conserved in these cytopathic viruses in that they selectively infect and kill OKT4[+] helper cells resulting in immune suppression. However, genetic comparison of HIV-1 and HIV-2 has revealed significant sequence divergence at both nucleotide and amino acid levels[3]. The RTs of HIV-1 and HIV-2 also show significant sequence divergence, having approximately 60% of their amino acids in common. Nevertheless, the DNA polymerase activities of both enzymes share similar biochemical properties such as ion and template-primer preferences[8,9]. Genetic analyses also suggest that the model previously proposed for HIV-1 RT[10] may not strictly apply to HIV-2 RT[11].

Given the limited amount of sequence homology, it is important to identify and compare the biological properties of these two enzymes. Preliminary studies have shown that despite differences at the molecular level, the HIV RTs are functionally similar[11]. It is critical from the viewpoint of AIDS therapy to determine if this functional similarity extends to the susceptibility of these enzymes to inhibitors. The total conservation of biological and functional properties would ultimately imply that strategies for HIV-1 antiviral therapy may be of value in treating diseases associated with HIV-2. Additionally, such a comparison could provide important information on the structure of both enzymes. In order to probe the structure and activity of these enzymes, it is appropriate to evaluate the effect of a diverse group of compounds. We have recently performed such a study with a variety of natural products. It was found that the two HIV-associated RTs do show small but significant differences in their sensitivities to certain natural product inhibitors[12].

STRUCTURE AND FUNCTION OF HIV-1 AND HIV-2 RT

RT converts the RNA genome of the virus into a duplex DNA intermediate that can be integrated into the genome of the host cell. This proviral DNA is then transcribed leading to the production of viral proteins. HIV RTs are multifunctional, as are other retroviral RTs. A single protein possesses DNA polymerase activity that can transcribe utilizing either RNA or DNA templates. In contrast, the ribonuclease H (RNase H) activity is capable of degrading RNA only when it is part of an RNA - DNA heteroduplex[13]. All these domain-mediated functional activities of the enzymes provide reasonable targets for the action of anti-HIV drugs.

Structural analysis has identified proteins of M_r 66 000 (p66) and 51 000 (p51) as the HIV-1 RT[14]. HIV-2 RT is also a heterodimer, consisting of M_r 68 000 (p68) and 55 000 (p55) proteins. Amino acid sequence analysis has shown that the p51 subunit of HIV-1 RT is generated by the carboxy-terminal processing of p66; the two subunits have a common sequence at the amino terminus[15]. By analogy, the p55 subunit of the HIV-2 RT is proposed to be a product of processive cleavage of the p68 subunit. Sequence homology studies also revealed that all retroviral RTs (including the HIV-1 RT) have the DNA polymerase domain as the amino-terminal portion of the molecule and the RNase H as the carboxy-terminal portion. The same was found with HIV-2 RT[11].

MECHANISMS OF ACTION OF RT INHIBITORS

RT inhibitors may function by several mechanisms of action. Substrate or template-primer analogs of RTs act by mimicking the natural nucleotides and nucleic acids with which the enzyme interacts. AZT and the 2',3'-dideoxynucleosides[16] such as ddI are substrate analogs of HIV RT, while template-primer analogs that inhibit RT activity include 5-mercapto-polycytidylic acid[17]. Phosphonoformate and other pyrophosphate[18] derivatives constitute product analogs of RTs. Template-primer binding agents such as the benzophenanthridine alkaloids[19] are thought to intercalate with nucleic acids and, hence, interfere with their template properties. Agents which bind metal ions, such as the thiosemicarbazones, either deplete the system of divalent cations necessary for enzyme activity, or they may interact with the metal ion which is complexed with the enzyme[20].

Compounds which bind selectively to the active site of the enzyme should yield useful responses. Unfortunately, such agents are not frequently encountered. Compounds that bind to proteins in a relatively nonspecific fashion such as polyphenolic compounds[21] are observed to have more generalized effects on a variety of enzyme systems. This property often contributes significantly to the toxicity of such compounds.

Similarly, compounds that exert their inhibitory action by complexing with synthetic template-primers are typically not very specific for the RT, and may be cytotoxic. The RNA genome of the virus (70S RNA), though a novel feature of retroviruses, does not appear to offer any uniqueness from a physical and chemical standpoint that would distinguish it from cellular nucleic acids. However, approaches based on specificity (such as antisense oligonucleotides) offer promise[22].

SCREENING OF NATURAL PRODUCTS FOR HIV-1 AND HIV-2 RT INHIBITORY ACTIVITY

The HIV RT Assay

The assay developed for the detection of RT activity in virions[23], involving polyadenylic acid [poly (rA)], oligodeoxythymidylic acid [oligo (dT)], and radiolabeled thymidine triphosphate ([³H]TTP), has been adopted in our laboratory as a simple method for screening the HIV-1 and HIV-2 RT inhibitory potential of natural products[12,24]. Although the optimal conditions for the assay of HIV-2 RT have not been systematically determined per se, identical assay conditions for both enzymes yielded good rates of substrate incorporation. This is expected because of the conservation of biological properties in both enzymes.

HIV-1 And HIV-2 RT

The efficient expression of enzymatically active HIV-1[25] and HIV-2 RT[26] in bacterial systems has made available large quantities of the enzymes for biochemical, genetic, immunological and proteolytic analyses. The 66 kDa polypeptide of HIV-1 RT, when expressed in *Escherichia coli*, has biochemical properties that closely resemble the dimeric p66/p51 form of the enzyme purified from virions[25]. Furthermore, the p66 enzyme was shown to be active in RT assays and exhibit inhibitory properties with several known antiretroviral agents (e.g., AZT and suramin) that are indistinguishable from the viral enzyme[27]. The purified recombinant enzyme was sufficiently similar to the viral enzyme that it can be substituted for the latter in drug screening assays[28]. Hence, the intercomparison of data obtained with monomeric and dimeric forms of HIV-1 and HIV-2 RT is not expected to commonly cause any discrepancies.

Solvent Selection

Adaptation of the RT assay for the evaluation of natural products necessitates the selection of appropriate solvents. Various solvents have been tested and found to exert different degrees of inhibitory effect on the polymerization reaction. In general, the monomeric forms of the RTs were found to be more susceptible to solvent inhibition than the dimeric forms, possibly due to the greater stability of the latter. This, however, does not influence the inhibitory potential of compounds and extracts determined with the system since the solvent effect is monitored as the negative control. DMSO at a final concentration of 10% (v/v) is well-tolerated in systems incorporating either structural form of HIV-RT. At this concentration, only slight inhibition of enzyme activity was noted. Satisfactory dissolution of test compounds and extracts may be expected, except for those which are very nonpolar in nature. The aqueous nature of the enzyme reaction mixture also precludes the use of nonpolar organic solvents and hence the testing of nonpolar compounds. Certain compounds which appear to dissolve completely in DMSO may precipitate when added to the final reaction mixture. A lower concentration of compound may be tested in these cases. Methanol and ethanol at final concentrations of 10% (v/v) are poorly tolerated resulting in approximately 55% and 100% inhibition of monomeric HIV-1 RT activity, respectively[24]. Much lower concentrations of these solvents [e.g., 2% (v/v)] may be used when DMSO is not suitable.

Evaluation of Experimental Parameters

The most fundamental and yet crucial part of the development of the HIV-RT assay involves the evaluation of experimental parameters. Attempts should be made to optimize all assay conditions with respect to the ratios and concentrations of template-primers, monovalent and divalent ion concentrations, and pH of the reaction buffer. The concentration of reagents used (especially that of the enzyme, template-primer and substrate) will determine the catalytic efficiency of the enzyme, and ultimately its susceptibility to inhibition by agents with varied mechanisms of action. For example, the use of saturating enzyme, template-primer or substrate concentrations may attenuate the inhibitory effect of compounds acting competitively with respect to the enzyme, template-primer or substrate, respectively. On the other hand, when higher template-primer concentrations are used, greater inhibition values will be obtained for a compound that acts by an uncompetitive mode of action with respect to the template-primer. Hence, it is clear that the inhibitory potency of a particular agent depends not only on its concentration, structure and mode of action, but also on the assay conditions.

In addition, the concentration of RT, template-primer and substrate must be suitable from a kinetic viewpoint. Typical kinetic information to be examined includes the time-course profile and the saturation curves for the template-primer, substrate and enzyme[24]. Substrate incorporation should increase linearly with time for the duration of the assay and, in general, due to the aforementioned reasons, subsaturation conditions should be employed preferably. Thus, the standardization of experimental conditions is especially important when the inhibitory potential of a single agent is to be compared in systems incorporating two different enzymes. Wherever possible, the effects of inhibitors should be evaluated at the same region on the saturation curves of the two enzymes in an attempt to standardize the polymerization activity of each system and hence their susceptibility to the inhibitors.

A question frequently raised concerning the *in vitro* evaluation of compounds as potential antiviral agents is the relevance of the template-primer employed in the enzyme assay. As a prerequisite, the template-primer used should be sensitive in order to achieve good substrate incorporation and hence, a system perceptive towards the presence of inhibitory compounds. However, inhibitors of nucleic acid-polymerizing enzymes frequently demonstrate template specificity. Such is the case with fagaronine chloride[29] and *O*-methylpsychotrine sulfate heptahydrate[30], which have a preference for A:T (adenine:thymine) rich templates. Ideally, the RNA genome of the virus in question should be used for assessing the RT inhibitory potential of prospective antiviral agents. This is often not practical. However, since the genomes of the HIV-1[31] and HIV-2[3] are A:T rich, the use of $(rA)_n \cdot (dT)_n$ as a sensitive template-primer in the screen may be considered appropriate and of physiological relevance. Other template-primers such as $(rC)_n \cdot (dG)_n$ may be examined during the mechanistic analysis of active inhibitors.

Plant constituents which bind nonspecifically to proteins (e.g., polyphenolic compounds) may be frequently encountered during a screening procedure. The inclusion of bovine serum albumin (BSA) as a protective agent in an enzyme assay is optional. Its presence, however, will decrease the inhibitory activity of these compounds toward the enzyme of interest, and thus, serves as an effective means of screening out interfering substances. It is also worthwhile to optimize the concentration of BSA in the assay from the viewpoint of enzyme activity.

Since HIV-1 RT demonstrates a preference for Mg^{2+} over Mn^{2+} ions for most template-primers[23], the former is normally used in an assay of this enzyme activity. However, on occasion the type of divalent ion used in the RT assay has been found to have significant effects on the inhibitory activity demonstrated by certain compounds[32].

Assay Protocol and Strategies for Screening

A strategic and economical method for conducting the assay that results in highest efficiency involves the prescreen of only one concentration of pure compound or plant extract at a concentration of 200 µg/ml[24]. A dose-response is then obtained for active compounds and extracts by testing at least five concentrations in duplicate. The IC_{50} values (concentration giving 50% inhibition of enzyme activity) may be calculated from linearly regressed dose-response plots of percent control activity versus concentration or log concentration of compound or extract. Any RT inhibitor [e.g., fagaronine chloride, $IC_{50} = 5$ µg/ml (13 µM)[33]] may be used as the positive control substance. Negative control assays are performed without the compounds or extracts, but an equivalent amount of solvent is added.

Assay Sensitivity

The HIV-1 RT assay developed was found to be equally sensitive in detecting selected positive control inhibitors (e.g., fagaronine chloride) in the absence and presence of up to 200 µg/ml of an arbitrarily chosen plant extract, which had previously been determined to be inactive in the HIV-1 RT system[24]. The lower limit of detection was found to be 0.01% w/w fagaronine chloride in the plant extract.

Concentration Limits for Activity

Inhibition data generated from different laboratories are difficult to compare critically due to variations in assay conditions and test protocols used. Therefore, a certain compound may appear significantly active in one assay while relatively inactive in another. Such inconsistencies necessitate the standardization of an enzyme assay when it is applied to natural product screening. Similarly, activity cut-offs should be individually tailored for each enzyme assay protocol. Thresholds for activity can usually be determined after a preliminary round of screening involving a few hundred natural product entities. Consequently, for the HIV-1 RT screen, the following thresholds were established. After the screening of approximately 200 pure compounds it was deduced that a pure natural product may be classified as active (IC_{50} < 50 µg/ml), moderately active (50 µg/ml < IC_{50} < 150 µg/ml) or weakly active (150 µg/ml < IC_{50} < 200 µg/ml)[24]. During the prescreen, however, compounds demonstrating 50-100% inhibition at 200 µg/ml were selected for further activity evaluation.

Screening of Plant Extracts

The screening of crude extracts may result in a significant proportion of actives (>50% inhibition at 200 µg/ml). A response of this magnitude is indicative of the presence of high concentrations of potent inhibitory compounds in plant extracts. Frequently, this may be ascribed to polyphenolic compounds which are abundant in the plant kingdom, and which have been reported to inhibit RTs from RNA tumor viruses[34]. While certain tannins have HIV RT inhibitory activity, in our laboratory non-tannin inhibitory compounds are of primary interest. Tannin-precipitating reagents offer a quick and convenient way to verify the presence of tannins. A dark blue to brown solution with or without a precipitate results with the use of ferric chloride, whereas a gelatin/sodium chloride mixture gives a white precipitate in the presence of tannins. Relative to tannins, most non-polyphenolic inhibitory compounds are present in much lower concentrations in plant extracts. It was observed that when potent HIV RT inhibitors such as fagaronine chloride or nitidine chloride were tested at concentrations between 0.002 and 0.4 µg/ml in the presence of an inert extract (simulating a 0.001 - 0.2% composition in a 200 µg/ml plant extract), only 15 to 20% inhibition was obtained[24]. Thus, in the absence of tannins, responses of this magnitude must be considered significant. However, because it is generally not known from the outset whether the inhibition mediated by a plant extract is due to tannins or other substances, chromatographic fractionation of active extracts is required, and inhibition obtained with concentrated fractions will serve as a better indication of the presence of non-tannin inhibitory compounds. Alternatively, tannins can be removed prior to testing of the extract in the preliminary screen.

Evaluation of Methods for Tannin Removal

Due to the interfering nature of polyphenolics, it becomes mandatory to find quick and simple procedures for their removal that do not significantly affect the efficiency of the screening process. Methods that have been evaluated included the use of precipitating agents such as gelatin/sodium chloride solution, caffeine solution, soluble and insoluble polyvinylpyrrolidone (PVP), in addition to polyamide column chromatography[24].

The interaction leading to the removal of tannins is the formation of hydrogen bonds between the tannin phenolic hydroxy groups and the amide link (CONH) of the precipitating agent which results in the formation of insoluble complexes. Therefore, these procedures also remove non-tannin inhibitory compounds with phenolic hydroxy groups (e.g., flavonoids), except those that have catechol groups which are internally hydrogen bonded. Due to this structural characteristic, hydrolyzable tannins have been reported to elude such tannin removal procedures. Furthermore, quinones are irreversibly bound to these agents due to a covalent interaction[35].

It was previously reported that chromatography on polyamide is effective for the selective removal of polyphenolics[36]. With slight modification, this procedure may be accomplished quickly and conveniently with only a few milligrams of plant extract. This procedure is advantageous in that bound non-tannin inhibitory compounds with two or three phenolic hydroxy groups can be eluted by flushing with methanol.

The use of PVP, on the hand, may result in the removal of all compounds with phenolic hydroxy groups. The insoluble form of PVP facilitates its removal by centrifugation. However, soluble PVP has been found to be equally efficient, even though an excess will result in an increase in viscosity of the enzyme reaction mixture. A slight stimulation of enzyme activity was observed in the presence of soluble PVP. Nonetheless, at this stimulatory concentration, it did not interfere with the inhibitory effect of test compounds as demonstrated by the identical inhibition values observed in its presence and absence[24].

All other procedures examined were found to be unsuitable for tannin removal. Difficulty was encountered in using such procedures quantitatively because the reagents themselves were added as aqueous solutions, and there was an uncertainty as to the amount to add to the plant extract for the complete precipitation of tannins. The viscosity of these reagents and excess amounts in solution also interfered with the enzyme assay.

Thus, no single quantitative procedure involving the addition of a precipitating reagent was found to be suitable for tannin removal from plant extracts in preparation for the HIV RT assay, and no generalizations could be made regarding the concentrations or type of reagents that should be adopted. Different plant species contain different types of polyphenolics at highly variable concentrations. In general, since tannins are frequently present in high concentrations in plant extracts, all precipitating agents should have a high capacity for adsorbing these polyphenolic compounds. The methods of choice are obviously polyamide column chromatography and precipitation with insoluble PVP, procedures which do not introduce any extraneous material to the plant extract to be tested.

Bioactivity-Guided Fractionation Procedure

The utility of a bioassay in the process of drug discovery from natural sources relies on its capacity to detect minute quantities of active constituents in plant extracts derived from various stages of a fractionation procedure, and hence guiding this often lengthy process to ultimately yield an active compound(s). When the HIV-1 RT assay was applied at various stages of the fractionation scheme of the bark of *Plumeria rubra* L. (Apocynaceae), known to contain the active iridoid fulvoplumierin, it proved capable of directing the isolation process to yield the active compound from the petroleum ether extract after silica gel column chromatography[24]. The ability of the assay to detect active constituents in plant extracts depends, in part, on the concentration of extract tested, in addition to the concentration and inhibitory potency of the active constituent(s) present. Therefore, in a bioactivity-guided fractionation scheme, *relative* % inhibition values should be of concern rather than the *actual* % inhibition. For example, in the case of *P. rubra*, a petroleum ether fraction demonstrated only 35% inhibition at 200 µg/ml, from which fulvoplumierin, a moderately active HIV-1 RT inhibitor, was later isolated. This % inhibition value was not outstanding in itself, but it warranted attention when compared with the aqueous and chloroform fractions which were almost devoid of detectable inhibitory activity[24].

Data Interpretation

A variety of interactions occurring in an *in vitro* enzyme system can modulate the catalytic rate of DNA synthesis and the effect of potential inhibitors. An awareness of such factors will facilitate the correct interpretation of enzyme inhibition data. Some examples of such interactions include the effect of the type of divalent cations (Mn^{2+} or Mg^{2+}) on DNA synthesis obtained with different template-primers, the role of chelating agents or cation binders, thiol reducing agents and BSA in the reaction buffer, and the effect of using different concentrations of enzyme, template-primer and substrate. Thiols may react with certain potential inhibitors and alter their inhibitory activity while the detergent effect on the activity of rifamycins is well documented[37]. In addition, certain natural products such as saponins may have surfactant properties, and their enzyme inhibitory effect may well be due to an alteration of the physicochemical properties of the reaction medium. Indeed, it was observed with the HIV-1 RT assay currently used in our laboratory[24] that the inhibitory effect of these compounds is quite irreproducible (unpublished observations). The interaction between certain compounds and nucleic acids results in the immediate precipitation of the complex. This artifact, if overlooked, may be misconstrued as a legitimate inhibition of enzyme activity. The effect of compounds which bind nonspecifically to proteins is worth reiterating. Thus, it is imperative to evaluate potential inhibitors of HIV RT under carefully controlled conditions.

NATURAL PRODUCT INHIBITORS OF HIV-1 RT

Benzophenanthridine Alkaloids

Benzophenanthridine alkaloids (Figure 1) are a class of isoquinoline alkaloids found in the plant families Rutaceae and Papaveraceae that possess interesting biological and pharmacological properties[38]. Fagaronine chloride (1), isolated from *Fagara xanthoxyloides* Lam. (Rutaceae), demonstrated potent antileukemic (P-388 and L1210 lymphocytic leukemias)[33] activity but was considered inactive in a few solid tumor models investigated[39]. Nitidine chloride (2), isolated from *Fagara macrophylla* (Oliv.) Engl. (Rutaceae), was found to inhibit Lewis lung carcinoma and murine L1210 and P388 leukemias[40]. Sanguinarine chloride (3) and chelerythrine chloride (4) have been shown to possess antimicrobial activity[41,42]. Enzyme systems affected by the benzophenanthridine alkaloids include catechol O-methyltransferase[43] and transfer RNA methyltransferase[43,44] of rat liver, NaK-ATPase of guinea pig brain[45,46] and protein kinase C of rat brain[47]. Benzophenanthridine alkaloidal salts such as fagaronine chloride, O-methylfagaronine fluorosulfonate, nitidine chloride, allonitidine methylsulfate (5) and 6-methoxy-5,6-dihydronitidine bromoacetate (6) were reported to be potent inhibitors of avian myeloblastosis virus (AMV) RT and Rauscher murine leukemia virus (RMLV) RT demonstrating IC_{50} values between 6-18 µg/ml for fagaronine chloride and O-methylfagaronine fluorosulfonate, and 40-60 µg/ml for the latter three alkaloids[19,48]. HIV-1 RT was later shown to be similarly affected, with fagaronine chloride ($IC_{50} = 10$ µg/ml) and nitidine chloride ($IC_{50} = 49$ µg/ml) being the most potent compounds tested[24].

The inhibitory effects of these alkaloids on several mammalian polymerase enzymes have also been described[49]. While reverse transcriptase and DNA polymerase α activities were strongly inhibited by these alkaloids, RNA polymerases I and II were only moderately affected. Polyadenylic acid (poly A) polymerase activity was only weakly inhibited. Fagaronine chloride and nitidine chloride gave IC_{50} values in the range of 20-45 µg/ml with both viral and mammalian DNA polymerases. Additionally, fagaronine chloride inhibited protein synthesis in cell-free systems, and caused irreversible inhibition of DNA, RNA and protein syntheses in KB cells[50]. An interaction with the ribosomal system was proposed.

A comparison of the effect of 15 structurally-related benzophenanthridine alkaloids revealed that fagaronine chloride possesses optimum structural features for AMV RT inhibitory activity[51]. Decreased activity was observed when the phenolic hydroxy group at position 2 was methylated or when the methoxy group at position 9 was replaced by a phenolic hydroxy group. Also, nitidine chloride and allonitidine methylsulfate which have a methylene dioxy group at positions 2 and 3 or 8 and 9, respectively, showed less inhibition of AMV RT. Saturation of ring B (e.g., 6-methoxy-5,6-dihydronitidine bromoacetate) or rings B and C [e.g., chelidonine (7)] with the concomitant loss of the charge on the nitrogen atom, was observed to produce a significant loss of activity. Likewise, demethylation of fagaronine chloride resulted in the loss of its antileukemic, bactericidal, cytotoxic and RT inhibitory effects[39,24]. Furthermore, substituents at positions 7 and 8 (e.g., chelerythrine chloride) or at any other positions [e.g., chelirubine chloride (8)] on the molecule reduced the RT inhibitory effect. These data indicate that the quaternary nitrogen and the methoxy groups at positions 8 and 9 are essential for optimum activity. These general considerations have also been shown to apply to the HIV-1 RT[24].

It has also been suggested that the RT inhibitory activity of these alkaloidal salts is in accordance with the postulated active site (iminium ion -C=N$^+$-CH$_3$) for antitumor and other biological activities[51]. In aqueous solution at suitable pH, benzophenanthridine alkaloids covalently take up hydroxide ions to form a pseudobase or alkanolamine adduct. The alkaloids have been shown to attain different iminium ion concentrations in solution, the equilibria of which is highly dependent on the pK_a of the molecule and the effect of substituent groups[52]. A dichotomy of biological activies was observed. While a high iminium ion concentration correlated with potent antitumor activity (fagaronine chloride and nitidine chloride), a low concentration of the ion was favorable for antimicrobial activity and NaK-ATPase inhibition (sanguinarine chloride and chelerythrine chloride). The carbon of the iminium function

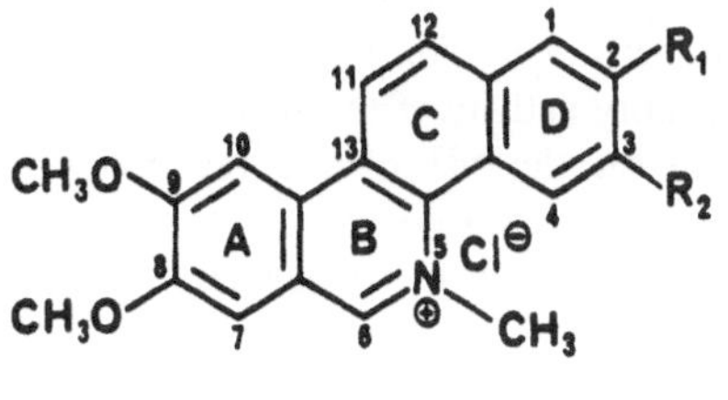

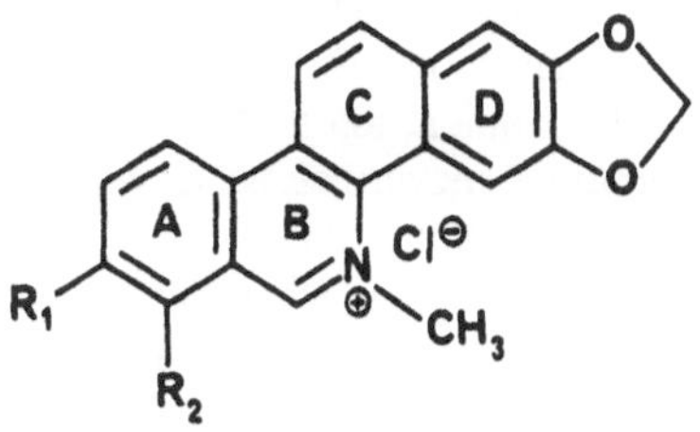

1 Fagaronine chloride (R₁ = OH; R₂ = OCH₃)
2 Nitidine chloride (R₁, R₂ = OCH₂O)
3 Sanguinarine chloride (R₁, R₂ = OCH₂O)
4 Chelerythrine chloride (R₁, R₂ = OCH₃)

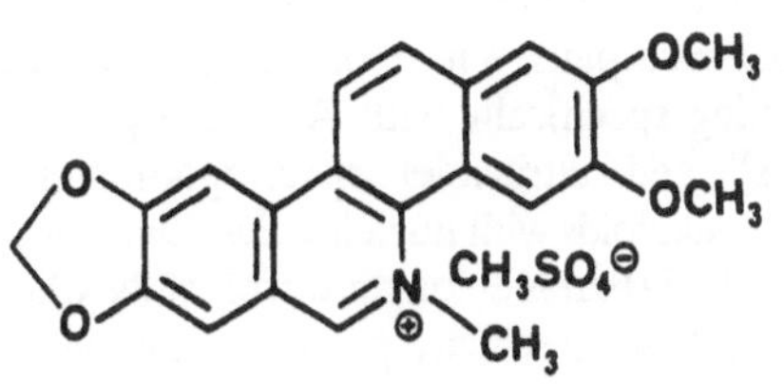

5 Allonitidine methylsulfate

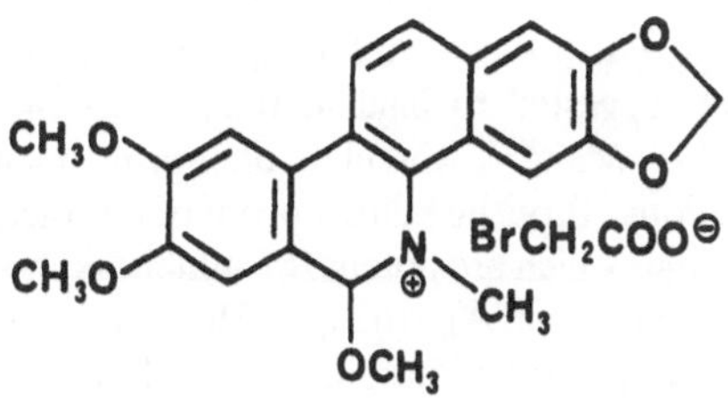

6 6-Methoxy-5,6-dihydronitidine bromoacetate

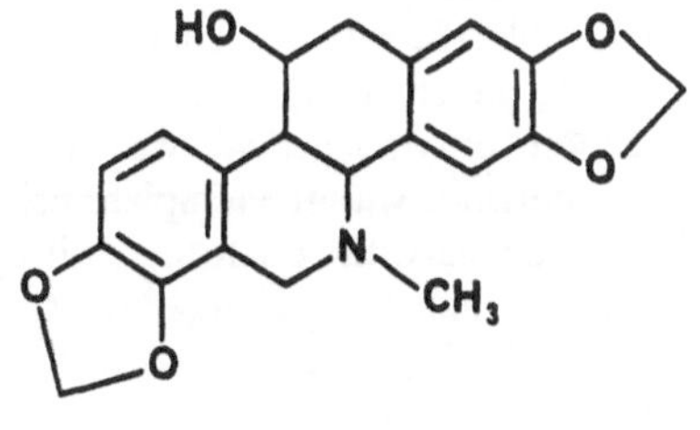

7 Chelidonine

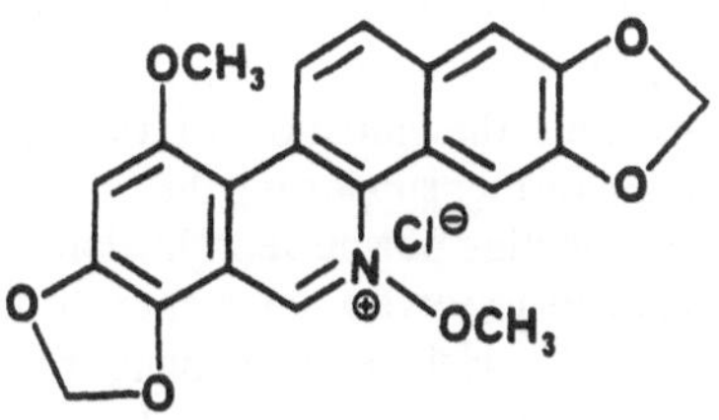

8 Chelirubine chloride

FIGURE 1. BENZOPHENANTHRIDINE ALKALOIDS

represents an effective alkylation site. The presence of substituents at positions 7 and 8 such as in sanguinarine chloride and chelerythrine chloride could sterically block nucleophilic attack at this iminium carbon, and may thus account for the absence of antitumor and antileukemic effects in these molecules[53]. A substituent at position 7 could also interact sterically with the peri-H of ring B and destabilize the iminium ion, shifting the equilibrium in favor of the alkanolamine ion[52]. From the perspective of structure-activity relationship (SAR), it would seem that the iminium ion is also necessary for RT inhibitory activity. The iminium-alkanolamine equilibrium, nevertheless, is very much dependent on the pH of the reaction medium. Sanguinarine chloride and chelerythrine chloride have lower pK_a values than the other members of the benzophenanthridine alkaloids. Consequently, there are significant amounts of pseudobase forms present at physiological pH[54].

It has also been reported that many nonalkylating antileukemic alkaloids possess two oxygen atoms and a nitrogen atom that define a triangle of certain optimum interatomic distances[55]. Any deviations from these specified interatomic distances is postulated to result in decreased antileukemic activity. This observation was found to apply to the benzophenanthridine alkaloids as well[53].

Marked template-primer preference was observed for the benzophenanthridine alkaloids. A significant degree of RT inhibition was obtained for activated DNA, 70S viral RNA and template-primers rich in A:T (adenine:thymine) or dA:T (deoxyadenine:thymine) base pairs compared to those containing G:C (guanine:cytosine) base pairs. Therefore, these alkaloids were suggested to bind to template-primers, interacting specifically with A:T base pairs of nucleic acids[19,29]. Circular dichroism measurements[56] and ultraviolet spectrophotometric titrations[57] furnished direct proof of interaction of these alkaloids with nucleic acids. Only those alkaloids which are potent RT inhibitors showed induced CD bands at longer wavelengths when admixed with $r(A)_n \cdot (dT)_{12-18}$. However, contrary to the base specificity proposed, ultraviolet spectral perturbations suggestive of an intercalation process were obtained when fagaronine chloride was admixed with calf thymus DNA, $(dA-dT)_n$, $(dG)_n \cdot (dC)_n$ or with other double-stranded nucleic acids. Hence, the binding of fagaronine chloride is not limited to polynucleotides containing only A and T. In contrast, DNase I footprinting studies have revealed that sanguinarine chloride and chelerythrine chloride failed to interact with $(AT)_n$ sequences on DNA due to the unique structure of the DNA helix in such a region[54]. Although the mode of action of fagaronine chloride has not been unequivocally characterized, interaction with the template-primer appears to be a major mechanism. Consistent with this suggestion, the kinetics of inhibition of RTs by the benzophenanthridine alkaloids was competitive with respect to the template-primer[56]. Interaction of fagaronine chloride with the enzyme protein itself has not been established. When these alkaloids were added to the enzyme reaction mixture after the initiation of the polymerization reaction, enzyme activity was abruptly decreased, further supporting the notion that the alkaloids interacted with the template-primer, thereby, affecting polynucleotide elongation instantly[19,49]. In summary, the quaternary nitrogen atom appears to serve as an anchor through interaction with the negatively charged phosphate backbone of nucleic acids enabling the fused coplanar aromatic ring system to intercalate[58,59]. This proposition is advocated by the fact that N-demethylfagaronine failed to interact effectively with double stranded nucleic acids[57]. Interactions of this type could also explain why poly A polymerase, which requires a single stranded RNA primer with a free 3' terminus as the template for activity, is the only polymerase enzyme relatively unaffected by these alkaloids[49]. Noteworthy is the fact that it is the iminium ion rather than a quaternary N atom, in general, that mediates the interaction with nucleic acids and, hence, RT inhibitory activity. Quaternary alkaloids of other structural classes, e.g., echitamine chloride (an indole alkaloid) and magnoflorine chloride (an aporphine), were virtually inactive in the HIV-1 RT system[24].

A good correlation was observed between the RT inhibitory activities of the benzophenanthridine alkaloids and their antileukemic effects, implying common molecular features responsible for activity. The role that RT inhibition plays in the antineoplastic activity of these alkaloids is uncertain at this point. In addition, the role of cellular metabolic activation cannot be overlooked where the *in vivo* effects of the benzophenanthridine alkaloids are concerned. For example, rat liver preparations have been shown to substantially enhance the bactericidal activity of fagaronine[57]. Unfortunately, the usefulness of these alkaloids as potential antiviral or anticancer agents is limited by their toxicity[53,57] and potent anti-HIV activity in cell culture remains to be demonstrated.

Protoberberine Alkaloids

Protoberberine alkaloids (Figure 2) which are widely distributed in plants of the Berberidaceae, Menispermaceae, Ranunculaceae and Rutaceae, are structurally related to the benzophenanthridine alkaloids. Berberine chloride (**9**), a typical representative of these alkaloids, has been extensively studied with respect to its chemical and pharmacological properties[60]. Much of the biological activities of the benzophenanthridine and protoberberine alkaloids are similar. Both berberine chloride and palmatine (**10**) were also effective against experimental tumors but lacked antileukemic activity against P-388 murine lymphocyte leukemia[61]. However, the synthetic dehydroprotoberberine coralyne chloride (**11**) exhibits inhibitory properties against both leukemias L1210 and P388 in mice[62,63]. Furthermore, protoberberine alkaloids were reported to inhibit a number of enzymes including horse liver[64] and yeast[65] alcohol dehydrogenase, tyrosine decarboxylase[66], tryptophanase[67] and catechol *O*-methyl transferase[62] by distinct mechanisms. The potent RT inhibitory effects of the benzophenanthridine alkaloids prompted an investigation on the related protoberberine alkaloids[68]. The IC_{50} values of palmatine and berberine chloride were in the range of 30-35 µg/ml and 100-105 µg/ml, respectively, for RTs from AMV, RMLV, and simian sarcoma virus type 1 (SSV-1). Similar IC_{50} values were observed for the HIV-1 RT[24].

The RT inhibitory properties of these alkaloids are identical to those of the benzophenanthridines. The use of template-primers containing only G:C base pairs resulted in weak inhibition of enzymatic activity. Based on this indirect evidence, a preference for A:T base pairs of nucleic acids was suggested[69]. The inhibition mediated by these alkaloids exhibited competitive type kinetics with respect to the template-primer but interaction with the template-primer binding site of the enzyme has not been conclusively demonstrated. A previous report related the antileukemic activity of the protoberberine alkaloids with molecular conformation and DNA binding properties of the molecules[61]. It became apparent that in berberine chloride and palmatine, the plane of ring A is twisted slightly out of that of the CD rings by the completely saturated ring B, thus lowering their ability to intercalate relative to the completely planar coralyne. A change in the absorption spectrum of coralyne acetosulfate when added to a calf thymus DNA solution has also been reported[70]. Other studies have concurred that the biological activities of berberine and coralyne are due to their ability to bind DNA probably by intercalation[71,72,73]. The concept of the triangular N-O-O pharmacophore[54] which was found among a number of antileukemic compounds may also be applied to coralyne and the protoberberine alkaloids.

SAR studies reveal that the presence of methoxyl groups at positions 2 and 3 of ring A and positions 9 and 10 or 10 and 11 of ring D was essential for optimum RT inhibitory activity[68]. Saturation of ring C as in tetrahydropalmatine (**12**) also yielded an inactive compound. Even though SARs are not as well defined in this series of alkaloids, they are observed to be weaker inhibitors of RT than the benzophenanthridines possibly due to differences in intercalating ability imposed by structural differences. While the benzophenanthridine molecule constitutes a fused planar aromatic ring system, that of the protoberberine does not[61].

Earlier studies indicated that the RT inhibitory property of benzophenanthridine alkaloids corresponded well with their antileukemic activity but this did

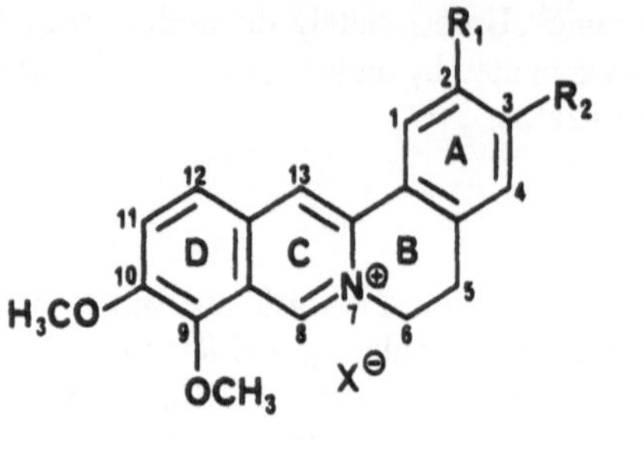

9 Berberine chloride (R_1, R_2 = OCH_2O ; X = Cl)
10 Palmatine (R_1, R_2 = OCH_3; X = OH)

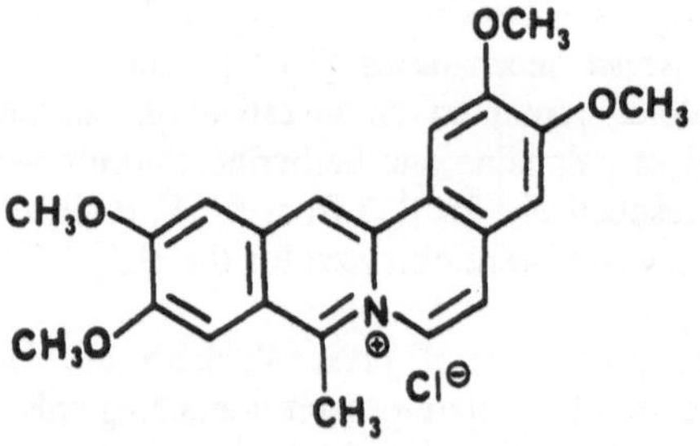

11 Coralyne chloride

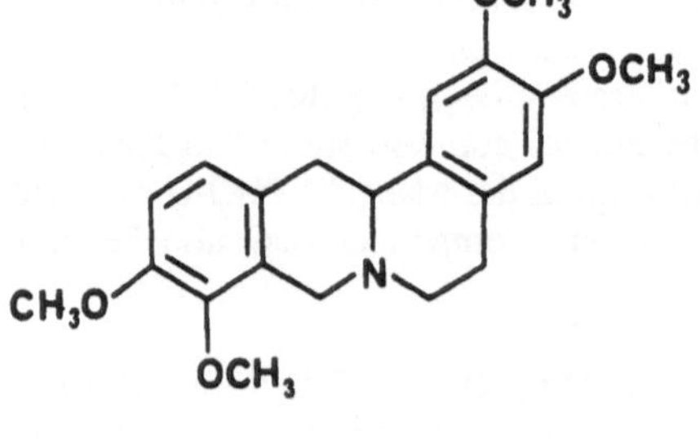

12 Tetrahydropalmatine

FIGURE 2. PROTOBERBERINE ALKALOIDS

not seem to be the case for the protoberberine alkaloids in general[74]. The mechanism of RT inhibition may well follow similar arguments, although additional mechanisms cannot be ruled out. Analogous to the benzophenanthridine alkaloids, the potential clinical utility of these compounds as antivirals or chemotherapeutic agents is hampered by their toxicity[75]. The anti-HIV activity of this group of compounds has yet to be determined.

Ipecac Alkaloids

Previous efforts in the screening of diverse structural classes of natural products[24] revealed two novel inhibitors from the ipecac class of alkaloids (Figure 3), O-methylpsychotrine sulfate heptahydrate (MP) (13) and psychotrine dihydrogen oxalate (PDO) (14). Relative to approximately 200 natural product drugs that were evaluated for inhibitory activity, MP and PDO were two of the most potent inhibitors uncovered. MP and PDO demonstrated IC_{50} values in the HIV-1 RT system of 10 μg/ml (14 μM) and 6 μg/ml (9 μM), respectively, relative to fagaronine chloride, which demonstrated an IC_{50} value of 5 μg/ml (13 μM). As a result, additional members of the ipecac alkaloids were tested for HIV-1 RT inhibitory activity with the aim of defining relevant structure-activity relationships[30].

MP is a minor alkaloid of ipecac, the dried rhizome and root of *Cephaelis ipecacuanha* (Brotero) A. Richard (Rubiaceae)[76]. Like the benzophenanthridine and protoberberine alkaloids, ipecac alkaloids are biosynthesized from isoquinoline intermediates. Ipecac yields more than 2% (w/w) of ether-soluble alkaloids, with the three principal ones being emetine, cephaeline and psychotrine[77]. These structurally-related alkaloids are prepared as their inorganic acid salts for the improvement of solubility and stability. The therapeutic properties of the ipecac alkaloids as emetics, expectorants and amebicides have long been recognized. The pharmacological effects of the two major alkaloids, emetine and cephaeline, include antiamebic[78], antitumor[79], and antiviral activities[80], in addition to the irreversible inhibition of DNA synthesis in cell culture[81,82]. They did not, however, affect cellular thymidine kinase or the incorporation of TTP into DNA with *in vitro* assay systems[81]. The synthesis of cellular RNA is unaffected by emetine but viral RNA synthesis in polio virus infected HeLa cells was strongly inhibited[81]. However, the toxicity of the drug, manifested primarily as cardiotoxicity when used for extended periods, precludes its therapeutic use as an antiviral.

Emetine and cephaeline are also potent inhibitors of protein synthesis in mammalian, yeast and plant cells, while bacterial cells are resistant to the effects of these alkaloids[81]. All biological effects of emetine and cephaeline, both therapeutic and toxic, are consistent with the inhibition of protein synthesis as the primary mode of action. Protein synthesis is thought to be inhibited at the stage of translocation[83,84], possibly by a direct effect of these compounds on an elongation factor. In contrast, O-methylpsychotrine and isoemetine were reported to have very weak effects on protein synthesis (<1% of the activity of emetine)[81].

An attempt has been made to characterize the mechanism of inhibition mediated by MP[30]. The inhibition was noncompetitive with respect to TTP and uncompetitive with respect to $(rA)_n$ and $(dT)_{12-18}$ (4:1) at low template-primer concentrations but competitive at high concentrations (> 200 μM). Identical non-Michaelis-type kinetics were observed when activated DNA was used as the template. The biphasic nature of the double-reciprocal plots and Hill coefficients of less than one indicated that MP functions as an allosteric inhibitor of the enzyme that appears to possess multiple active sites which interact in a cooperative (negative) fashion in the presence of the inhibitor. MP was selective for recombinant HIV-1 RT (p66) utilizing $(rA)_n$ and $(dT)_{12-18}$ (4:1) as template-primer. Greater inhibition was observed with this template-primer as compared to other natural and synthetic template-primers tested. Even though A:T or dA:T systems were strongly inhibited, spectroscopic studies did not reveal changes in the absorption spectra of MP in the presence of various concentrations of A or T rich homopolymers or activated DNA. Hence, it appears that MP does not interact with nucleic

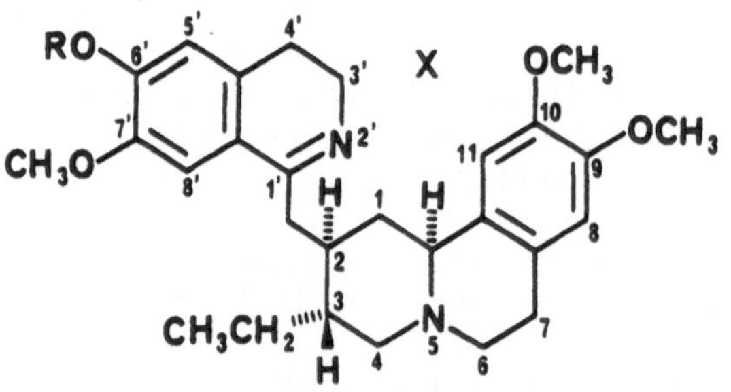

13 *O*-Methylpsychotrine sulfate heptahydrate (R = CH₃; X = H₂SO₄)
14 Psychotrine dihydrogen oxalate [R = H; X = 2(COOH)₂]

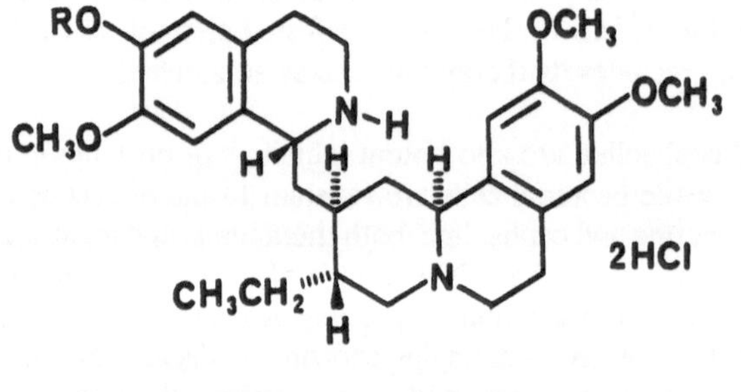

15 Emetine hydrochloride (R=CH₃)
16 Cephaeline hydrochloride (R=H)

FIGURE 3. IPECAC ALKALOIDS

acids and, therefore, does not interfere with their template properties. Since the p66 monomeric enzyme was used for the assay it can be concluded that a putative binding site(s) for MP resides on this catalytic subunit of the HIV-1 RT even though the recombinant p66/p51 heterodimer was equally susceptible to inhibition. Data further supported the conclusion that MP binds to a site other than the substrate or template-primer binding site of the enzyme possibly through hydrophobic or hydrogen bonding interactions.

MP had significantly less effect on AMV RT as well as mammalian or bacterial DNA and RNA polymerases. Other members of the ipecac class of alkaloids, e.g., emetine dihydrochloride (15) and cephaeline dihydrochloride (16), were inactive against all of these enzymes, including HIV-1 RT. Conversely, MP did not inhibit *in vitro* protein synthesis, a property manifested by all the other ipecac alkaloids tested. The inhibition of protein synthesis was observed to bear an apparent inverse correlation with RT inhibition. The imine functionality at positions 1' and 2' of MP and PDO appeared to be the key structural requirement for activity. It is also of necessity that these compounds be in the salt form in order to demonstrate HIV-1 RT inhibitory activity. MP and PDO demonstrated similar potency in the HIV-1 RT system suggesting that the presence of the methoxy group at position 6' of MP had no consequence on the binding of the compound to its putative binding site on the enzyme. The salt form of the active molecules may impart a positive charge on the N atoms in solution and hence facilitate interaction with the enzyme in some unknown way. It is uncertain at this point if the nature of the compounds (salt or free base) will influence the molecular conformation adopted in solution.

The fact that they do not inhibit protein synthesis implies that psychotrine, MP and PDO may be devoid of some toxic effects mediated by other biologically active ipecac alkaloids. However, while the inhibition of protein synthesis is structurally specific, the emetic action of these compounds does not correlate with their ability to inhibit protein synthesis[81]. *O*-Methylpsychotrine is very bitter and induces nausea when ingested. If compounds of this type are to find therapeutic use it will be necessary to explore chemical modifications that may reduce bitter and emetic properties while retaining or even improving HIV-1 RT inhibitory properties.

Flavonoids

Flavonoids (Figure 4), 2-phenyl benzo-γ-pyrone derivatives, are of wide occurrence in vascular plants including fruits, vegetables and grains. Flavonoids are classified according to ring substitution patterns and the degree of benzopyrone ring saturation as flavones, flavanones, isoflavones, isoflavanones, anthocyanins, chalcones and aurones, to name a few. These groups in turn show varying degrees of hydroxylation, methoxylation and glycosylation[85].

This large group of low molecular weight compounds has a wide spectrum of pharmacological properties[86], the mechanisms of which remain to a large extent unknown. Hence, flavonoids exert diverse effects on mammalian enzyme systems which contribute to their potential use in the treatment of a wide variety of human ailments[85]. These natural products are potent inhibitors of enzymes such as iodothyronine deiodinase[87], cyclic AMP phosphodiesterase[88], protein kinase C[89] and catechol-*O*-methyl-transferase[90]. The enzyme inhibitory properties of flavonoids were observed to show stringent conformational requirements. No unifying mechanism explains the effect of this rather homogenous group of molecules on such a perplexing number and variety of enzymes[91]. Flavonoids structurally resemble nucleosides, isoalloxazine and folic acid, and this similarity may be the basis of many of the current hypotheses of their physiological action[86].

Progress in the realms of molecular modelling and conformational analysis has permitted the visualization of the crystal and molecular conformation of flavonoids[92,93]. This facilitates the study of SAR involved in the host of biological activities mediated by

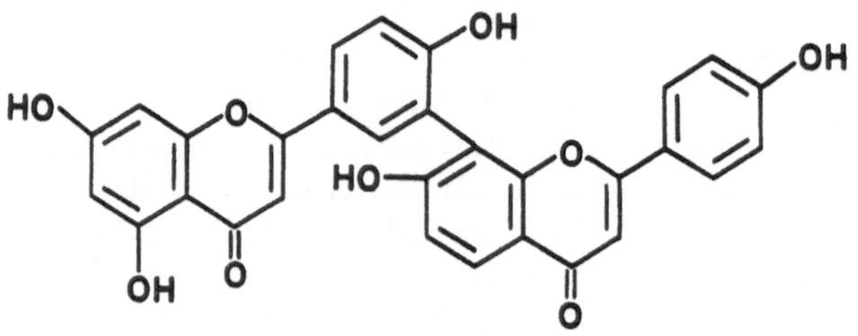

17 Amentoflavone

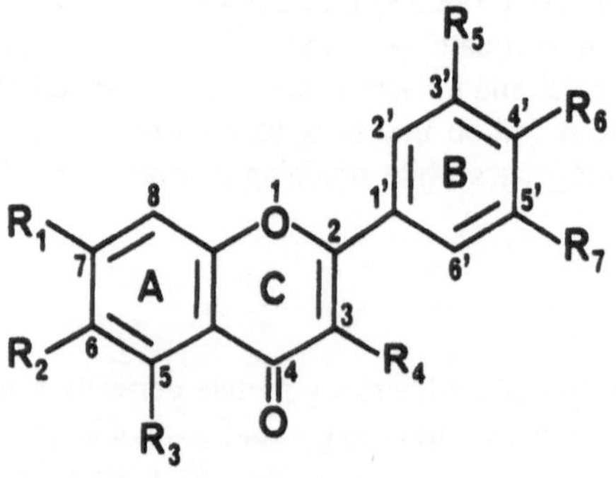

18 Scutellarein (R_1, R_2, R_3, R_6 = OH; R_4, R_5, R_7 = H)
19 Quercetin (R_1, R_3, R_4, R_5, R_6 = OH; R_2, R_7 = H)
20 Baicalein (R_1, R_2, R_3 = OH; R_4, R_5, R_6, R_7 = H)
21 Myricetin (R_1, R_3, R_4, R_5, R_6, R_7 = OH; R_2 = H)
22 Quercetagetin (R_1, R_2, R_3, R_4, R_5, R_6 = OH; R_7 = H)
23 Kaempferol (R_1, R_3, R_6 = OH; R_2, R_5, R_7 = H)

FIGURE 4. FLAVONOIDS

these molecules. Conformational analyses have indicated that the basic ring system of flavones, aurones and chalcones are planar. In contrast, saturation of the C_2-C_3 bond causes the pyran ring of flavanones, anthocyanins and isoflavanones to be puckered, thereby significantly affecting the conformation of these molecules. However, when a double bond is present, an extended conjugated resonance system is created which may play a role in their interaction with enzymes.

Numerous flavonoids have demonstrated promising antiviral activity[94-96] via mechanisms of action ranging from binding to virus capsid proteins (i.e., antiinfective)[97] to interference with viral RNA or protein synthesis (i.e., antireplicative)[98]. Preliminary results were published on the oncornavirus RT inhibitory activity of flavonoids[99]. It was later confirmed that these compounds are potent *in vitro* RT inhibitors. A systematic study of the potential RT inhibitory effects of 18 flavonoids representing seven different chemical classes has been conducted[100]. Amentoflavone (17), scutellarein (18) and quercetin (19) were found to inhibit AMV RT, Rous-associated virus-2 (RAV-2) RT and Moloney murine leukemia virus (MMLV) RT in a concentration-dependent manner. For each of these flavonoids, the IC_{50} value was approximately 10 μM. The effect of these compounds on MMLV RT was observed to be much greater than that on the avian RTs. In another study, baicalain (5,6,7-trihydroxyflavone) (20) inhibited the activity of RTs from RMLV, MMLV and HIV-1 at concentrations as low as 2 μg/ml, while the activities of mammalian DNA polymerase α and β were not affected in this concentration range[32]. Moreover, the degree of inhibition was observed to be greater when Mn^{2+} was used in the assay mixture instead of Mg^{2+}. Quercetin, baicalein, myricetin (21) and quercetagetin (22) were also found to inhibit the RTs from RMLV and HIV, demonstrating IC_{50} values of less than 0.5 μg/ml[101]. All three flavonoids demonstrated differential effects on DNA polymerases α, β, and γ purified from KB-III cells, terminal deoxynucleotidal transferase obtained from calf thymus, and DNA and RNA polymerases from *Escherichia coli*. The activity of flavonoids on nucleotide polymerizing enzymes[102,103] may account for their cytotoxicity[104]. In addition, quercetin was observed to inhibit the activity of murine RNA polymerase II[105]. During a screen aimed at identifying natural product HIV-1 RT inhibitors specifically[24], flavonoids such as quercetin and kaempferol (23) emerged as weak inhibitors of the enzyme (IC_{50} = 150-200 μg/ml).

When inhibition kinetics were analyzed, flavonoids exhibited varied modes of action. With few exceptions, the inhibition of RT was competitive or mixed with respect to the template-primer $[r(A)_n \cdot (dT)_{12-18}]$ used and noncompetitive with respect to the TTP substrate. In the case of RMLV RT, for example, quercetin, baicalein, myricetin and quercetagetin competed with the polynucleotide template-primer but not with the triphosphate substrate[101,103]. However, when HIV RT was used, the mode of inhibition was of the mixed type with respect to the template-primer. These results indicate that flavonoids act by binding to the template-primer binding site of the enzyme or to the template-primer itself. The molecular planarity of flavones such as quercetin and myricetin is thought to facilitate the intercalation of these molecules with the template-primer. It has been suggested that polar flavonoids have this intercalative ability[106]. The planarity of flavonoids has also been correlated with their ability to produce open circular DNA from a covalently closed cyclic plasmid, an observation which may be used as a measure of intercalating ability[102].

A comparative study of numerous flavonoid molecules was conducted to define SARs involved in RMLV and HIV RT inhibition[101]. It was concluded that the presence of both the unsaturated double bond between positions 2 and 3 of the flavonoid pyrone ring, and the hydroxy groups at positions 5, 6 and 7 (i.e., baicalein) were prerequisites for RT inhibition. Neither flavone nor various mono- and dihydroxyflavones (except 6,7-dihydroxyflavone) showed appreciable inhibitory activity up to 10 μg/ml. The fact that phenolic hydroxy groups contribute to the activity of flavonoids was clearly demonstrated by glycosidation of the hydroxyl groups at position 7 of baicalein and position 3 of quercetin which greatly reduced the inhibitory effects of their respective parent compounds. No clear-cut SARs have emerged from the present state of knowledge on the RT inhibitory activity of flavonoids. It can be postulated, however, that in addition to the requirement of structural planarity, the phenolic hydroxy groups on these molecules, which are often present in numbers greater than two, have a profound effect on the RT inhibitory potential of flavonoids.

Several flavonoids including quercetin have been reported to show *in vivo* antiviral effects[107,108]. Controversy still shrouds the effect of flavonoids on the common cold[109]. Preliminary results have shown some of these flavonoids to be toxic to cultured cells at concentrations that did not inhibit HIV-1 induced cytopathogenicity[101]. Nevertheless, the toxicity of flavonoids is typically low in humans and it is possible that pharmacologically significant concentrations could be reached in tissues and influence the outcome of viral infections[86]. Structural modifications to minimize toxicity and enhance activity may further resolve these problems.

Miscellaneous Compounds

Numerous other natural products have been shown to inhibit the RT of various retroviruses. Selected examples of these compounds are shown in Figure 5.

Even though the inhibitory effect of polyanionic substances on the replication of enveloped viruses were reported more than two decades ago, it did not generate much interest until dextran sulfate and heparin were found to be highly inhibitory to the replication of HIV-1 *in vitro*[110]. Until now, this antiviral effect was considered to be largely nonspecific. Pentosan polysulfate[111] also emerged as a potent anti-HIV compound which was associated with low toxicity and anticoagulant activity[112]. Consistent with previous findings[113], anionic polysaccharides have been shown to inhibit virus adsorption to cells. Pentosan polysulfate also acts as a competitive inhibitor of HIV-1 RT with respect to the template-primer, but whether this compound is actually taken up by cells to enable RT inhibition to play a role in its anti-HIV effects, remains subject for further study. However, it was noted that the concentrations of sulfated polysaccharides required to inhibit HIV-1 RT *in vitro* were considerably higher than those required for inhibition of HIV-1 replication in cell culture[112]. The IC_{50}s for HIV-1 RT inhibition for pentosan polysulfate, fucoidan, dextran sulfate and heparin were 19.1, 29.5, 32.9, and 410 μg/ml, respectively[112]. HIV-1 and AMV RT inhibitory carrageenans have also been isolated from sea algae[114].

All the polysaccharides that inhibit RT activity have sulfate residues. Those which do not (e.g., dextran and lentinan), were essentially devoid of RT inhibitory activity. Thus, sulfate residues play a key role in RT inhibition. Indeed, the addition of sulfate groups to glycyrrhizin and other nonsulfated polysaccharides resulted in the endowment of RT inhibitory activity to the substances involved[115]. It was also reported that the sulfated side chains of polyvinyl sulfate competed with the template-primer for the active site of RT[116]. It was observed that the HIV-1 RT inhibitory effect of these compounds was negated if bovine serum albumin was added to the reaction mixture, indicating nonspecific binding to protein[112].

Avarol (29) is a sesquiterpenoid hydroquinone present in the marine sponge *Dysidea avara*. Both avarol and its quinone derivative avarone (24) exhibit a dose-dependent inhibition of HIV-1 replication in human H9 cells *in vitro*[117]. Avarol and avarone, however, were devoid of any significant inhibitory effect on HIV-1 RT as well as mammalian DNA polymerases[118]. Subsequently, several novel secondary metabolite derivatives of avarol and avarone, namely avarone A (25), B (26) and E (28), and avarol F (31) were found to inhibit all the three catalytic activities associated with the HIV-1 RT (DNA-dependent and RNA-dependent DNA polymerase and ribonuclease H activities)[119]. Kinetic experiments performed with avarone E revealed that the compound binds RT molecules at sites different from the binding sites of the template-primer or the substrate TTP. Avarol C (30) and avarone D (27) which have 6'-acetyl substituents were devoid of any inhibitory activity against the HIV-1 RT. Taken together, these observations indicate that the hydroxyl group at the *ortho* position to the carbonyl group of the quinone ring (as in the case of derivatives A, B, E and F) is required for potent HIV-1 RT inhibitory activity.

Analogous to avarone, various quinones have also been reported to inhibit retroviral RTs including the HIV-1 RT[120-123]. Quinone compounds interfere with the binding of the template-primer to the enzyme, most probably by interaction with the enzyme. Structure-activity studies of a diverse group of quinone molecules revealed a common feature. Hydroxy groups critically situated play a pivotal role in the activity of these compounds. Most often activity is attributed to hydroxy groups situated *ortho* to the quinone carbonyl. Studies have also shown that quinones function as electron acceptors and, therefore, may oxidize sulfydryl groups in the active site of the enzyme[124]. Furthermore, it was demonstrated that hypericin (32), an anthraquinone dimer, binds to proteins in a nonspecific manner[123]. This property may be partially responsible for its HIV-1 RT inhibitory activity.

In an extensive survey of plant extracts for HIV-1 RT inhibitory activity, four phloroglucinol derivatives, mallotojaponin, mallotochromene (33), mallotophenone and mallotolerin isolated from the pericarps of *Mallotus japonicus* Muell. Arg. (Euphorbiaceae), were found to be active[125]. Mallotojaponin and mallotochromene were particularly active with IC_{50} values of less than 10 µg/ml. Differences in activity were attributed to the nature of the side chains present. Mallotojaponin was also found to interfere with the binding of the template-primer to the enzyme.

Taspine (34) is a dilactone tertiary alkaloidal base isolated from *Croton lechleri* M. Arg. (Euphorbiaceae). Taspine hydrochloride inhibited RMLV, SSV-1 and AMV RT with IC_{50} values falling in the range of 70-98 µg/ml[126]. The alkaloid was more specific for RT than for cellular DNA polymerases or *E. coli* DNA or RNA polymerases. Preliminary experiments involving UV spectrophotometric titrations also indicated that the alkaloid interacted with calf thymus DNA.

Pretazettine (35), an alkaloid isolated from the bulbs of *Narcissus tazetta* L. "Sacred lily" was claimed to be the first RT inhibitor discovered that binds to the AMV RT and affects steps subsequent to the formation of the enzyme-template-primer complex[127]. The immediate cessation of polymerization upon the addition of the alkaloid to an ongoing reaction suggested that the inhibitor interfered with initiation or elongation of the polymerization process, rather than with the initial binding of the template to the enzyme.

Many antibiotics such as the antitumor anthracyclines adriamycin and daunomycin are known inhibitors of retroviral RTs[128]. The predominant mode of action is via an interaction with the template-primer, hence, these compounds have received relatively little attention as antiviral agents due to their toxicity. In contrast, quinoline quinone antibiotics such as streptonigrin (36)[129] and the ansamysin antibiotic rifamycin SV (37)[130] have been shown to interact with the enzyme molecule itself. Where quinone compounds such as streptonigrin are concerned, a specific site of interaction on the RT molecule, referred to as the "quinone pocket", has been proposed[131]. A comparative study of the inhibitory properties of antibiotics on HIV RT, AMV RT and cellular DNA polymerases has also been published[132].

As already mentioned, the application of the HIV-1 RT assay recently developed to the bioactivity guided fractionation scheme of the bark of *Plumeria rubra*, yielded the active iridoid fulvoplumierin (38) (IC_{50} = 45 µg/ml in the HIV-1 RT system)[24]. Additional iridoids evaluated such as plumericin and plumieride were inactive.

CONCLUSION

Therefore, the standard reverse transcriptase assay is a specific, sensitive, simple and

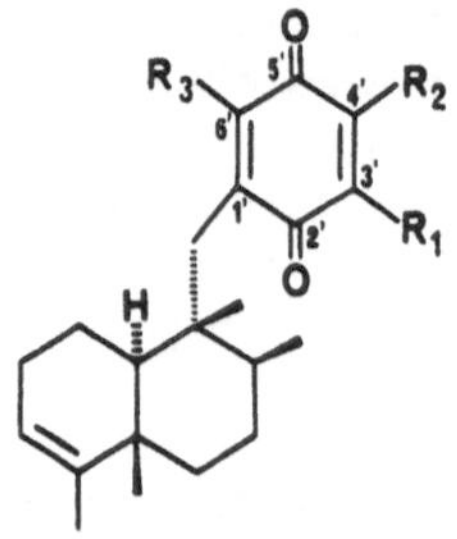

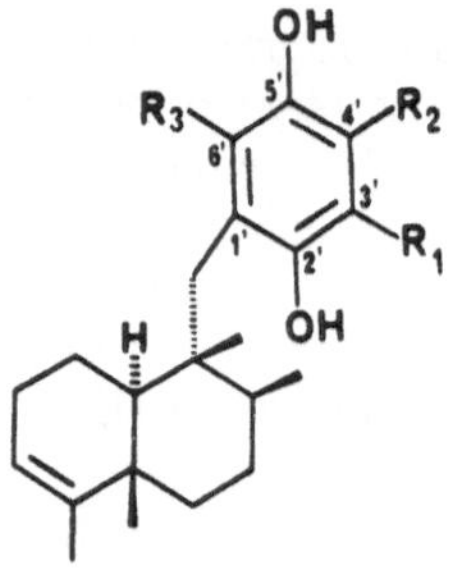

24 Avarone (R₁, R₂, R₃ = H)
25 Avarone A (R₁ = OH; R₂, R₃ = H)
26 Avarone B (R₁ = H; R₂, R₃ = OH)
27 Avarone D (R₁, R₂ = H; R₃ = CH₃COO)
28 Avarone E (R₁ = H; R₂ = OCH₃; R₃ = OH)

29 Avarol (R₁, R₂, R₃ = H)
30 Avarol C (R₁, R₂ = H; R₃ = CH₃COO)
31 Avarol F (R₁, R₂ = H; R₃ = OH)

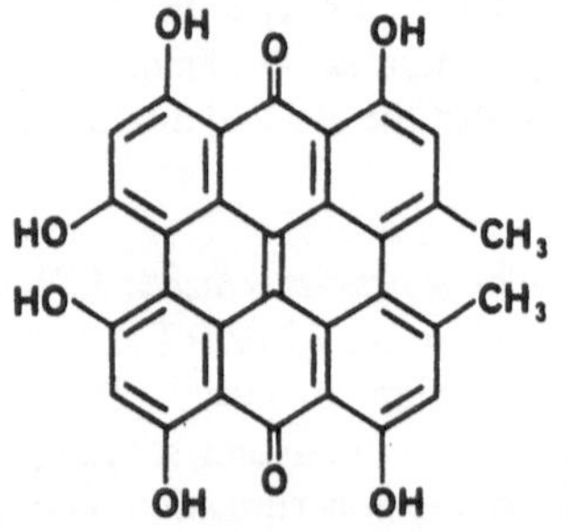

32 Hypericin

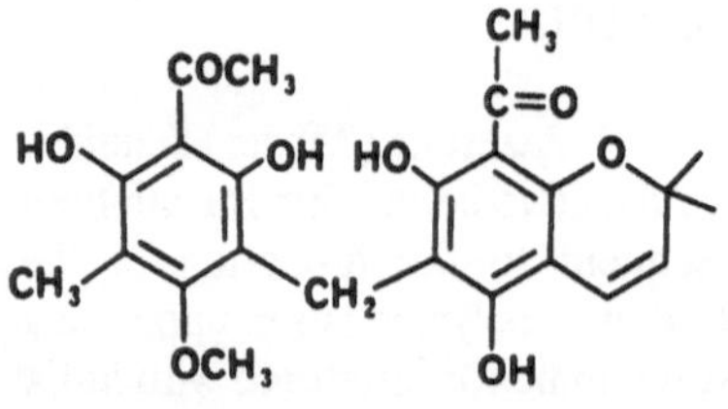

33 Mallotochromene

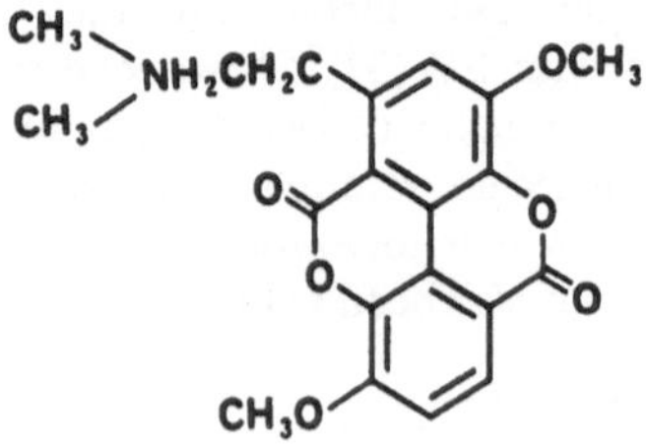

34 Taspine

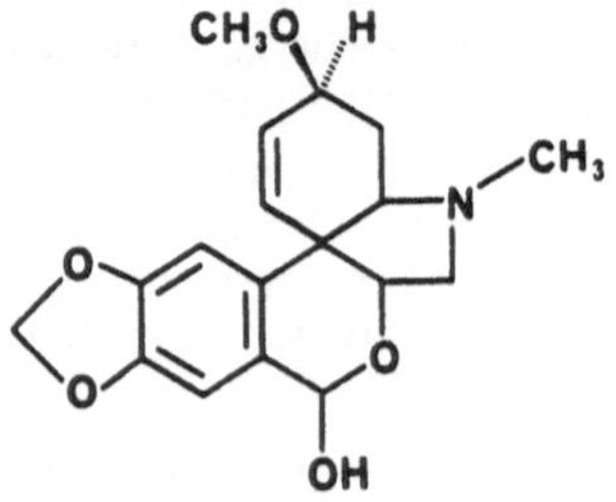

35 Pretazettine

FIGURE 5. MISCELLANEOUS COMPOUNDS

36 Streptonigrin

37 Rifamycin SV

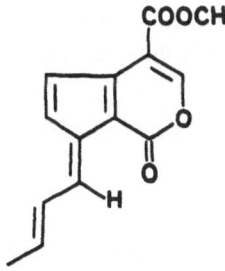

38 Fulvoplumierin

FIGURE 5. MISCELLANEOUS COMPOUNDS, CONT.

reliable method for screening the HIV-1 and HIV-2 RT inhibitory potential of natural products. Results are reproducible, and the procedure is applicable to both pure natural products and plant extracts. Moreover, the method has proven to be useful in a bioactivity-guided fractionation procedure. However, tannins are active inhibitors of reverse transcriptase and, in general, only non-tannin inhibitory compounds are of interest. Since no clear-cut threshold for activity could be established for plant extracts containing tannins, these substances are preferably removed before the assay is performed.

Since retroviruses are thought to play an important role in carcinogenesis, a variety of inhibitors of the retroviral RT has been uncovered from the viewpoint of understanding the mechanism of malignant transformation of virus-infected cells. Interest in these RT inhibitors rekindled with the discovery of AIDS. The clinical utilization of AZT has established the potential value of RT inhibitors in the management of AIDS. In spite of our knowledge of a great variety of RT inhibitors, however, an absolutely specific HIV-1 or HIV-2 RT inhibitor is not yet available. The potential of higher plants to provide novel prototype drugs is considerable. The present review clearly demonstrates that a diverse group of natural products possesses RT inhibitory activity. The empirical screening of structurally diverse materials may be the most efficient way to identify novel HIV-1 and HIV-2 RT inhibitors, until such time as the structure of the enzymes is completely understood, and alternative methods are defined.

Once active leads are identified, further screening of structural analogs is generally warranted. Mechanistic studies are in order to determine the selectivity of the mode of action manifested. One very important approach to demonstrate selectivity of compounds is to compare their inhibitory effects against various cellular DNA polymerases. Finally, it will be necessary to examine their properties as potential inhibitors of HIV infection and replication in cell culture followed by more advanced testing in animal models.

Although complete HIV eradication and a cure of AIDS cannot be achieved with agents that disrupt the replication of the viral genome, they are considered extremely important for improving the quality of life and survival of the host, and they are of value in combination regimens of chemotherapy. Chemotherapeutic intervention of the viral replicative cycle is also a prerequisite for immunotherapy aimed at reconstruction of the immune system. In addition, substances found to function as HIV RT inhibitors may serve as inhibitors of other types of retroviruses.

ACKNOWLEDGMENT

The authors wish to thank Dr. S. H. Hughes of the Development Center, ABL, Inc., Basic Research Program, Frederick, Maryland 21701, for kindly supplying HIV-1 and HIV-2 RT.

REFERENCES

1. F. Barre-Sinoussi, J. C. Chermann, J. Rey, M. T. Nugebe, S. Chamaret, J. Gruest, C. Dauguet, C. Axler-Blin, F. Vezinek-Brun, C. Rouzioux, W. Rozenbaum, and L. Montagnier, Isolation of a T-lymphotropic retrovirus from a patient at risk for acquired immune deficiency syndrome, *Science* 220:868 (1983).
2. F. Clavel, D. Guetard, F. Brun-Vezinet, S. Chamaret, M. A. Rey, M. O. Santos-Ferreira, A. G. Laurent, C. Dauguet, C. Katlama, C. Rouzioux, D. Klatzmann, J. L. Champalimaud, and L. Montagnier, Isolation of a new human retrovirus from West African patients with AIDS, *Science* 233:343 (1986).
3. M. Guyader, M. Emerman, P. Sonigo, F. Clavel, L. Montagnier, and M. Alizon, Genome organization and transactivation of the human immunodeficiency virus type 2, *Nature (London)* 326:662 (1987).
4. P. S. Sarin, Molecular pharmacologic approaches to the treatment of AIDS, *Ann. Rev. Pharmacol.* 28:411 (1988).
5. M. S. Hirsch, Chemotherapy of human immunodeficiency virus infections: Current practice and future prospects, *J. Infect. Dis.* 161:845 (1990).

6. K. Ono, H. Nakane, F. Barre-Sinoussi, and J. C. Chermann, Differential inhibition of various mammalian DNA polymerase activities by ammonium-21-tungsto-9-antimoniate (HPA 23), *Eur. J. Biochem.* 176:305 (1988).

7. K. Ono, H. Nakane, and M. Fukushima, Differential inhibition of various deoxyribonucleic and ribonucleic acid polymerases by suramin, *Eur. J. Biochem.* 172:349 (1988).

8. S. F. Le Grice, R. Zehnle, and J. Mous, A single 66-kilodalton polypeptide processed from the human immunodeficiency virus type 2 *pol* polyprotein in *Escherichia coli* displays reverse transcriptase activity, *J. Virol.* 62:2525 (1988).

9. A. L. DeVico, T. D. Copeland, F. Di Marzo Veronese, S. Oroszlan, R. C. Gallo, and M. G. Sarngadharan, Purification and partial characterization of human immunodeficiency virus type 2 reverse transcriptase, *AIDS Res. Hum. Retroviruses* 5:51 (1989).

10. A. M. Barber, A. Hizi, J. V. Maizel, Jr., and S. H. Hughes, HIV-1 reverse transcriptase: Structure predictions for the polymerase domain, *AIDS Res. Hum. Retroviruses* 6:1061 (1990).

11. A. Hizi, R. Tal, and S. H. Hughes, Mutational analysis of the DNA polymerase and ribonuclease H activities of Human Immunodeficiency Virus type 2 reverse transcriptase expressed in *Escherichia coli, Virology* 180:339 (1991).

12. Manuscript in preparation.

13. H. Varmus, Retroviruses, *Science* 240:1427 (1988).

14. F. D. Di Marzo Veronese, T. D. Copeland, A. L. DeVico, R. Rahman, S. Oroszlan, R. C. Gallo, and M. G. Sarngadharan, Characterization of highly immunogenic p66/p51 as the reverse transcriptase of HTLV-III/LAV, *Science* 231:1289 (1986).

15. W. G. Farmerie, D. D. Loeb, N. C. Casavant, C. A. Hutchinson, III, M. H. Edgell, and R. Swanstrom, Expression and processing of the AIDS virus reverse transcriptase in *Escherichia coli, Science* 236:305 (1987).

16. H. Mitsuya, and S. Broder, Inhibition of the *in vitro* infectivity and cytopathic effect of human T-lymphotropic virus type-III/lymphadenopathy associated virus (HTLV-III/LAV) by 2',3'-dideoxynucleosides, *Proc. Natl. Acad. Sci. USA* 83:1911 (1986).

17. P. Chandra, U. Ebener, and A. Götz, Inhibition of oncornaviral DNA polymerase by 5-mercapto polycytidylic acid: mode of action, *FEBS Lett.* 53:10 (1975).

18. L. Vrang, and B. Öberg, PPi analogs as inhibitors of human T-lymphotropic virus type III reverse transcriptase, *Antimicrob. Agents Chemother.* 29:867 (1986).

19. M. L. Sethi, Inhibition of reverse transcriptase activity by benzophenanthridine alkaloids, *J. Nat. Prod.* 42:187 (1979).

20. H. M. Temin, and D. Baltimore, RNA-directed DNA synthesis and RNA tumor viruses, *Adv. Virus Res.* 17:129 (1972).

21. E. Haslam, T. H. Lilley, Y. Cai, R. Martin, and D. Magnolato, Traditional herbal medicines: The role of polyphenols, *Planta Med.* 55:1 (1989).

22. P. C. Zamecnik, J. Goodchild, Y. Taguchi, and P. S. Sarin, Inhibition of replication and expression of human T-cell lymphotropic virus type-III in cultured cells by exogenous synthetic oligonucleotides complementary to viral RNA, *Proc. Natl. Acad. Sci.* 83:4143 (1986).

23. A. D. Hoffman, B. Banapour, and J. A. Levy, Characterization of the AIDS-associated retrovirus reverse transcriptase and optimal conditions for its detection in virions, *Virology* 147:326 (1985).

24. G. T. Tan, J. M. Pezzuto, A. D. Kinghorn and S. H. Hughes, Evaluation of natural products as inhibitors of human immunodeficiency virus type-1 (HIV-1) reverse transcriptase, *J. Nat. Prod.* 54:143 (1991).

25. A. Hizi, C. McGill, and S. H. Hughes, Expression of soluble, enzymatically active, human immunodeficiency virus reverse transcriptase in *E. coli* and analysis of mutants, *Proc. Natl. Acad. Sci. USA* 85:1218 (1988).

26. A. Hizi, R. Tal, and S. H. Hughes, Mutational analysis of the DNA polymerase and ribonuclease H activities of Human Immunodeficiency Virus type 2 reverse transcriptase expressed in *Escherichia coli, Virology* 180:339 (1991).

27. R. F. Schinazi, B. F. H. Eriksson, and S.H. Hughes, Comparison of inhibitory activities of various antiretroviral agents against particle-derived and recombinant human immunodeficiency virus type 1 reverse transcriptases, *Antimicrob. Agents Chemother.* 33:115 (1989).

28. P. K. Clark, A. L. Ferris, D. A. Miller, A. Hizi, K. -W. Kim, S. M. Deringer-Boyer, M. L. Mellini, A. D. Clark, Jr., G. F. Arnold, W. B. Lebherz, III, E. Arnold, G. M. Muschik, and S. H. Hughes, HIV-1 reverse transcriptase purified from a recombinant strain of *Escherichia coli, AIDS Res. Human Retroviruses* 6:753 (1990).

29. V. S. Sethi, Base specificity in the inhibition of oncornavirus reverse transcriptase and cellular nucleic acid polymerases by antitumor drugs, *Ann. N. Y. Acad. Sci.* 284:508 (1977).

30. G. T. Tan, A. D. Kinghorn, S. H. Hughes, and J. M. Pezzuto, Psychotrine and its O-methyl ether are selective inhibitors of human immunodeficiency virus-1 reverse transcriptase, *J. Biol. Chem.* 35:23529 (1991).

31. L. Ratner, W. Haseltine, R. Patarca, K. J. Livak, B. Starcich, S. F. Josephs, E. R. Doran, J. A. Rafalski, E. A. Whitehorn, K. Baumeister, L. Ivanoff, S. R. Petteway, Jr., M. L. Pearson, J. A. Lautenberger, T. S. Papas, J. Ghrayeb, N. T. Chang, R. C. Gallo, and F. Wong-Staal, Complete nucleotide sequence of the AIDS virus, HTLV-III, *Nature* 313:277 (1985).

32. K. Ono, H. Nakane, M. Fukushima, J.-C. Chermann, and F. Barré-Sinoussi, Inhibition of reverse transcriptase activity by a flavonoid compound, 5,6,7-trihydroxyflavone, *Biochem. Biophys. Res. Commun.* 160:982 (1989).

33. W. M. Messmer, M. Tin-Wa, H. H. S. Fong, C. Bevelle, N. R. Farnsworth, D. J. Abraham, and J. Trojanek, Fagaronine, a new tumor inhibitor isolated from *Fagara zanthoxyloides* Lam. (Rutaceae), *J. Pharm. Sci.* 61:1858 (1972).

34. N. Kakiuchi, M. Hattori, T. Namba, M. Nishizawa, T. Yamagishi, and T. Okuda, Inhibitory efect of tannins on reverse transcriptase from RNA tumor viruses, *J. Nat. Prod.* 48:614 (1985).

35. W. D. Loomis and J. Battaile, Plant phenolic compounds and the isolation of plant enzymes, *Phytochemistry* 5:423 (1966).

36. M. E. Wall, H. Taylor, L. Ambrosio, and K. Davis, Plant antitumor agents III: A convenient separation of tannins from other plant constituents, *J. Pharm. Sci.* 58:839 (1969).

37. F. M. Thompson, L. J. Libertini, U. R. Joss, and M. Calvin, Detergent effects on reverse transcriptase activity and on inhibition by rifamycin derivatives, *Science* 178:505 (1972).

38. S. Simeon, J. L. Rios, and A. Villar, Pharmacological activities of benzophenanthridine and phenanthrene alkaloids, *Pharmazie* 44:593 (1989).

39. M. Arisawa, J. M. Pezzuto, C. Bevelle, and G. A. Cordell, Potential anticancer agents XXXI. N-Demethylation of fagaronine, *J. Nat. Prod.* 47:453 (1984).

40. R. K. Y. Zee-Cheng, and C. C. Cheng, Preparation and antileukemic activity of some alkoxybenzo[c]phenanthridinium salts and corresponding dihydro derivatives, *J. Med. Chem.* 18:66 (1975).

41. L. A. Mitscher, Y. H. Park, D. Clark, G. W. Clark, III, P. D. Hammesfahr, W. N. Wu, and J. L. Beal, Antimicrobial agents from higher plants. An investigation of *Hunnemannia fumariaefolia* pseudoalcoholates of sanguinarine and chelerythrine, *Lloydia* 41:145 (1978).

42. F. R. Stermitz, J. P. Gillespie, L. G. Amoros, R. Romero, T. A. Stermitz, K. L. Larson, S. Earl, and J. E. Ogg, Synthesis and biological activity of some antitumor benzophenanthridinium salts, *J. Med. Chem.* 18:708 (1975).

43. J. W. Lee, J. O. MacFarlane, R. K. Y. Zee-Cheng, and C. C. Cheng, Inhibition of catechol O-methyltransferase and transfer RNA methyltransferases by coralyne, nitidine, and related compounds, *J. Pharm. Sci.* 66:986 (1977).

44. C. C. Cheng, Inhibitors of tRNA O-methyltransferase as possible antineoplastic agents, *J. Pharm. Sci.* 61:645 (1972).

45. K. D. Straub, and P. Carver, Sanguinarine, inhibitor of Na-K-dependent ATPase, *Biochem. Biophys. Res. Commun.* 62:913 (1975).

46. H. G. Cohen, E. E. Seifen, K. D. Straub, C. Tiefenback, and F. R. Stermitz, Structural specificity of the NaK-ATPase inhibition by sanguinarine, an isoquinoline benzophenanthridine alkaloid, *Biochem. Pharmacol.* 27:2555 (1978).

47. J. M. Herbert, J. M. Augereau, J. Gleye, and J. P. Maffrand, Chelerythrine is a potent and specific inhibitor of protein kinase C, *Biochem. Biophys. Res. Commun.* 172:993 (1990).

48. V. S. Sethi, and M. L. Sethi, Inhibition of reverse transcriptase activity of RNA-tumor viruses by fagaronine, *Biochem. Biophys. Res. Commun.* 63:1070 (1975).

49. V. S. Sethi, Inhibition of mammalian and oncornavirus nucleic acid polymerase activities by alkoxybenzophenanthridine alkaloids, *Cancer Res.* 36:2390 (1976).

50. C. A. C. Torres, and A. Baez, Effects of the antitumor drugs 3-nitrobenzothiazolo[3,2α]quinolinium and fagaronine on nucleic acid and protein synthesis, *Biochem. Pharmacol.* 35:679 (1986).

51. M. L. Sethi, Screening of benzophenanthridine alkaloids for their inhibition of reverse transcriptase activity and preliminary report on the structure-activity relationships, *Can. J. Pharm. Sci.* 16:29 (1981).

52. M. A. Caolo, and F. R. Stermitz, Benzophenanthridinium salt equilibria, *Heterocycles* 12:11 (1979).

53. F. R. Stermitz, K. A. Larson, and D. K. Kim, Some structural relationships among cytotoxic and antitumor benzophenanthridine alkaloid derivatives, *J. Med. Chem.* 16:939 (1973).

54. N. P. S. Bajaj, M. J. McLean, M. J. Waring, and E. Smekal, Sequence-selective, pH-dependent binding to DNA of benzophenanthridine alkaloids, *J. Mol. Recognit.* 3:48 (1990).

55. K. Y. Zee-Cheng, and C. C. Cheng, Common receptor-complement feature among some antileukemic compounds, *J. Pharm. Sci.* 59:1630 (1970).

56. N. Kakiuchi, M. Hattori, H. Ishii, and T. Namba, Effect of benzo(c)phenanthridine alkaloids on reverse transcriptase and their binding property to nucleic acids, *Planta Medica* 53:22 (1987).

57. J. M. Pezzuto, S. K. Antosiak, W. M. Messmer, M. B. Slaytor, and G. R. Honig, Interaction of the antileukemic alkaloid, 2-hydroxy-3,8,9-trimethoxy-5-methylbenzo[c]phenanthridine (fagaronine), with nucleic acids, *Chem.- Biol. Interact.* 43:323 (1983).

58. R. Nandi, and M. Maiti, Binding of sanguinarine to deoxyribonucleic acids of differing base compositions, *Biochem. Pharmacol.* 34:321 (1985).

59. R. Nandi, K. Chaudhuri, and M. Maiti. Effects of ionic strength and pH on the binding of sanguinarine to deoxyribonucleic acid, *Photochem. Photobiol.* 42:497 (1985).

60. Y. Kondo, Organic and biological aspects of berberine alkaloids, *Heterocycles* 4:197 (1976).

61. M. Cushman, F. W. Dekow, and L. B. Jacobsen, Conformations, DNA binding parameters, and antileukemic activity of certain cytotoxic protoberberine alkaloids, *J. Med. Chem.* 22:331 (1979).

62. R. K. Y. Zee-Cheng, K. D. Paull, and C. C. Cheng, Experimental antileukemic compounds: Coralyne, analogs and related compounds, *J. Med. Chem.* 17:347 (1974).

63. R. K. Y. Zee-Cheng, and C. C. Cheng, Tetramethoxydibenzoquinolizinium salts. Preparation and antileukemic activity of some positional and structural isomers of coralyne, *J. Med. Chem.* 19:882 (1976).

64. J. Kovar, and S. Pavelka, Characterization of binding site of horse liver alcohol dehydrogenase for berberines and auramine O, *Collect. Czech. Chem. Commun.* 41:1081 (1976).

65. J. Kovar, J. Stejskal, and L. Matyska, Interaction of yeast alcohol dehydrogenase with protoberberine alkaloids, *J. Enzym. Inhib.* 1:35 (1985).

66. S. Kuwano, and K. Yamauchi, Effect of berberine on tyrosine decarboxylase activity of *Streptococcus faecalis*, *Chem. Pharm. Bull.* 8:491 (1960).

67. S. Kuwano, and K. Yamauchi, Competition of berberine with pyridoxal phosphate in the tryptophanase system of *Escherichia coli*, *Chem. Pharm. Bull.* 8:497 (1960).

68. M. L. Sethi, Enzyme inhibition VI: Inhibition of reverse transcriptase activity by protoberberine alkaloids and structure-activity relationships, *J. Pharm. Sci.* 72:538 (1983).

69. M. L. Sethi, Enzyme inhibition VIII: Mode of inhibition of reverse transcriptase activity by analogues, isomers, and related alkaloids of coralyne, *J. Pharm. Sci.* 74:889 (1985).

70. R. K. Y. Zee-Cheng, and C. C. Cheng, Interaction between DNA and coralyne acetosulfate, an antileukemic compound, *J. Pharm. Sci.* 62:1572 (1973).

71. W. D. Wilson, A. N. Gough, J. J. Doyle, and M. W. Davidson, Coralyne. Intercalation with DNA as a possible mechanism of antileukemic action, *J. Med. Chem.* 19:1261 (1976).

72. A. K. Krey, and F. E. Hahn, Berberine: complex with DNA, *Science* 166:755 (1969).

73. M. W. Davidson, I. Lopp, S. Alexander, and W. D. Wilson, The interaction of plant alkaloids with DNA. II. Berberinium chloride, *Nucleic Acids Res.* 4:2697 (1977).

74. M. L. Sethi, Comparison of inhibition of reverse transcriptase and antileukemic activities exhibited by protoberberine and benzophenanthridine alkaloids and structure-activity relationships, *Phytochemistry* 24:447 (1985).

75. Y. C. Wu, Y. F. Liou, S. T. Lu, C. H. Chen, J. J. Chang, and K. H. Lee, Cytotoxicity of isoquinoline alkaloids and their N-oxides, *Planta Med.* 55:163 (1989).

76. G. A. Cordell, "Introduction to Alkaloids: A Biogenetic Approach," John Wiley & Sons, Inc., New York, NY, pp. 560-567, (1981).

77. V. E. Tyler, L. R. Brady, and J. E. Robbers, "Pharmacognosy," 9th edn., Lea & Febiger, Philadelphia, PA, pp. 209-212, (1988).

78. E. B. Vedder, A preliminary account of some experiments undertaken to test the efficacy of the Ipecacuanha treatment of dysentery, *J. Trop. Med.* 14:149 (1911).

79. A. I. Grollman, Emetine in the treatment of intra-abdominal and retroperitoneal non-specific granulomas, *Surg. Gynecol. Obstet.* 120:792 (1965).

80. E. Grunberg, and H. N. Prince, Antiviral activity of emetine, 2-dehydroemetine and 2-dehydro-3-noremetine, *Antimicrob. Agents Chemother.* 6:527 (1966).

81. A. P. Grollman, Inhibitors of protein biosynthesis, *J. Biol. Chem.* 243:4089 (1968).

82. A. P. Grollman, Structural basis for inhibition of protein synthesis by emetine and cycloheximide based on an analogy between ipecac alkaloids and glutarimide antibiotics, *Proc. Natl. Acad. Sci. U.S.A.* 56:1867 (1966).

83. A. P. Grollman, Emetine: New uses for an old drug, *Ohio State Medical J.* 66:257 (1970).

84. A. P. Grollman, Mode of action of emetine: The inhibition of protein synthesis, *Pharmacologist* 8:190 (1966).

85. V. Cody, E. Middleton, Jr., and J. B. Harborne, eds., "Plant Flavonoids in Biology and Medicine: Biochemical, Pharmacological and Structure-activity Relationships," Alan R. Liss, Inc., New York (1986).

86. B. Havsteen, Flavonoids, a class of natural products of high pharmacological potency, *Biochem. Pharmacol.* 32:1141 (1983).

87. M. Auf'mkolk, J. Koehrle, R. D. Hesch, and V. Cody, Inhibition of rat liver iodothyronine deiodinase: Interaction of aurones with the iodothyronine ligand-binding site, *J. Biol. Chem.* 261:11623 (1986).

88. J. E. Ferrell, P. D. G. Chang-Sing, G. Loew, R. King, J. M. Mansour, and T. E. Mansour, Structure-activity studies of flavonoids as inhibitors of cyclic AMP phosphodiesterase and relationship to quantum chemical indices, *Molec. Pharmacol.* 16:556 (1979).

89. E. Middleton, Jr., and P. Ferriola, Effect of flavonoids on protein kinase C: Relationship to inhibition of human basophil histamine release, *in*: "Plant Flavonoids in Biology and Medicine II: Biochemical, Cellular, and Medicinal Properties," V. Cody, E. Middleton, Jr., J. B. Harborne, and A. Beretz, eds., Alan R. Liss, Inc., New York (1988).

90. R. T. Borchardt, and J. A. Huber, Catechol-O-methyl-transferase. 5. Structure-activity relationships for inhibition by flavonoids, *J. Med. Chem.* 18:120 (1975).

91. M. Rossi, L. F. Rickles, and W. A. Halpin, The crystal and molecular structure of quercetin: A biologically active and naturally occurring flavonoid, *Bioorganic. Chem.* 14:55 (1986).

92. V. Cody, Crystal and molecular structures of flavonoids, *in*: "Plant Flavonoids in Biology and Medicine II: Biochemical, Cellular, and Medicinal Properties," V. Cody, E. Middleton, Jr., J. B. Harborne, and A. Beretz, eds., Alan R. Liss, Inc., New York, pp. 29-44 (1988).

93. J. S. Cantrell, Crystal structures, bonding, and hydrogen bonding in flavonoid compounds, *in*: "Plant Flavonoids in Biology and Medicine: Biochemical, Pharmacological and Structure-activity Relationships," V. Cody, E. Middleton, Jr., and J. B. Harborne, eds., Alan R. Liss, Inc., New York, pp. 391-394 (1986).

94. J. W. T. Selway, Antiviral activity of flavones and flavans, *in*: "Plant Flavonoids in Biology and Medicine: Biochemical, Pharmacological and Structure-activity Relationships," V. Cody, E. Middleton, Jr., and J. B. Harborne, eds., Alan R. Liss, Inc., New York, pp 521-536 (1986).

95. T. N. Kaul, E. Middleton, Jr., and P. L. Ogra, Antiviral effect of flavonoids on human viruses, *J. Med. Virol.* 15:71 (1985).

96. Y. Tsuchiya, M. Shimizu, Y. Hiyama, K. Itoh, Y. Hashimoto, M. Nakayama, T. Horie, and N. Morita, Antiviral activity of naturally occurring flavonoids *in vitro*, *Chem. Pharm. Bull.* 33:3881 (1985).

97. Y. Ninomiya, C. Uhsawa, M. Aoyama, I. Umeda, Y. Suhara, and H. Ishitsuka, Antivirus agents, Ro 09-0410, binds to rhinovirus specifically and stabilizes the virus conformation, *Virology* 134:269 (1984).

98. J. L. Castrillo, D. Vanden Berghe, and L. Carrasco, 3-Methyl-quercetin is a potent and selective inhibitor of poliovirus RNA synthesis, *Virology* 152:219 (1986).

99. P. H. Fischer, M. A. Apple, and I. Strokon, Quantitative structure-activity analysis of flavonol inhibition of RNA directed vs DNA directed polynucleotide polymerases, *Pharmacologist* 17:229 (1975).

100. G. Spedding, A. Ratty, and E. Middleton, Jr. Inhibition of reverse transcriptases by flavonoids, *Antiviral Res.* 12:99 (1989).

101. K. Ono, H. Nakane, M. Fukushima, J.-C. Chermann, and F. Barré-Sinoussi, Differential inhibitory effects of various flavonoids on the activities of reverse transcriptase and cellular DNA and RNA polymerases, *Eur. J. Biochem.* 190:469 (1990).

102. K. Shinozuka, Y. Kikuchi, C. Nishino, A. Mori, and S. Tawata, Inhibitory effect of flavonoids on DNA-dependent DNA and RNA polymerases, *Experientia* 44:882 (1988).

103. K. Ono, and H. Nakane, Mechanisms of inhibition of various cellular DNA and RNA polymerases by several flavonoids, *J. Biochem. (Tokyo)* 108:609 (1990).

104. A. Mori, C. Nishino, N. Enoki, and S. Tawata, Cytotoxicity of plant flavonoids against HeLa cells, *Phytochemistry* 27:1017 (1988).

105. K. Nose, Inhibition by flavonoids of RNA synthesis in permeable WI-38 cells and of transcription by RNA polymerase II, *Biochem. Pharmacol.* 33:3823 (1984).

106. A. Beretz, R. Anton, and J. C. Stoclet, Flavonoid compounds are potent inhibitors of cyclic AMP phosphodiesterase, *Experientia* 34:1054 (1978).

107. A. Veckenstedt, I. Béladi, and I. Mucsi, Effect of treatment with certain flavonoids on Mengo virus-induced encephalitis in mice, *Arch. Virol.* 57:255 (1978).

108. J. Guttner, A. Veckenstedt, H. Heinecke, and R. Pusztai, Effect of quercetin on the course of Mengo virus infection in immunodeficient and normal mice. A histologic study, *Acta Virol.* 26:148 (1982).

109. I. M. Baird, R. E. Hughes, H. K. Wilson, J. E. W. Davies, and A. N. Howard, The effects of ascorbic acid and flavonoids on the occurrence of symptoms normally associated with the common cold, *Am. J. Clin. Nutr.* 32:1686 (1979).

110. M. Ito, M. Baba, A. Sato, R. Pauwels, E. De Clercq, and S. Shigetta, Inhibitory effect of dextran sulfate and heparin on the replication of human immunodeficiency virus (HIV) *in vitro*, *Antiviral Res.* 7:361 (1987).

111. G. Sydow, and H.-P. Krocking, Effect of pentosan polysulfate (SP54) on the reverse transcriptase activity of several retroviruses, *Biomed. Biochem. Acta* 46:527 (1987).

112. M. Baba, M. Nakajima, D. Schols, R. Pauwels, J. Balzarini, and E. DeClercq, Pentosan polysulfate, a sulfated oligosaccharide, is a potent and selective anti-HIV agent *in vitro*, *Antiviral Res.* 9:335 (1988).

113. P. De Somer, E. De Clercq, A. Billiau, E. Schonne, and M. Claesen, Antiviral activity of polyacrylic and polymethacrylic acids. I. Mode of action *in vitro*, *J. Virol.* 2:878 (1968).

114. H. Nakashima, Y. Kido, N. Kobayashi, Y. Motoki, M. Neushul, and N. Yamamoto, Purification and characterization of an avian myeloblastosis and human immunodeficiency virus reverse transcriptase inhibitor, sulfated polysaccharides extracted from sea algae, *Antimicrob. Agents Chemother.* 31:1524 (1987).

115. H. Nakashima, T. Matsui, O. Yoshida, Y. Isowa, Y. Kido, Y. Motoki, M. Ito, S. Shigeta, T. Mori, and N. Yamamoto, A new anti-human immunodeficiency virus substance, glycyrrhizin sulfate; endowment of glycyrrhizin with reverse transcriptase inhibitory activity by chemical modification, *Jpn. J. Cancer Res. (Gann)* 78:767 (1987).

116. F. M. Hallinan, S. H. S. Lee, and K. R. Rozee, Inhibition of reverse transcriptase by polyvinyl sulfate (PVS), *Cancer Biochem. Biophys.* 98:97 (1981).

117. P. S. Sarin, D. Sun, A. Thornton, and W. E. G. Muller, Inhibition of replication of the etiologic agent of acquired immune deficiency syndrome (human T-lymphotropic retrovirus/lymphadenopathy associated virus) by avarol and avarone, *J. Natl. Cancer Inst.* 78:663 (1987).

118. Y. Kuchino, S. Nishimura, H. C. Schroder, M. Rottmann, and W. E. G. Muller, Selective inhibition of formation of suppressor glutamine tRNA in Moloney murine leukemia virus-infected NIH-3T3 cells by avarol, *Virology* 165:518 (1988).

119. S. Loya, and A. Hizi, The inhibition of human immunodeficiency virus type 1 reverse transcriptase by avarol and avarone derivatives, *FEBS Lett.* 269:131 (1990).

120. K. Ono, H. Nakane, S. Shimizu, and S. Koshimura, Inhibition of HIV reverse transcriptase activity by asterriquinone and its analogs, *Biochem. Biophys. Res. Commun.* 174:56 (1991).

121. H. Higuchi, K. Mori, A. Kato, T. Ohkuma, T. Endo, H. Kaji, and A. Kaji, Antiretroviral activities of anthraquinones and their inhibitory effects on reverse transcriptase, *Antiviral Res.* 15:205 (1991).

122. S. Loya, R. Tal, Y. Kashman, and A. Hizi, Illimaquinone, a selective inhibitor of the RNase H activity of human immunodeficiency virus type 1 reverse transcriptase, *Antimicrob. Agents Chemother.* 34:2009 (1990).

123. R. F. Schinazi, C. K. Chu, J. R. Babu, B. J. Oswald, V. Saalmann, D. L. Cannon, B. F. H. Eriksson, and M. Nasr, Anthraquinones as a new class of antiviral agents against human immunodeficiency virus, *Antiviral Res.* 13:265 (1990).

124. M. M. Wick, and G. Fitzgerald, Inhibition of reverse transcriptase by tyrosinase-generated quinones related to levodopa and dopamine, *Chem.-Biol. Interact.* 38:99 (1981).

125. H. Nakane, M. Arisawa, A. Fujita, S. Koshimura, and K. Ono, Inhibition of HIV-reverse transcriptase activity by some phloroglucinol derivatives, *FEBS Lett.* 286:83 (1991).

126. M. L. Sethi, Inhibition of RNA-directed DNA polymerase activity of RNA tumor viruses by taspine, *Can. J. Pharm. Sci.* 12:7 (1977).

127. T. S. Papas, L. Sandhaus, M. A. Chirigos, and E. Furusawa, Inhibition of DNA polymerase of avian myeloblastosis virus by an alkaloid extract from *Narcissus tazetta* L., *Biochem. Biophys. Res. Commun.* 52:88 (1973).

128. K. V. R. Dhananjaya, and A. Antony, Inhibition of avian myeloblastosis virus reverse transcriptase and its associated activities by daunomycin and adriamycin, *Ind. J. Biochem. Biophys.* 24:265 (1987).

129. M. A. Chirigos, J. W. Pearson, T. S. Papas, W. A. Woods, H. B. Wood, Jr., and G. Spahn, Effect of streptonigrin (NSC-45383) and analogs on oncornavirus replication and DNA polymerase activity, *Cancer Chemother. Rep.* 57:305 (1973).
130. C. Gurgo, and P. Grandgenett, Different modes of inhibition of purified ribonucleic acid directed deoxyribonucleic acid polymerase of avian myeloblastosis virus by rifamycin SV derivatives, *Biochemistry* 16:786 (1977).
131. Y. Hafuri, E. Takemori, K. Oogose, Y. Inouye, S. Nakamura, Y. Kitahara, S. Nakahara, and A. Kubo, Mechanism of inhibition of reverse transcriptase by quinone antibiotics. II. Dependence on putative quinone pocket on the enzyme molecule, *J. Antibiot.* 41:1471 (1988).
132. Y. Take, Y. Inouye, S. Nakamura, H. S. Allaudeen, and A. Kubo, Comparative studies of the inhibitory properties of antibiotics on human immunodeficiency virus and avian myeloblastosis virus reverse transcriptases and cellular DNA polymerases, *J. Antibiot.* 42:107 (1989).

NATURAL PRODUCTS AS ANTI-HIV AGENTS

Daniel Lednicer and Ven L. Narayanan

Drug Synthesis and Chemistry Branch
Developmental Therapeutics Program
Division of Cancer Treatment
National Cancer Institute
Bethesda, Maryland 20892

INTRODUCTION

The publication, in 1981, of a report announcing the occurrence of a cluster of diagnoses of a rare form of pneumonia due to the microorganism *Pneumocystis carinii* in a group of male homosexuals in San Francisco constitutes the first notice of the advent of a new disease[1]. It was quickly recognized, subsequent to that report, that the occurrence of this then rare infection was secondary to the profound depletion of the immune system caused by the new disease, which soon acquired the name of Acquired Immune Deficiency Syndrome, or more familiarly, AIDS. The causative organism was identified as a retrovirus, named Human Immunodeficiency Virus (HIV), which attacks the T4 lymphocytes of the immune system[2,3].

With the recognition of the varied symptomology presented by AIDS at its various stages, came the realization that the incidence of the disease was increasing at a rapid and alarming rate. The new availability of serum assays for detecting infection by HIV prior to the onset of full blown disease confirmed this impression. The serious nature of AIDS was underscored by developing statistics which seemed to point to the inevitable fatal outcome of AIDS. The National Cancer Institute became involved in research on AIDS almost from the beginning; one of the first laboratories to recognize the retroviral causative agent was, in fact, located within the NCI[3]. A report from Broder's group at NCI, barely one year after that which announced the identification of HIV as the cause of AIDS, recorded the anti-HIV activity of AZT[4].

In 1987 the NIH was given the charge to undertake a National program to develop treatments for AIDS.

Natural Products as Antiviral Agents, Edited by C.K. Chu
and H.G. Cutler, Plenum Press, New York, 1992

Responsibility for the drug discovery phase of the effort was delegated to the National Cancer Institute because of its long term involvement in therapeutic development. NCI has had in operation, for many years, the various activities required to take chemical compounds from the very initial investigative stages to drugs approved for clinical use in humans. Most important to this new effort was the long NCI involvement in developing and operating large throughput screens targeted at new cancer chemotherapy agents[5].

It was anticipated from the outset that a high throughput screen would be required which would be used to test compounds from the broadest possible selection of chemical structures. This is due to the fact that viruses differ in a very fundamental way from other infectious agents. As virtually inert organized collections of nucleotides, viruses have no metabolic existence independent of infected host cells. The ultimate therapeutic target thus consists of infected cells rather than the virions per se. The fine discrimination between infected and uninfected cells presents a difficult target. It is of note that the Physicians Desk Reference Lists just over a half dozen approved antiviral compounds. All but one of those are nucleosides with side effects typical for that class.

NCI ANTIVIRAL SCREEN

An in vitro screen was clearly indicated for a number of reasons. First and foremost was the fact that no good animal model existed for HIV infection at that time. It might be added parenthetically that this is still very much the case today. In addition an in vitro screen would have the additional advantages of high throughput capacity and modest sample size requirements. A whole infected cell model was selected since this would involve less mechanistic bias than assays based on inhibition of viral enzymes as, for example, reverse transcriptase or HIV protease. An added advantage to the whole cell assay is the fact that such a test also measures cell wall penetration of potentially active compounds. (Cell wall viral receptor blockers, of course, excepted).

The first published protocol for the assay used for the screen[6] starts with the seeding of human CEM T4-lymphocytes from stock into the wells of 96 well microtiter plates. Solutions of test compounds are then added in duplicate to two sets of wells at half log dilutions over a four decade order of magnitude. HIV was added to the wells one hour later. The current protocol omits that brief preincubation[7]. The plates are then incubated at 37^0C for 6 days.

A variety of methods, such as reverse transcriptase levels, virus antigen titer, and the like, have been used to determine the effect on cells of test substances. The present assay directly addresses the desired therapeutic outcome by measuring cell survival by using a vital stain, soluble version (XTT; 2,3-bis[methoxy-4-nitro-5-sulfonylphenyl]-5-[(phenylamino)carbonyl]-2H-tetrazolium

hydroxide)[8] of the stain MTT. The reduction of the
tetrazolium rings in these colorless reagents to orange
formazans is proportional to the number of viable cells,
permitting quantitative determination of the number of
surviving cells. The assay is thus completed by adding a
solution of XTT containing the coupler, methylphenazonium
methylsulfate, at the end of the incubation period. Dose
response data is then produced automatically from the optical
densities obtained in an automatic plate reader using
especially developed software.

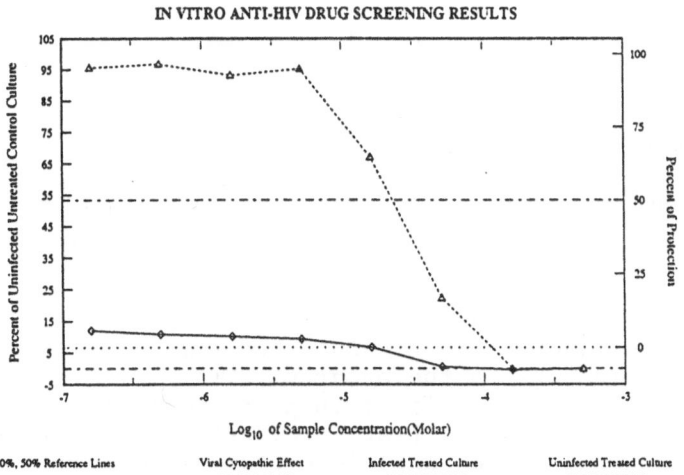

Figure 1

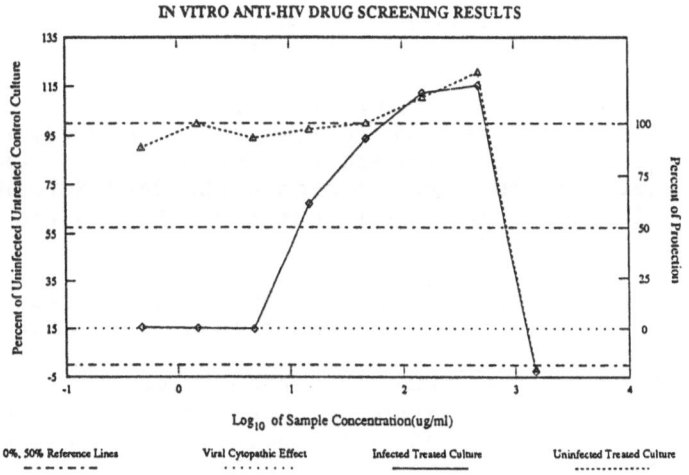

Figure 2

   Typical screening data are displayed in Figures 1 and 2.
Turning our attention first to a compound devoid of anti-HIV
activity (Figure 1), the dotted upper line represents
viability of uninfected CEM T-lymphocytes to increasing
concentrations of the test substance: cell survival declines
in dose related fashion as the concentration of compound
increases beyond 10^{-5} molar. Cells infected with HIV show only
very low survival at any level of drug as shown by the solid
lower line. Figure 2 shows typical results for an active

compound. The response for uninfected cells is rather similar to that above with cytotoxicity manifested rather abruptly at 100 μg/mL. The infected cells on the other hand show increasing levels of survival in the presence of test agent. The interval between 10 and 100 μg/mL describes an almost classical sigmoid dose response curve. Once the upper concentration is reached the cytotoxicity of the test agent is manifested and limits the therapeutic interval. The therapeutic index of two decade orders of magnitude (ratio of the ED_{50} to the IC_{50}) would normally qualify such an agent for further investigation. This particular compound was not pursued further since it consists of a sulfated polysaccharide of indeterminate molecular weight (hence the switch from molarity to weight concentration; *vide infra*).

PURE NATURAL PRODUCTS ACQUISITION PROGRAM

The National Cancer Institute has long been involved in acquiring natural products as test candidates for its cancer screens[9]. In order to test organic compounds from as diverse a group of structures as possible, the NCI depends largely on outside investigators as sources of test substances. Many investigators involved in natural products isolation are regular contributors to the program. This continuing relationship is based at least in part on NCI policy which allows donors free use of test data; acknowledgement of the source is requested when data is quoted in publications. In addition, donors retain the rights in those cases where the activity of donated compounds warrant filing patents. A significant number of donations come from commercial laboratories. Those collaborations are facilitated by the fact that the NCI will, on request, treat submissions as confidential. In those cases a compound is assigned a D (discreet) prefixed NSC number; that designation means that the structure as well as its source cannot be divulged to personnel outside NCI. Commercial sponsors and occasional academic institutions chose to sign a standard Screening Agreement with NCI which spells out the detailed rights of the parties. The natural product literature is also monitored for new structures of interest to the program; particularly interesting compounds are requested for the tests when a sufficient amount seems to be available.

The fact that the anti-HIV screen is conducted *in vitro* has served to make the amount of compound required for the screen more commensurate with the quantities involved in natural product chemistry. We currently request 10 mg for routine testing in the screen; this will permit initial screening at the level of about 100 μg/mL. If larger amounts are supplied testing can begin at correspondingly higher levels while lesser amounts will allow only a single test without the possibility of confirmatory assays. A minimum of 2 mg is, however, required.

NCI has recently brought into operation a large scale program for identification of potential AIDS antiviral and antitumor compounds from various natural product sources. This program obviously generates a large number of crude extracts and consequently consumes a large portion of the capacity of the screens. We have, thus, been constrained to restrict the input of outside samples to those which consist of characterized compounds.

Figures 3 and 4 graphically depict the international scope of the acquisitions program. As noted in the first chart the largest single group of contributors who have donated pure natural products to the program since March of 1985 (the date NSC 600000 was assigned), not unexpectedly, come from the U.S.. Close to two thirds of the 517 donors, however list their addresses outside the U.S.. The largest single group of those (126) have European addresses with no single country showing predominance. It should, however, be noted that the second largest group from a single country, after the U.S., is from Japan. That group interestingly accounts for almost twice the number of donors compared to all other Asian countries. The number of contributors from the Americas, other than the U.S. (27), is surprisingly small particularly in view of the rich sources of novel natural products available in the tropical and sub-tropical parts of the continent.

NATURAL PRODUCTS ACQUISITION
Geographic Distribution of Donors

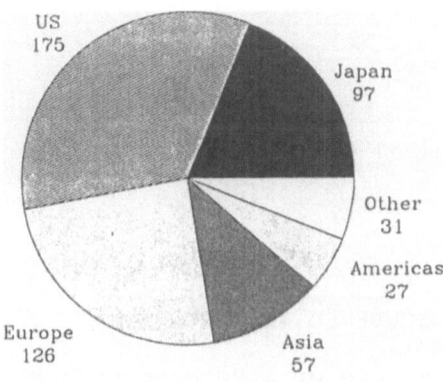

Number of Donors Since 3/85

Figure 3

Figure 4 shows the geographic origin of the 2511 compounds acquired during the same period, with origin related to donor rather than the source of the material from which the material was isolated. The rank order of the origins of those natural products is similar to that of the donors though the proportions differ significantly. In this case for example, close to half the compounds (1131) were sent by donors with addresses in the U.S. Investigators from Europe and Japan are nearly even numerically in terms of number of compounds submitted. These numbers point to an interesting difference in the number of compounds per donor; this ranges from 7 compounds per donor for the "other" category to 3 for Europe. The significance of this observation is not yet clear.

NATURAL PRODUCTS ACQUISITION
Geographic Origin of Compounds

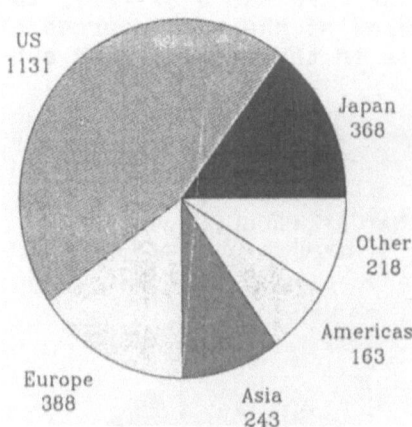

US
1131

Japan
368

Other
218

Americas
163

Asia
243

Europe
388

Number of Compounds Since 3/85

Figure 4

NATURAL PRODUCTS SCREENED AGAINST HIV

Somewhat in excess of 1620 compounds classified as natural products have been tested in the AIDS antiviral screen since it came into operation in late 1987. Chemical and biological data on all those compounds is available on-line to authorized NCI personnel through several data bases resident on a VAX 9210 computer. In addition, a rough structural index of the substances which have been tested is maintained to insure that no single structural class is over represented in the screen. Structures are grouped by ring number and then by natural product or structural class within the group. Tables 1,2 and 3 are representative examples of that index for respectively mono-, bi-, and tetracyclic compounds.

TABLE 1

NATURAL PRODUCTS SCREENED AGAINST HIV

Monocyclic compounds	
Structure Class	Number
Cyclobutanes	7
Cyclopentanes	5
Cyclohexanes and cyclohexenes	16
Cyclohexanes with a spiro substituent	19
Cyclohexanones	8
Monocyclic aromatic	39
Styrene derivatives	17
Furans, furanones etc..	48
N and S 5 membered heterocycles	29
Other monocyclic heterocycles	13
Sulfoglycolipids	4
Cyclitols & Aminocyclitols	18
Aminoglycosides	13
Nucleosides	17

TABLE 2

NATURAL PRODUCTS SCREENED AGAINST AIDS

Fused bicyclic compounds	
Structure Class	Number
Naphthalenes, decalins and indans	78
Hydroazulenes	25
Dolabellines	7
Flavones	39
Benzofurans	9
Benzopyrans	10
Hydro-furanofurans & pyranopyrans	19
Coumarins	13
Pyrrolizidines	37
Indolizidines	8
Indoles	20
Isoquinolines	40
Quinolines	7
Quinine derivatives	5
Miscellaneous	39

TABLE 3

NATURAL PRODUCT SCREENED AGAINST HIV

Fused tetracyclic compounds

Structure Class	Number
Anthracyclines	51
Nogalomycins	6
Steroids (& jervines)	75
Quassinoids	12
Azabenzphenanthrenes	8
Elipticines	7
Phorbols	10
Ibogaines & Yohimbines	37
Bridged tetracyclic	6
Miscellaneous	52

Figure 5 below represents a summary of distribution of fused ring numbers in the 1620 compounds tested in the screen to date. It should be noted that this sort includes the basic nucleus only and disregards pendant rings. A compound which contains two furans connected by a chain will be counted as monocyclic. The ring count at 4 and beyond is to some extent arbitrary since no special provision has been made for bridged ring systems in the interest of simplicity.

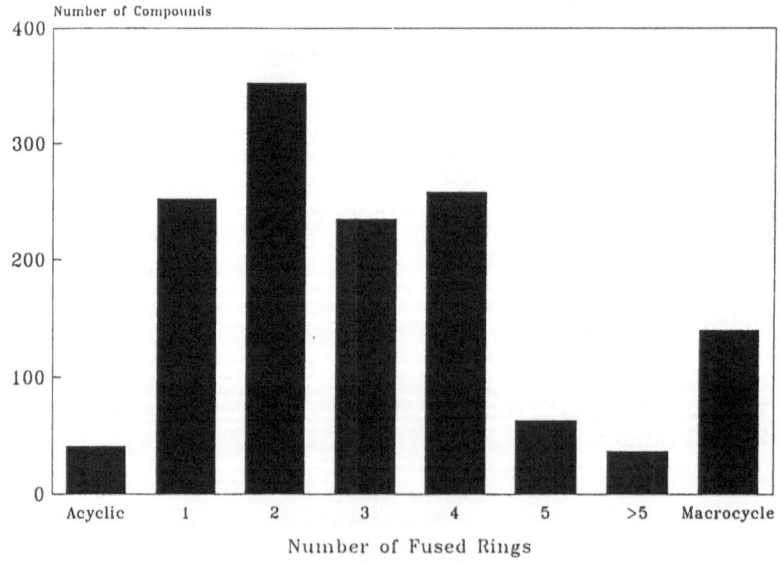

Natural Products Screened Against HIV
Compounds Sorted by Ring Number

Figure 5

SCREENING RESULTS

Close to 14,000 synthetic compounds and, as noted above, just in excess of 1620 characterized natural products have been tested in the NCI AIDS antiviral screen as of this writing. As is the case in any screen, the vast majority of compounds have little effect on the cytopathic effect of HIV on T-lymphocytes. Compounds which do spare those cells are ranked as either "Moderate" or "Active".

"Moderate" compounds comprise those agents which cause at least 50% of the infected cells to be viable at the end of the experiment but which achieve less than 90% protection. This rating is also given to those compounds which afford greater than 90% protection at a single concentration. This typically includes those compounds whose cytotoxic effects are manifested at concentrations only slightly higher than those required for greater than 90% protection from the cytopathic effect.

"Active" compounds are those which provide at least a 90% protection from the cytopathic effect over at least a decade log interval in replicated experiments. In many cases, this activity has been confirmed in secondary testing using a similar protocol with p24 protein and syncytium formation as end points insead of the XTT vital stain.

TABLE 4

SCREENING RESULTS
Synthetics and Natural Products

	Synthetics	Natural Products
Number Screened	25,000	1620
Moderately Active	497 2.0%	50 3.1%
Confirmed Active	488 2.0%	24 1.5%

The proportion of active compounds to the total number tested from the synthetic and natural product collection (Table 4) is more complicated than might appear at first sight. Several active series have already been identified on the synthetic side. At least two of these leads have been followed up by the donors using extensive synthetic programs aimed at optimizing the activity. Those series of targeted compounds distort the "moderate" and "active" columns for the synthetics since many of those analogues display at least moderate activity. In the case of synthetics the finding of an active compound thus leads to a proliferation of related compounds with a far better than random chance of showing activity in the screens.

Though the rate of finding moderates is somewhat higher

for natural products than for synthetics, Table 4 suggests that the rate of finding actives is lower. This is attributable, at least in part, to the structural complexity of natural products. The consequent difficulty in obtaining the analogues required for optimizing activity makes optimization of a lead a slower process. It should, however, be noted that four of the entries in the natural product "confirmed active" column represent derivatives prepared from an active lead in order to optimize potency. To address this , the NCI has just put into place a contract synthesis program specifically aimed at optimizing natural product leads.

TABLE 5

Natural Products With
Confirmed Anti-HIV Activity

Polyethers and derivatives	5
Sulfated polysaccharides	4
Tannins	4
Protamine derivatives	2
Lignans	2
Naphthoanthraquinone	1
Antibiotic O-sulfate	1
Phorbol ester	1
Plant protein derivative	1
Castanospermine	1
Unknown structures	2

The 24 natural products with confirmed AIDS antiviral activity are distributed among a variety of structural classes as noted in Table 5.

A discreet polyether and its four targeted analogues form the largest single structural class with 5 entries; the confidential nature of the submission precludes disclosure of the structures at this time. Four members of this group are, as noted above, targeted analogues and thus artificially inflate the group. It is of note however that 3 additional polyether antibiotics, lasalocid A, nigericin and inostamycin show moderate activity in the screen. Each of these 3 compounds shows an EC50 at very low concentration; the activity is, however, interrupted by the onset of cytotoxicity at only slightly higher concentration (Figure 6). Other investigators have reported that the polyether antibiotics monensin[10] and kijimycin[11] similarly exert anti-HIV activity. The protocol used to test kijimycin differs from that used by NCI precluding direct comparison of activities. The authors do comment that this antibiotic inhibits HIV replication in the absence of cytotoxicity over a wide range. Detailed studies suggested that the compound may act on the internalization step in HIV replication. Support is lent to this hypothesis from the fact that monensin has been shown to inhibit the entry process of picornaviruses.[12].

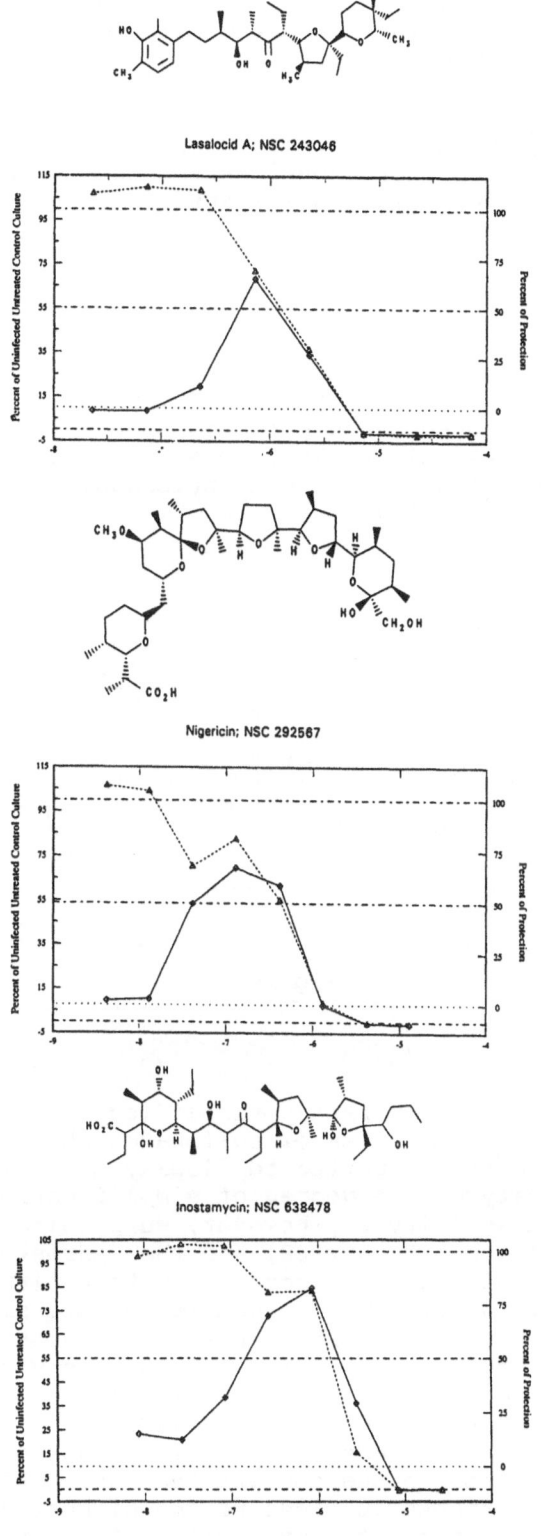

Lasalocid A; NSC 243046

Nigericin; NSC 292567

Inostamycin; NSC 638478

Figure 6

Polyether antibiotics which show moderate anti-HIV activity

Another large group of active compounds consists of polyanionic compounds including the sulfated polysaccharides and the tannins. The antiviral activity of polymeric polyanions has been recently reviewed [13,14]. The sulfated polysaccharides are thought to act by interacting with the T-cell CD-4 receptor and thus inhibit binding of the viral particle, on the basis of experiments conducted with the semisynthetic product dextran sulfate.

Dose response plots for two representative sulfated polysaccharides from unstated plant sources are shown in Figure 7. Both substance effect 100% protection of HIV infected T-lymphocytes starting at concentrations in the range of 1 μg/mL with no apparent cytotoxicity up to the top dose tested (100 μg/mL). This intriguing activity may be of little practical consequence since compounds from this class are typically poorly bioavailable when administered orally. As a class, they tend to manifest anticoagulant effects when administered parenterally.

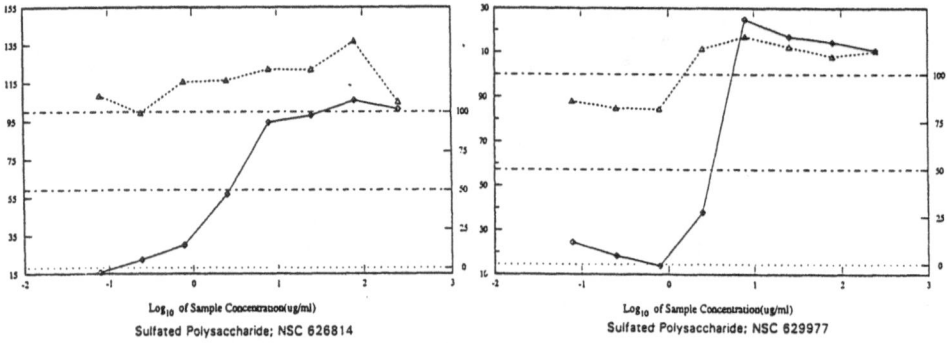

Figure 7

Sulfonates can be markedly simplified yet still retain anti-HIV activity. A series of sulfonated cholic acid derivatives have been reported to display selective AIDS antiviral activity[15]. The degree of simplification which is consistent with activity is, however, surprising. Thus both cholesterol sulfate (NSC 632988) and the androstene-3,17-diol derivative, NSC 628320, show some moderate anti-HIV activity. The activity of the former is restricted to a single data point before cytotoxicity intervenes; the androstenediol is active at only the top dose. The SAR in this very limited series is far from clear. The very closely related steroid sulfonates NSC 632987, NSC 636777 and NSC 636778, show no activity up to the top dose tested. This lack of activity should be considered in the context that the active compounds show protective effects at close to the highest dose. The activity, and lack thereof, may simply reflect relatively small differences in potency.

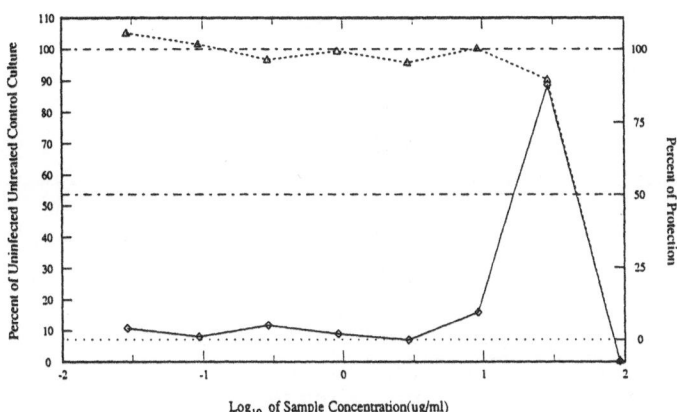

NSC 628320

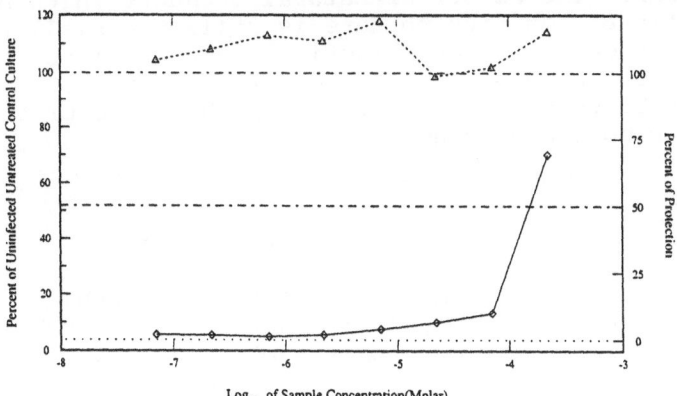

NSC 632988

Figure 8
Active Steroid Sulfates

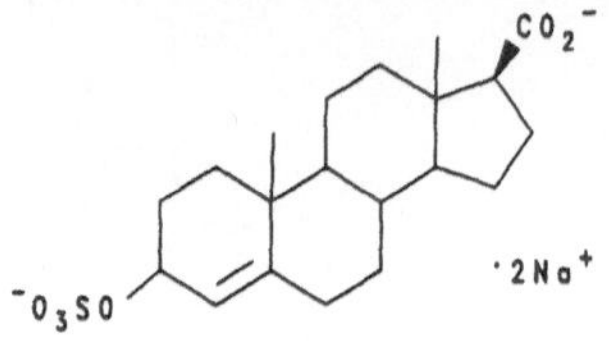

NSC 632987

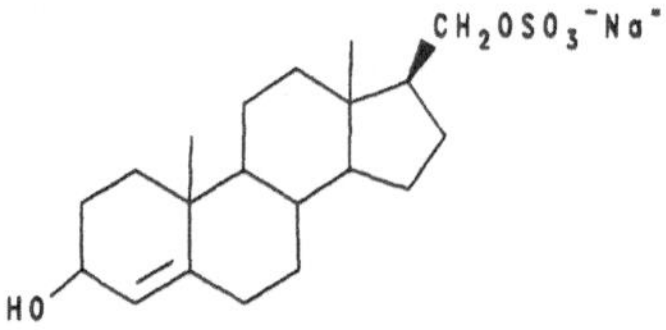

NSC 636777

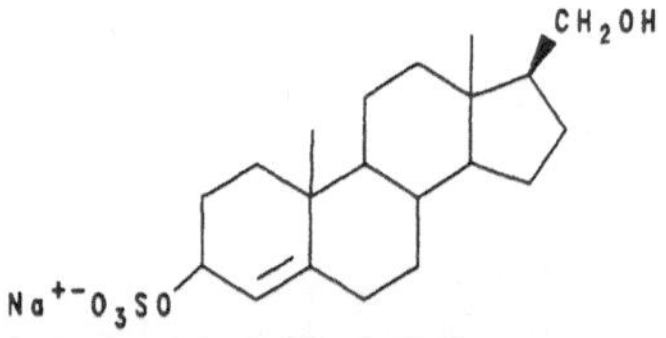

NSC 636778

Figure 9
Inactive Steroid Sulfates

The NCI's AIDS antiviral screen has now been in full
scale operation for close to four years. The assay has proven
very applicable to high turnover, providing quite reproducible
results. It is encouraging to note that all well
authenticated anti-HIV compounds show good activity in the *in
vitro* screen. The number of natural products which have
shown confirmed activity to date is still too small to permit
the formulation of structure-activity correlations. The
proportion of compounds which show at least moderate activity
suggests that this is a potentially rich source of compounds
for further screening candidates.

REFERENCES

1.   M.S.Gottlieb, H.M.Shanker, P.T.Fan, A.Saxon,
      J.D.Weisman and I.Pozalski. *Pneumocystis* Pneumonia-
      Los Angeles. Mort.Morb.Wkly.Repts., 30:250 (1981).

2.   F. Barre-Sinoussi, J.C.Chermann, F.Rey,
      M.T.Nugeyre, S.Chamaret, J.Gruest, C.Dauguet,
      C.Axler-Blin, F.Vezinet-Brun, C.Rouzioux,
      W.Rozenbaum and L. Montagnier. Isolation of a T-
      Lymphotropic retrovirus from a patient at risk from
      acquired immune deficiency syndrome (AIDS).
      Science, 20:868 (1983).

3. S.K.Arya, R.C.Gallo, B.H. Hahn, G.M.Shaw, M.Popovic, S.Z.Salahudin and F.Wong-Staal. Homology of genome of Aids-associated virus with genomes of human T-cell leukemia viruses. <u>Science</u>, 31:927 (1984).

4. H.Mitsuya, K.J.Weihold, P.A.Furman, M.H.St.Clair, S.N.Lehrman, R.C.Gallo, D.Bolognesi, D.W.Barry and S.Broder. 3'-Azido-3'deoxythymidine (BW A509U): an antiviral agent that inhibits the infectivity and cytopathic effect of human T-lymphotropic virus type III/lymphadenopathy-associated virus in vitro. <u>Proc.Natl.Acad.Sci.USA</u>, 82:7096 (1985).

5. J.Driscoll. The preclinical new drug research program of the National Cancer Institute. <u>Cancer Treat.Repts.</u> 68:69 (1984).

6. O.S.Weislow, R.Kiser, D.L.Fine, J.Bader, R.H.Shoemaker and M.R.Boyd. New soluble-formazan assay for HIV cytopathic effects: application to high-flux screening of synthetic and natural products for AIDS-antiviral activity. <u>J.Natl.Cancer Inst.</u> 81:577 (1989).

7. R.J.Gulakowski, J.B.McMahon, P.G.Staley, R.A.Moran and M.R.Boyd. A semiautomated multiparameter approach for anti-HIV drug screening. <u>J.Virolog.Meth.</u>, 00:0000 (1991).

8. K.D.Paull, R.H.Shoemaker, M.R.Boyd, D.C.Baker, J.L.Parsons, P.A.Risbood, W.A.Barbera, M.N.Sharma, E.Hand, D.C.Scudiero, A.Monks, M.C.Alley and M.Grote. The synthesis of XTT- a new tetrazolium reagent that is bioreducible to a soluble formazan. <u>J.Heterocyclic Chem.</u>, 25:911 (1988).

9. M.Suffness. Development of Antitumor Natural Substances at the National Cancer Institute, <u>in</u> "Antitumor Natural Products", Gann Monograph on Cancer Research No.36, T.Takeuchi, K.Nitta, N.Tanaka eds., Japan Scientific Societies Press, Tokyo (1989).

10. M.Nakamura, T.Ohno, S.Kunimoto, H.Nakagawa, T.Takeuchi. Kijimycin: an Inhibitor of human immunodefeciency virus in acutely and chronically infected cells. <u>J.Antibiotics</u>, 44:569 (1991).

11. R.L.Dewar, M.B.Vasudevachari, V.Natarajan, N.P.Salzman. Biosynthesis and processing of human immunodeficiency virus type 1 envelope glyproteins: effects of monensin on glycosylation and transport. <u>J.Virol.</u>, 63:2452 (1989).

12. S.I.Olsnes, K.Madhus, K. Sandvig. Entry mechanisms of picornoviruses; virus atachment and entry into cells. <u>Am.Soc.Microb.(Proc.ASM.Conf.)</u>, 171 (1986).

13. D. Lednicer, K.M.Snader. Plants and other organisms as sources of anti-HIV drugs, _in_ "Economic and Medicinal Plant Research", Vol.6.,N.R.Farnsworth ed., Academic Press, London (1991).

14. M.Baba, D.Schols, H.Nakashima R.Pauwels G.Parmentier, D.K.F.Meijer, E.DeClercq. Selective activity of several cholic acid derivatives against human immunodeficiency virus replication in vitro. J.Acquired Immune Defic.Syndr., 2:264 (1989).

15. M.Baba, E. DeClercq, Sulfated polymers as inhibitors of HIV replication. Pharmacochem.Libr., 14:85 (1990)

ANTIVIRAL ACTIVITY OF NATURAL PHLOROGLUCINOLS AND

THEIR ANALOGUES

Masahiro TADA[*], Kazuhiro CHIBA and Takao YOSHII[+]

Tokyo University of Agriculture and Technology
Laboratory of Bio-organic Chemistry, Fuchu
Tokyo, 183
+National Institute of Health, Central Virus
Diagnostic Laboratory, 4-7-1 Gakuen
Musashimurayama, Tokyo, 190-12

INTRODUCTION

There are many reports concerning antimicrobial constituents in higher plants. Schönbeck investigated the antibiotic properties of individual parts of a plant separately and found that the flowers of many plants showed especially high antibiotic activities, followed by leaves, roots and lastly branches (Schönbeck, 1967). So far as we know, there are few reports on biologically active compounds from flowers (Mitchell et al., 1970). In the course of our chemical investigation on the self-defensive substances of plants, we examined antibacterial (Escherichia coli and Bacillus subtilis) and antiviral activities (vesicular stomatitis virus [VSV] and herpes simplex virus [HSV]) of methanol extracts from various plants (especially flowers), and found that almost all plant extracts showed anti-Bacillus activities and antiviral activities against both, or either, of VSV and/or HSV (Tada et al., unpublished data). Various constituents of plants, including phenols (Sakagami et al., 1989; Van Hoof et al., 1989), lignoid (MacRae et al., 1989), flavones (Nagai et al., 1990; Gonzalez et al., 1990), coumarins (Reusser et al., 1989), quinones (Konoshima et al., 1989), phloroglucinols (Chan et al, 1989; Tada et al, 1990), sesquiterpenes (De Tommasi et al., 1990), diterpenes (Koehn et al., 1991), and tannins (Nonaka et al., 1990) have been reported with respect to antiviral activity. The flowers of Hypericum chinense L. (Guttiferae) (Japanese name: byouyanagi) exhibited extremely strong antibacterial activity. The plants belonging to the Guttiferae family are well known folk medicines for external wounds with anodyne, stanching and antiphlogistic properties, in Japan. So far we have isolated antimicrobial compounds, chinesin I and II from flowers of Hypericum chinense L (Nagai and Tada, 1987), otogirin and otogirone from Hypericum erectum (Tada et al., 1991). The structures of these constituents attracted our attention as their partial structure is similar to the 1,3-dioxo-2-carboxamide structure of tetracycline (Boothe et al., 1953). Several antimicrobial acylphloroglucinols were isolated from higher plants, e.g. aspidin from Proiobteris austricaca (Riedl and Mitteldorf, 1956),

Natural Products as Antiviral Agents, Edited by C.K. Chu
and H.G. Cutler, Plenum Press, New York, 1992

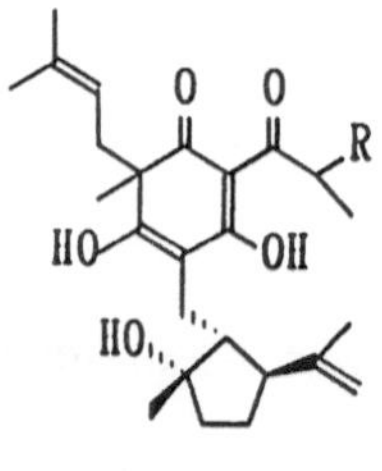

chinesin I: R=Et
chinesin II: R=Me

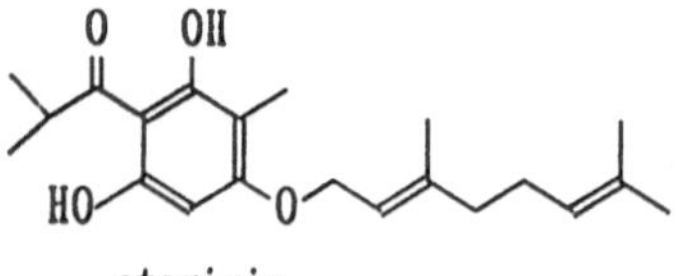

otogirin

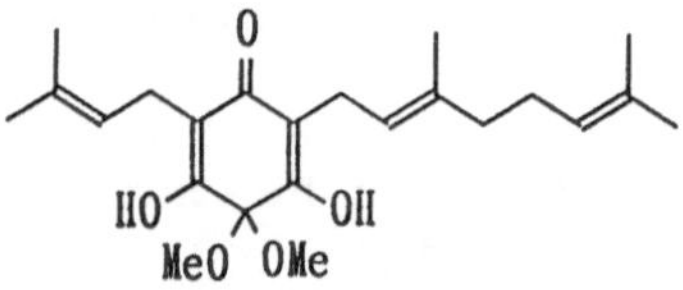

otogirone

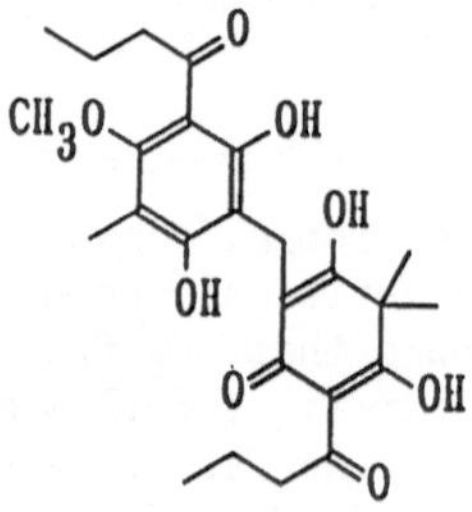

aspidin

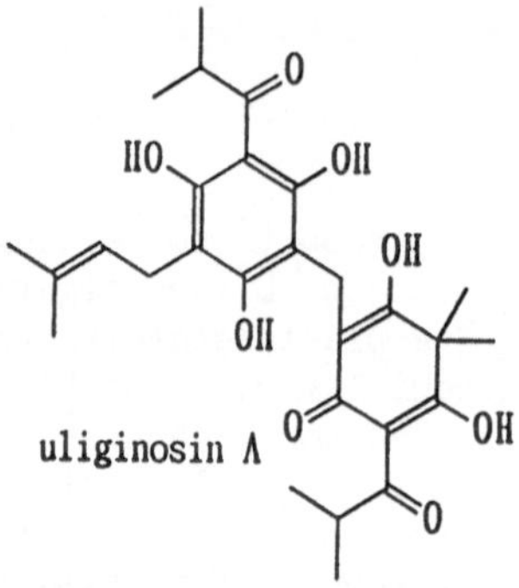

uliginosin A

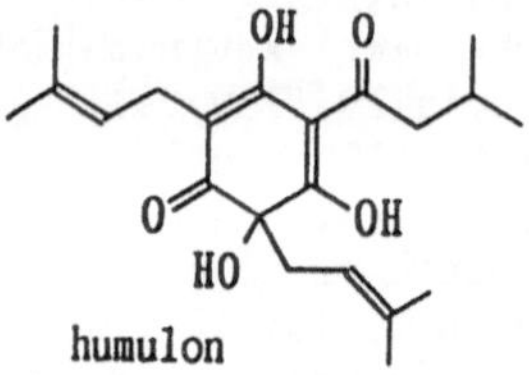

humulon

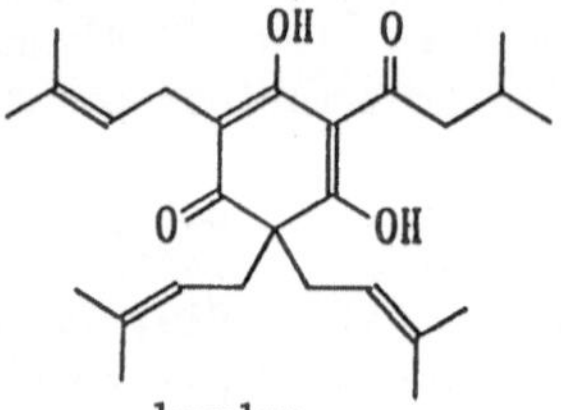

lupulon

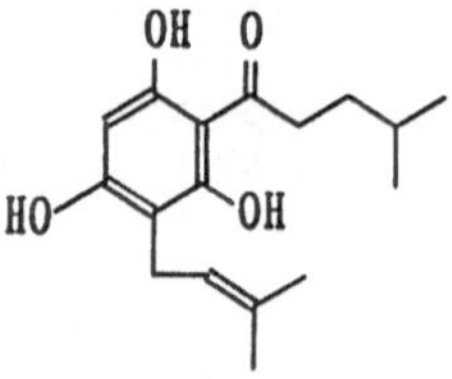

caespitin

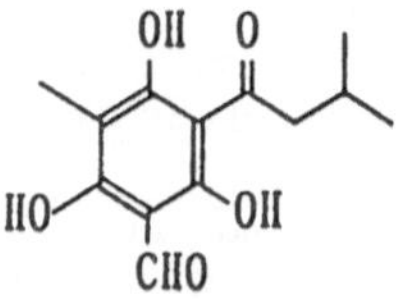

grandinol

uliginosin A from Hypericum uliginocum (Parker and Johnson, 1968), humulon (De Keukeleire and Verzele, 1970) and lupulon from Humulus lupulus (Ashurst, 1967). Antimicrobial activities of caespitin and its analogues from the indigenous South African plant, Helichrysum caespititium have been reported (Van der Schye et al., 1986). Acylphloroglucinols, grandinol and its analogues, are also known as plant growth regulators which inhibit photosynthetic electron transfer in plants (Bolte et al., 1984; Yoshida et al., 1988; Yoneyama et al., 1989). These fragments of information prompted our research into the structure activity relationships of 2,4-diacylphloroglucinols, 2-acylcyclohexane-1,3-diones and 2-carboxamidocyclohexane-1,3-dione. Various phloroglucinol derivatives and the related compounds were synthesized to examine the antimicrobial and antiviral activities. This structure-activity relationship is discussed. The correlation between the inhibition of virus replication and the alkyl chain length of the substitutents on the phloroglucinol ring was observed. A prestudy of the mode of phloroglucinols suggested that a certain derivative inhibits the synthesis of RNA and proteins of VSV.

ANTIVIRAL ACTIVITY OF NATURAL PHLOROGLUCINOLS

So far, a small number of natural antiviral phloroglucinols have been reported. Sessiliflorene, sessiliflorol A and B are acylphloroglucinols isolated from Melicope sessiliflora with inhibitory activity against herpes simplex virus (Chan et al., 1989). Isomallotochroman, butyrylmallotochromanol and related compounds from Mallotus japonicus, also inhibit the replication of HSV-1 (Arisawa et al., 1990). Mallotojaponin and mallotochromene, isolated from Mallotus japonicus exhibit strong inhibition of human immunodeficiency virus (HIV)-reverse transcriptase activity (Nakane et al., 1991). Recently, the inhibitory effects of natural phloroglucinols, euglobal-G1, G2 and G3 from Eucalyptus grandis (Takasaki et al., 1990) and syzygiol from Syzygium polycephaloides (Nishizawa et al, 1991) on Epstein-Barr virus activation, have also been reported. Chinesin I and II were isolated from flowers of Hypericum chinense and found to have strong activities against Gram-positive bacteria among plant constituents (Table 1), but no activity against Gram-negative bacteria (Nagai and Tada, 1987). Chinesins are also effective against both RNA and DNA viruses with envelope (VSV and HSV-1 and HSV-2), but not effective against poliovirus type 1 (LSc, 2ab) (polio-1) without envelope (Table 2) (Tada et al., 1990).

Table 1. Antibacterial Activities of Chinesin I

Gram-positive bacteria	MIC. (ug/ml)
Staphylococcus aureus	3.13
Staphylococcus epidermidis	>100
Micrococcus luteus	6.25
Bacillus subtilis	3.13

Table 2. Antiviral Activities of Chinesins (I/II 3:1 mixture)

	virus yield ($\log_{10}$ P.F.U./ml)			
conc.(ug/ml)	VSV	HSV-1	HSV-2	Polio-1
1.25	9.0	X	X	X
2.5	6.3	7.7	5.3	X
5.0	5.2	6.5	2.9	8.7
10.0	4.2	4.7	2.0	8.3
control	9.6	8.7	6.9	8.7

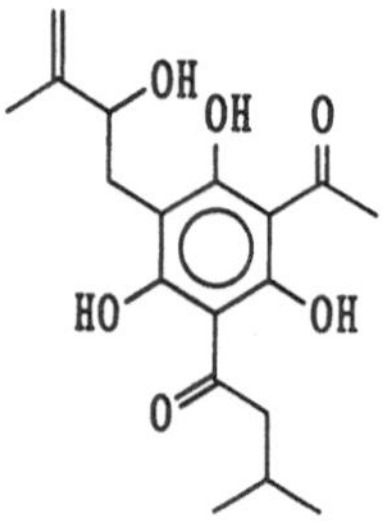

sessiliflorene 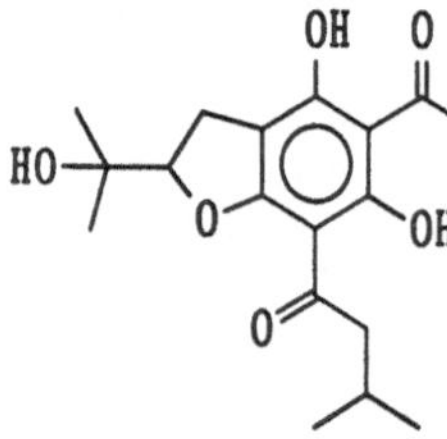   sessiliflorol A 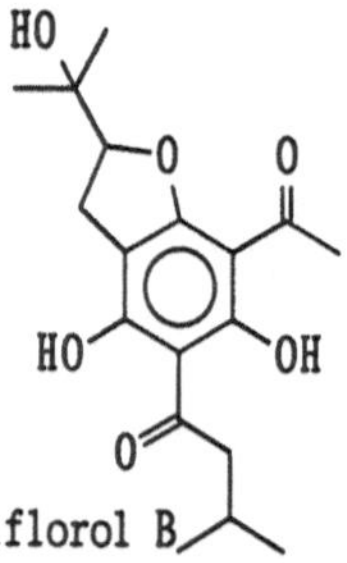   sessiliflorol B

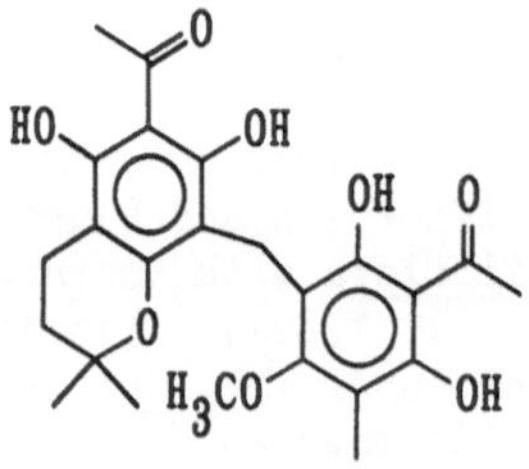

isomallotochroman

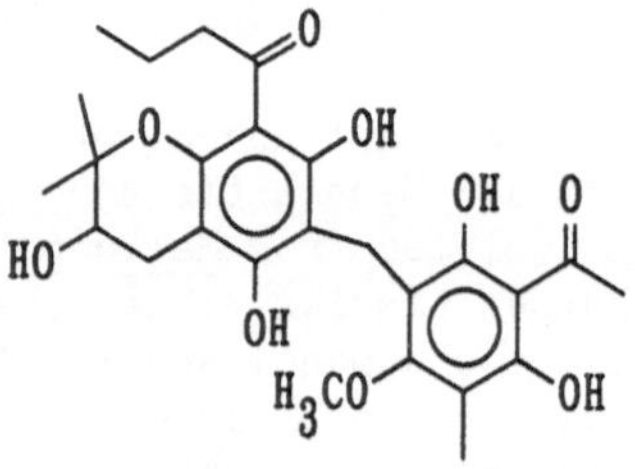

butyrylmallotochromanol

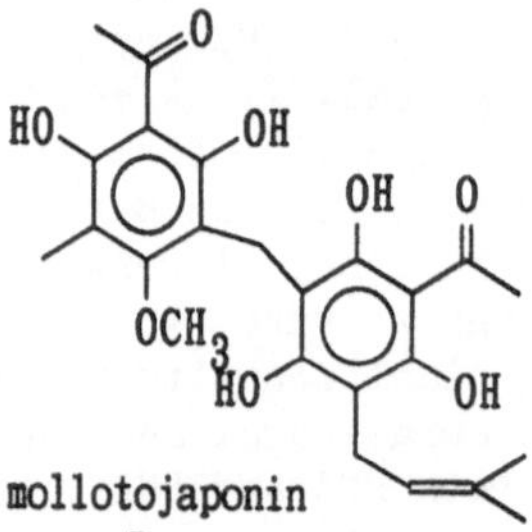

mollotojaponin

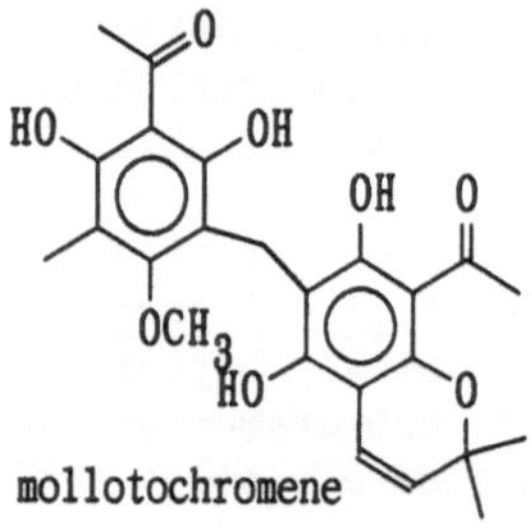

mollotochromene

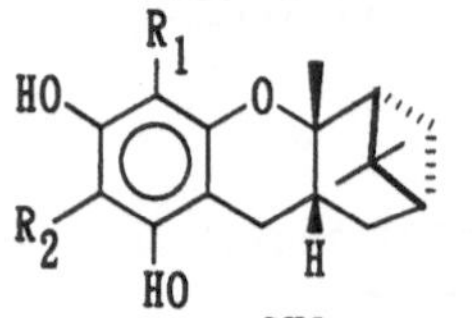

euglobal-G1:R1=COCH2CH(CH3)2 R2=CHO
euglobal-G2:R1=CHO R2=COCH2CH(CH3)2

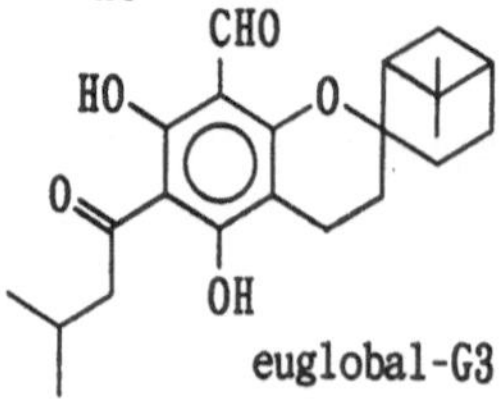

euglobal-G3

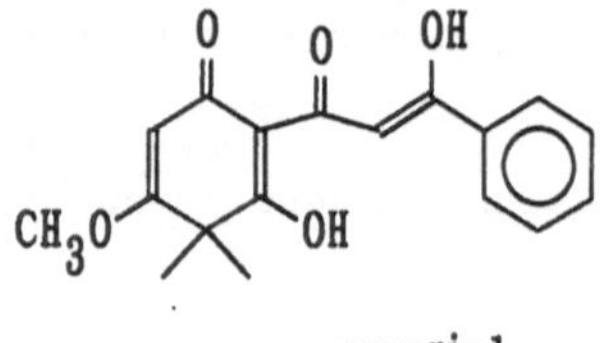

syzygiol

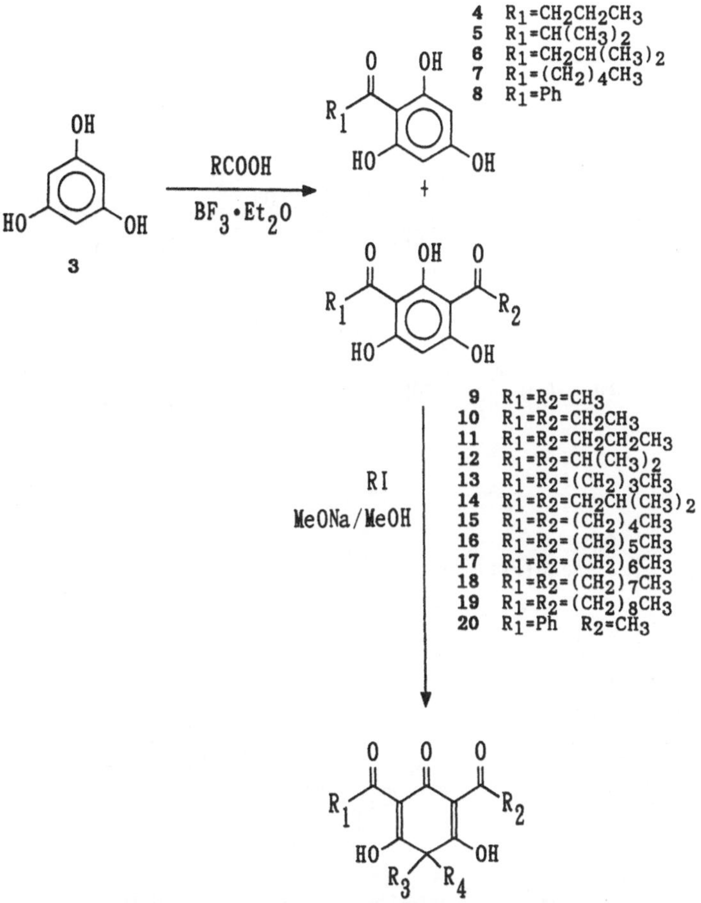

4 $R_1=CH_2CH_2CH_3$
5 $R_1=CH(CH_3)_2$
6 $R_1=CH_2CH(CH_3)_2$
7 $R_1=(CH_2)_4CH_3$
8 $R_1=Ph$

$\dfrac{RCOOH}{BF_3 \cdot Et_2O}$

9   $R_1=R_2=CH_3$
10 $R_1=R_2=CH_2CH_3$
11 $R_1=R_2=CH_2CH_2CH_3$
12 $R_1=R_2=CH(CH_3)_2$
13 $R_1=R_2=(CH_2)_3CH_3$
14 $R_1=R_2=CH_2CH(CH_3)_2$
15 $R_1=R_2=(CH_2)_4CH_3$
16 $R_1=R_2=(CH_2)_5CH_3$
17 $R_1=R_2=(CH_2)_6CH_3$
18 $R_1=R_2=(CH_2)_7CH_3$
19 $R_1=R_2=(CH_2)_8CH_3$
20 $R_1=Ph$ $R_2=CH_3$

RI
MeONa/MeOH

21 $R_1=R_2=CH(CH_3)_2$ $R_3=R_4=CH_3$
22 $R_1=R_2=CH(CH_3)_2$ $R_3=R_4=CH_2CH_3$
23 $R_1=R_2=CH(CH_3)_2$ $R_3=R_4=CH_2CH_2CH_3$
24 $R_1=R_2=CH_2CH(CH_3)_2$ $R_3=R_4=CH_3$
25 $R_1=R_2=CH_2CH(CH_3)_2$ $R_3=R_4=CH_2CH_3$
26 $R_1=R_2=CH_2CH(CH_3)_2$ $R_3=R_4=CH_2CH_2CH_3$
27 $R_1=R_2=CH_2CH(CH_3)_2$ $R_3=R_4=(CH_2)_3CH_3$
28 $R_1=R_2=CH_2CH(CH_3)_2$ $R_3=R_4=(CH_2)_4CH_3$

Scheme I

SYNTHESIS OF ACYLPHLOROGLUCINOLS

Among antimicrobial natural compounds from the plants, chinesin I **1** and II **2**, which were isolated from flowers of <u>Hypericum</u> <u>chinense</u> L., attracted our attention because of the potent activities against both bacteria and viruses. Chinesins possess one acyl chain and three alkyl chains at benzene carbons, and have no aromaticity by di-substitution of methyl and isopentenyl groups on the same carbon of the ring. These findings promoted us to synthesize the following phloroglucinol derivatives and evaluate the activities.

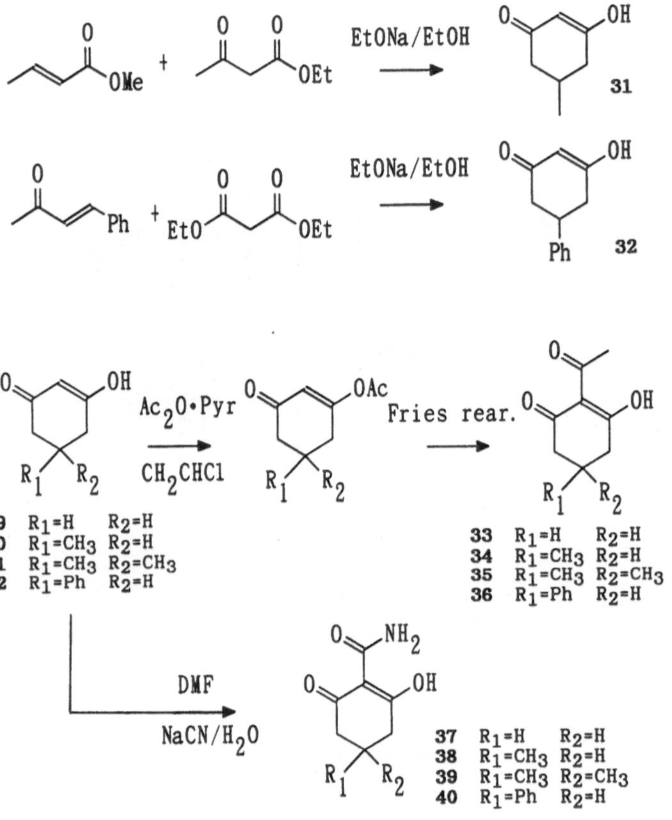

29	R_1=H	R_2=H
30	R_1=CH$_3$	R_2=H
31	R_1=CH$_3$	R_2=CH$_3$
32	R_1=Ph	R_2=H

33	R_1=H	R_2=H
34	R_1=CH$_3$	R_2=H
35	R_1=CH$_3$	R_2=CH$_3$
36	R_1=Ph	R_2=H

37	R_1=H	R_2=H
38	R_1=CH$_3$	R_2=H
39	R_1=CH$_3$	R_2=CH$_3$
40	R_1=Ph	R_2=H

Scheme II

2-Acyl-, 2,4-diacylphloroglucinols and 2,6-diacyl-4,4-dialkylcylohexane-1,3,5-triones were prepared according to Scheme I. 2-Acylphloroglucinols were synthesized by treating phloroglucinol with corresponding carboxylic acid in the presence of BF$_3$-Et$_2$O (boron trifluoride diethyl ether) complex. 2,4-Diacylphloroglucinols were synthesized in the same way, and product ratios of the monoacyl and the diacyl derivatives were controlled by the amount of carboxylic acids.

A diacylphloroglucinol was alkylated in a solution of sodium methoxide and methanol with the corresponding alkyl halides to yield 2,6-diacyl-4,4-dialkylcyclohexane-1,3,5-triones together with the by-product monoalkylated compounds.

2-Acetylcyclohexane-1,3-dione 33 was synthesized by Fries rearrangement of 3-acetoxy-2-cyclohexen-1-one, which was prepared from cyclohexane-1,3-dione 29 and acetic anhydride in pyridine-ethylene chloride (40.2 %) according to Scheme II. Methyl, dimethyl and phenyl analogues (34, 35, 36) of 33 were synthesized from corresponding cyclohexane-1,3-diones (30, 31, 32), respectively (26.0 - 53.4 %). 5-Methylcyclohexane-1,3-dione 31 was obtained by condensation of methyl crotonate and ethyl acetoacetate catalyzed with sodium ethoxide in ethanol (53.4%). 5-Phenylcyclohexane-1,3-dione 32 was prepared by the similar condensation from diethyl malonate and 4-phenyl-3-buten-2-one (97.1 %).

2-Carboxamidocyclohexane-1,3-diones (37, 38, 39, 40) were synthesized from corresponding cyclohexane-1,3-diones (29, 30, 31, 32) with sodium cyanate in N,N-dimethylformamide-water at 100 $^{\circ}$C (15.5 - 23.4 %).

ANTIMICROBIAL ACTIVITIES OF THE SYNTHESIZED PHLOROGLUCINOL DERIVATIVES

All the synthesized compounds showed antimicrobial activity against B. subtilis, but weak activity, or none, against E. coli.

Table 3. Antimicrobial Activity of Chinesin Analogues

Compound	Activity	
	B. subtilis	E. coli
	100ug/disk	100ug/disk
3	ND	20.6
8	28.4	50.3
9	91.1	29.2
10	91.3	27.3
11	63.9	ND
12	94.8	51.4
14	74.1	ND
	10ug/disk	100ug/disk
29	3.5	ND
30	19.5	ND
31	29.9	3.1
32	32.6	ND
33	28.2	10.3
34	26.9	ND
35	32.4	ND
36	50.5	12.2
37	59.9	ND
38	71.7	20.4
39	53.3	ND
40	60.4	ND
	10ug/disk	10ug/disk
Tetracycline	100.0	100.0
Chinesins(1+2)	81.8	ND

ND: not done

Antimicrobial activity was tested by paper-disk method (Mitchell et al., 1970) against Bacillus subtilis (IFO 3734) and Escherichia coli (IFO 3301).   The antimicrobial activities were determined by the following formula (Table 3).

Activity = (Y - 0.6)/(X - 0.6) X 100
X (mm) : Diameter of inhibition zone by tetracycline
Y (mm) : Diameter of inhibition zone by sample
Diameter of the paper disk is 0.6 mm.

The activity of dicarbonyl compounds (29, 30, 31, 32) are weaker than corresponding 2-acetylcyclohexane-1,3-diones (33, 34, 35, 36). 2-Carboxamidocyclohexane-1,3-dione (37, 38, 39, 40) showed stronger activity than the corresponding 2-acetylcyclohexane-1,3-dione (33, 34, 35, 36). Clearly, 2,4-diacylphloroglucinols (9, 10, 11, 12, 14, 20) are more active than 2-acetylcyclohexane-1,3-dione and 2-carboxamidocyclohexane-1,3-dione. The antimicrobial activity of chinesins and 2,4-diacylphloroglucinols are comparable to tetracycline. These show that dicarbonyl groups on cyclohexane are a necessary structural unit for this activity.

A marked tendency to rise antimicrobial activity was observed together with the increase of number of carbonyl groups or enols in a molecule. The number of carbons and branching on acyl groups influence the activity. As 2,4-diisobutyryl-phloroglucinol 12 showed the strongest antimicrobial activity by the primary test, the MIC was measured against various microorganisms comparing the activity with cefazolin (Kariyone et al., 1970) (Table 4).   Compound 12 was also effective against Candida ablicans (NIH3147).

Table 4. MIC of 12 against Various Microorganisms

microorganism	MIC (ug/ml)	
	12	cefazolin
Staphylococus aureus	0.78	0.39
S. aureus ATCC 10537	3.13	0.1
S. epidermidis	100	1.56
Micrococcus luteus	6.25	1.56
Bacillus subtilis	3.13	0.1
Candida albicans NIH3147	50	100
Mycobacterium smegmatis 607	100	

ANTIVIRAL ACTIVITIES OF THE SYNTHESIZED PHLOROGLUCINOL DERIVATIVES

Antiviral activity of the dione 32, 2-carboxamidocyclohexane-1,3-diones (37, 39) and 2,4-diacylphloroglucinols (9, 10, 12) were tested at the highest non-toxic dose of the compounds by the inhibition of cytopathic effect (c.p,e.) development of vesicular stomatitis virus (VSV), herpes simplex virus type I (HSV-1) and polio-1 in HEL, RL-33 and FL cells, respectively. Furthermore, VSV yield in HEL cells was measured in the presence of compounds (9, 10, 12). Infectivity titration of VSV was done by plaque method in RL-33 cells. No activity was observed for the dione 32 and 2-carboxamidocyclohexane-1,3-diones (37, 39) whereas strong activities were observed for 2,4-diacylphloroglucinols (9, 10, 12), chinesin I and chinesin II against VSV and HSV-1. These suggest that the compounds which have acylphloroglucinol structure should be generally active against VSV and HSV-1. 2,4-Diacylphloroglucinols and chinesins showed no activity against polio-1. It is interesting that these acylphloroglucinols are effective to both RNA virus with envelope (VSV) and DNA virus with envelope (HSV-1), but not effective to RNA virus without envelopes

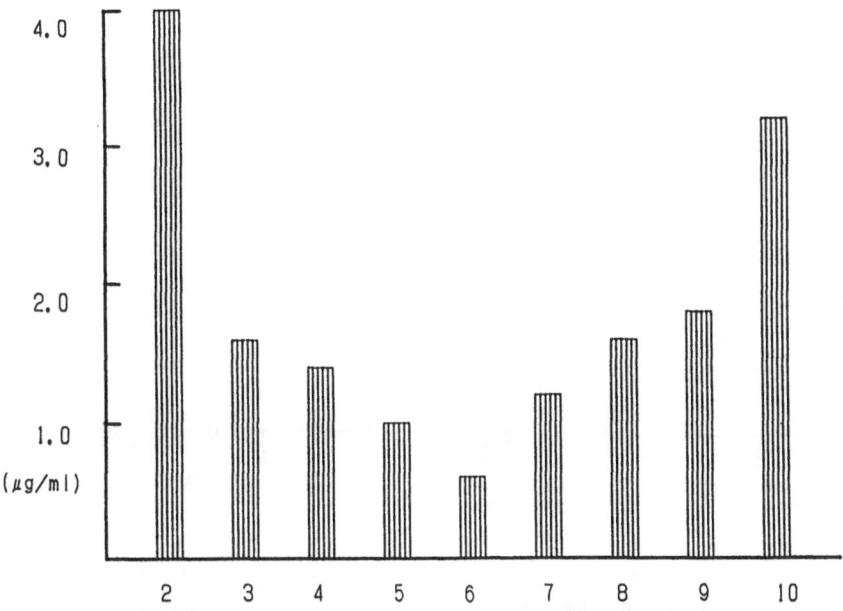

Carbon number of the acyl chains of diacylphloroglucinols.

Figure 1. Correlation between antiviral activity and acyl chain length of diacylphloroglucinols. Microgram per milliliter of the spindle was determined by the minimum concentrations of the compounds giving 10^{-3} yield reduction of VSV in HEL cells. RL-33 cells (Yoshii et al., 1977; Yoshii and Kono, 1978) derived from rabbit lung were grown in Eagle's minimum essential medium (MEM) containing 10% calf serum, 0.11% $NaHCO_3$ and antibiotics (penicillin 100 U/ml, streptomycin 100 ug/ml). Maintenance medium was MEM supplemented with 0.15% $NaHCO_3$, 2% fetal bovine serum and antibiotics. For the growth of HEL cells, the concentration of $NaHCO_3$ was reduced to 0.075%. The Indiana serotype of VSV was propagated in HEL cells with maintenance medium and stored at -70°C until use. Plaque titration was performed in RL-33 cells. Briefly, confluent monolayers of the cells in 2 oz prescription bottles were inoculated with appropriately diluted virus in 0.2 ml amounts. After virus adsorption for 1 hr. at 37°C, the infected cell cultures were overlaid with 5 ml amounts of agar medium, and incubated at 37°C. Agar overlay medium consisted of MEM containing 0.8% BiTek-agar (Difco), 0.15% $NaHCO_3$, 2% fetal bovine serum, and antibiotics (as in the growth medium). The second overlay medium containing 0.006% neutral red was added in 4 ml amounts to the cultures on the second day post infection (p.i.). Plaques were counted on day 3 p.i. A multicycle growth was employed in order to observe the growth characteristics of VSV. Confluent monolayers of HEL cells grown in 2 oz prescription bottles were infected with VSV at an input m.o.i. of 0.01 p.f.u. (plaque forming unit)/cell and the virus was allowed to adsorb for 1 hr. at 37°C. The infected cell cultures were then washed 3 times with MEM to remove unadsorbed virus, and 5 ml amounts of maintenance medium in the presence or absence of various concentrations of phloroglucinol derivatives were added immediately. Thereafter, cell cultures were incubated at 37°C for 24 hr. Cell cultures were disrupted by 3 cycles of freezing and thawing and the supernatant fluid obtained by low speed centrifugation was tested for infectivity assay.

(polio-1). It is well known that tannic acid inactivates enveloped viruses such as HSV, VSV. However, chinesins and these acylphloroglucinols did not show any virus inactivating activity.

In order to study the structure-activity relationship, chinesins and twenty-three synthesized phloroglucinol derivatives were evaluated with respect to antiviral activity against VSV.

Fig. 2 shows the effect of chinesins and acylphloroglucials on the replication of VSV in human embryonic lung fibroblast (HEL) cells. Activity of chinesin I and II were measured as a 3:1 mixture without separation. The antiviral activity of monoacylphloroglucinols was weaker than that of chinesins, however, potent inhibitory activity was found in the diacylphloroglucinols. The concentrations of the compounds **14** and **15** showing 10^{-3} virus yield reduction were 0.4 to 0.6ug/ml and

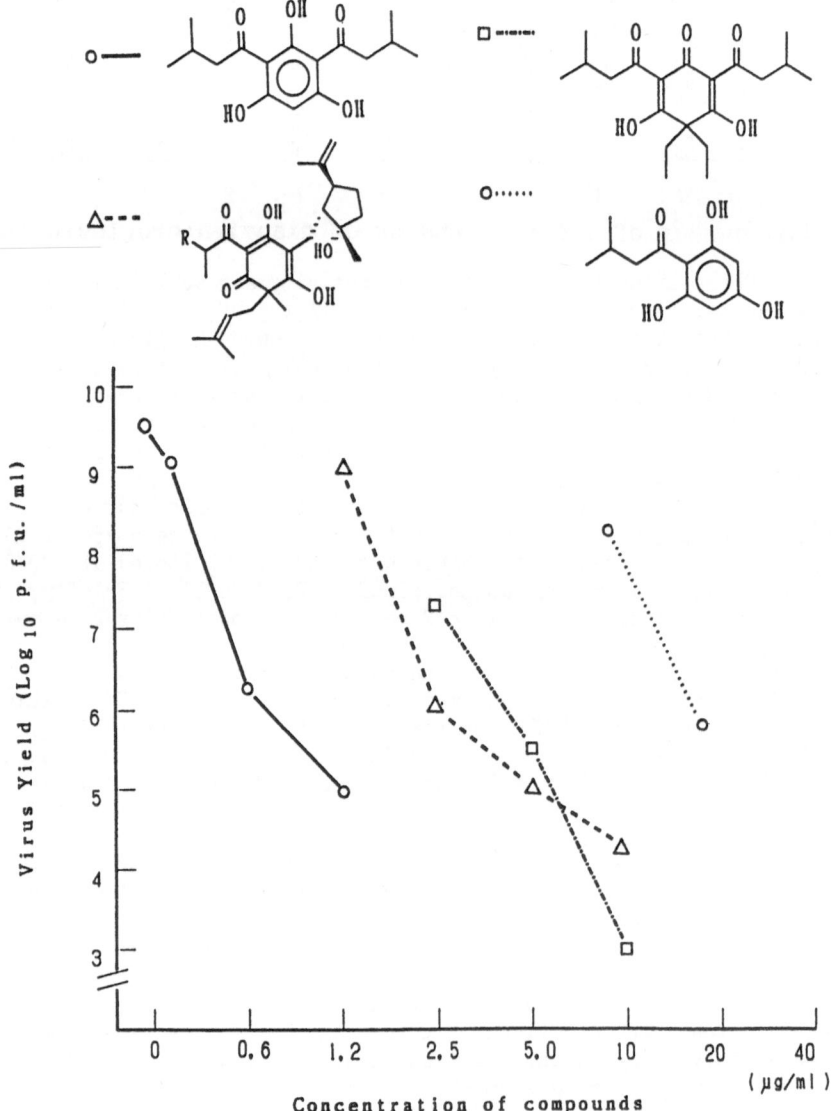

Figure 2. Effect of chinesins and related compounds on the replication of VSV in HEL cell cultures.

0.8 to 1.2ug/ml for 10^{-5} virus yield reduction respectively. Cytotoxicity was not observed for **4, 5,** and **6** with less than 50ug/ml and for chinesins, **7** was less than 25ug/ml. On the other hand, compound **11, 12, 14, 15** showed cytotoxicity at 5-10ug/ml. Monoacylphloroglucinols, which possessed an acyl chain composed of four carbons showed lower antiviral activity in comparison with those having an acyl chain of five or six carbons. The activity of diacylphloroglucinols also changed with the carbon number of the acyl chains.

The change of the concentrations of each derivatives for 10^{-3} yield reduction of VSV are shown in Fig. 1 with the acyl chain length of diacylphloroglucinols. These acyl groups were composed of straight chains. Compound **15** showed the strongest activity among diacylphloroglucinols compared with those straight chains so far tested.

The antiviral activity of the compounds decreased by the addition or reduction of the number of acyl chain carbons. Moreover, compounds **12** and **14** which possess branched acyl chains showed 10^{-3} yield reduction in a lower concentration than that of **11** and **13** (Fig. 1 and 2). The result suggested that the chain length and branching of acyl groups were closely correlated with the antiviral activity. The acyl chain of chinesins, which is composed of branched acyl chains of four or five carbons may, therefore, play an important role in the inhibition of VSV replication.

The antiviral activity of 2,6-diacyl-4,4-dialkylcyclohexane-1,3,5-triones which were derived from potent antiviral diacyl-phloroglucinols is shown in Table 5. Introduction of geminal methyl groups at C-4 of the benzene ring decreased the antiviral activity, but the activity was higher than the corresponding monoacylphloroglucinol. Furthermore, the activity became stronger when the alkyl chain length was increased, and compound **26, 27, 28** showed almost the same activity as that of the chinesins. Fig.2 shows the effect of the substituents of phloroglucinols on the virus yield reduction curves and the result was compared with that of chinesins. Similarity was found between compound **25** and the chinesins; both of them possessed at least one branched acyl chain composed of four or five carbons and disubstituted alkyl groups on the same carbon of the ring. Furthermore, cytotoxicity was not observed for 2,6-diacyl-4,4-dialkylcyclohexa-1,3,5-trienes in less than 50ug/ml. Selectivity between the antiviral activity and cytotoxicity became greater by introducing geminal alkyl groups on the benzene ring.

Table 5. The Effect of 2,6-Diacyl-4,4-dialkylcyclohexane-1,3,5-triones on the Replication of Vesicular Stomatitis Virus in HEL Cell Cultures[*]

Compounds	ug/ml	Virus Yield Log_{10}P.F.U./ml	Cytotoxicity (24 h)
21	10.0	4.6	(-)
22	5.0	4.4	(-)
23	3.0	4.6	(-)
VSV-control		9.6	

* HEL cells were seeded in 2-oz prescription bottles and infected at an input m.o.i of 0.01 p.f.u./cell. After adsorption for 1 h at 37 $^{\circ}$C, the infected cell monolayers were washed 3 times with PBS, 5 ml of maintenance medium containing various concentration of compound was added to the bottles and infected cultures were incubated at 37 $^{\circ}$C for 24 h after infection.

The results confirming the relations between structures of the compounds and antiviral activity including cytotoxicity give us important information for the development of suitable drugs for clinical use.

In conclusion The correlation between the antiviral activity on VSV and acyl chain length of diacylphloroglucinols was observed. It can be presumed that the hydrophile-lipophile balance of diacylphloroglucinols is closely correlated with the affinity for bio-membranes (Hansch and Fujita, 1964; Fujita et al., 1964). On the other hand, although 2,6-diacyl-4,4-dialkylcyclohexane-1,3,5-trienes were lipophiles, in comparison with the corresponding diacylphloroglucinol, their antiviral activity became stronger with the increase of the alkyl chain

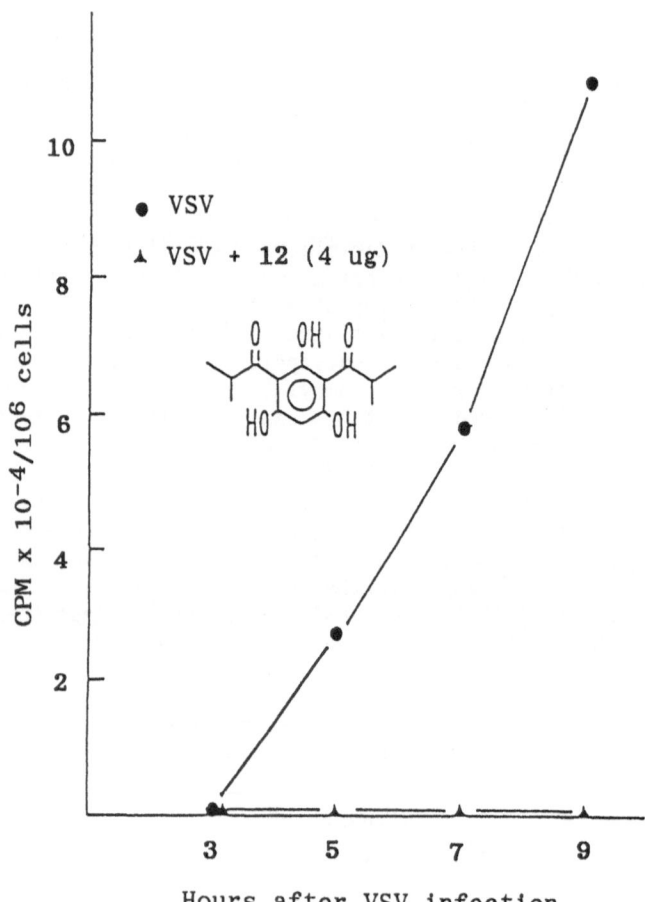

Hours after VSV infection

Figure 3. Effect of compound **12** on VSV RNA synthesis in FL cells. Confluent monolayers of FL cells grown in 35-mm plastic Petri-dish were infected with VSV at a m.o.i. of 20 pfu/cell. After virus adsorption for 1 hr at 37°C, the inoculum was removed and the cultures incubated with medium containing 1.5 ug/ml of actinomycin D and 5 uCi/ml of [^{3}H]uridine in the presence or absence of 4 ug/ml of **12**. At the indicated times after infection, the amount of [^{3}H]uridine incorporated into RNA was determined by deducing the average background of mock-infected cultures with actinomycin D.

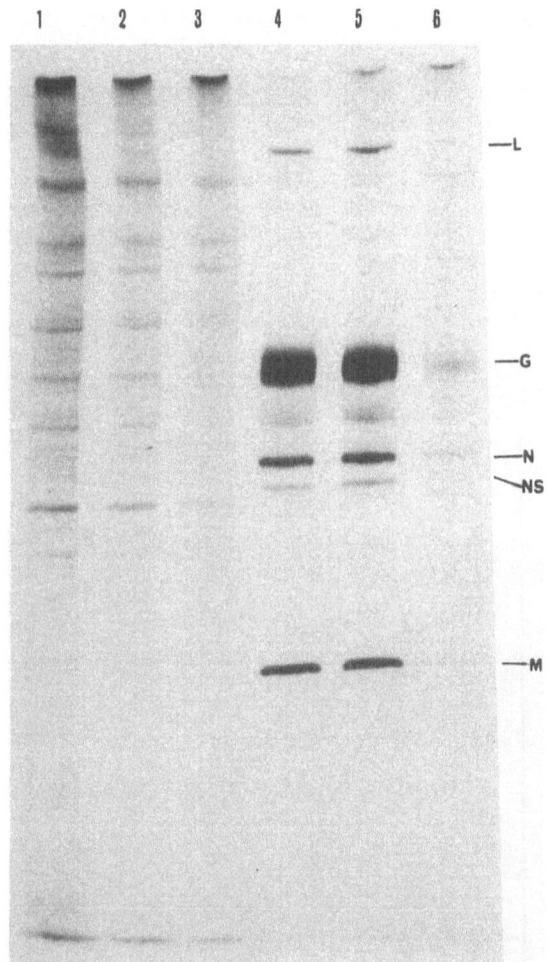

Figure 4. Inhibition of VSV protein synthesis in FL cells.

1	Uninfected	Cell control
2	Uninfected	+ 2 ug/ml of **12**
3	Uninfected	+ 4 ug/ml of **12**
4	VSV infected	Virus control
5	VSV infected	+ 2 ug/ml of **12**
6	VSV infected	+ 4 ug/ml of **12**

Confluent monolayers of FL cells grown in 35-mm plastic Petri-dish were infected with VSV at a m.o.i. of 20 pfu/cell. After virus adsorption for 1 hr at 37°C, unadsorbed virus was removed by washing with MEM, and the cultures were incubated in the presece or absence of the compound as indicated above. At 7 hr after virus infection, the cultures were pulse-labelled with 25 uCi/ml of [^{35}S]methionine for 1 hr. The cell lysates were analysed by SDS-polyacrylamide gel electrophoresis.

length.   The difference between the diacylphloroglucinols and 2,6-diacyl-4,4-dialkylcyclohexane-1,3,5-trione for affinity to bio-membranes is interesting.

MECHANISM OF ACTION

The investigations on the antiviral effects of phloroglucinol derivatives have quite recently been launched. Mechanisms of action are not yet clearly understood.

The RNA synthesis of VSV in FL cells was markedly inhibited in the presence of 4 ug/ml of **12** (Fig.3), whereas cellular RNA synthesis was not so affected (data not shown), and this was assumed from the amount of [^{3}H]uridine incorporation into TCA-insoluble fractions. In addition, reduction of VSV protein synthesis was evident at 4 ug/ml of **12** as shown

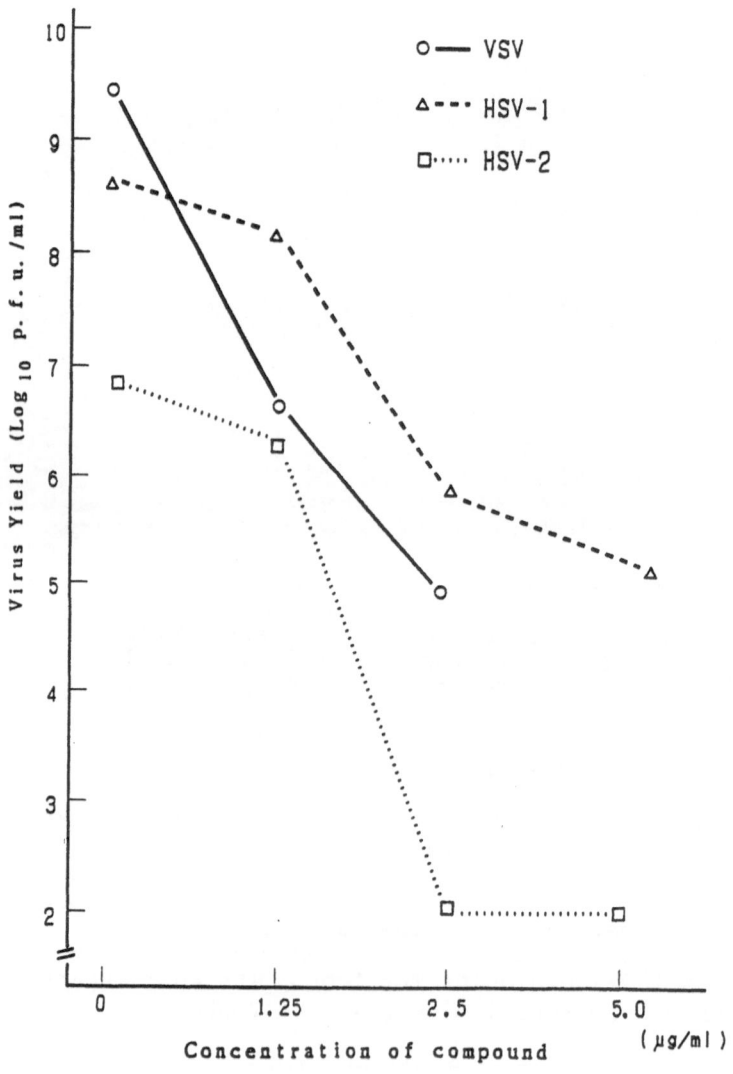

Figure 5. Inhibitory effect of diacylphloroglucinol **12** on the replication of VSV, HSV-1 and HSV-2 in HEL cells.

in Fig 4. Compound **12** did not show any inactivating activity to both VSV and HSV-1 at 5 ug/ml and 10 ug/ml, respectively, when incubated for 1 hr. at 37°C (Fig. 5). Based upon the above information, it seems likely that the main action of phloroglucinol derivatives on VSV replication is due to the inhibition of viral RNA and protein synthesis, but not to its activity of inactivation.

9-(2-Hydroxyethoxymethyl)guanine (acyclguanosine) and (E)-5-(2-bromovinyl)-2'-deoxyuridine (BVDU) are nucleoside analogues and are well known as potent antiherpetic agents. The modes of action of these compounds are similar to those reported on acyclguanosine (Elion et al., 1977; Fyfe et al., 1989) and BVDU (Clercq et al., 1980); i.g., phosphorylation by virus coded-thymidine (dThd) kinase is required for their antiviral effects. Phloroglucinol derivatives, however, inhibited the multiplication of dThd kinase deficient HSV-2 as illustrated in Fig. 5. This indicates that the compounds have no effect on the HSV coded-dThd kinase. Similar results were reported by Chan et al.(1989). On the other hand, while EB virus does not induce dThd kinase, its induction of early antigen was inhibited by euglobals. These compounds are novel acylphloroglucinols which are coupled with a monoterpene or a sesquiterpene structure (Takasaki et al., 1990).

HIV encoded-reverse transcriptase was also inhibited by some phloroglucinol derivatives, named mallotophenone and its analogues (Nakane et al., 1991). As reverse transcriptase is essential for the replication of retroviruses, it is reasonable to choose HIV-reverse transcriptase for selection of antiviral agents (Mitsuya et al., 1985; Furman et al., 1986). Although it was shown from the above study that inhibitory effects of natural and synthetic phloroglucinol derivatives displayed a fairly wide spectrum against both enveloped RNA and DNA viruses, further investigations for the detailed mechanism are needed.

REFERENCES

Arisawa, M., Fujita. A., Hayashi, T., Hayashi, K., Ochiai, H. and Morita, N., 1990, "Cytotoxic and Antiherpetic Activity of Phloroglucinol Derivatives from Mallotus japonicus (Euphorbiaceae)", Chem. Pharm. Bull. **38**: 1624.
Ashurst, P. R., 1967, "The chemistry of the hop resins", Fortschr. Chem. Org. Naturstoffe, **25**: 63.
Bolte, M. O., Bowers, J., Crow, W. D., Paton, D. M., Sakurai, A., Takahashi, N., Uji-ie, M. and Yoshida, S., 1984, "Germination Inhibitor from Eucalyptus pulverulenta", Agric. Biol. Chem., **48**: 373.
Boothe, J. H., Morton, J., Petisi, J. P., Wilkinson, R. G., Williams, J. H., 1953, "Tetracycline", J. Am. Chem. Soc., **75**: 4621.
Chan, J. A., Shultis, E. A., Carr, S. A., DeBrosse, C. W., Eggleton, D. S., Francis, T. A., Hyland, L. J., Johnson, W. P., Killmer, L. B., Staiger, D. B. and Westley, J. W., 1989, "Novel Phloroglucinols from the Plant Melicope sessiliflora (Rutaceae)", J. Org. Chem., **54**: 2098.
De Clercq, E., Descamps, J., Verhelst, G., Walder, R.T., Jones, A.S., Torrence, P. F., and Shugar, D., 1980, "Comparative efficacy of anti-herpes drugs against different strains of herpes simplex virus", J. Infec. Dise., **141**: 563.
De Keukeleire and Verzele, M., 1970, "Structure and the absolute configuration of (-)-humulon", Tetrahedron, **26**: 385.

De Tommasi, N., Pizza, C., Conti, C., Orsi, N. and Stein, M. L., 1990, "Structure and in vitro antiviral activity of sesquiterpene glycosides from <u>Calendula</u> <u>arvensis</u> ", <u>J</u>. <u>Nat</u>. <u>Prod</u>., **53**: 830.

Elion, G. B., Furman, P. A., Fyfe, J. A., De Miranda, P., Beauchamp, L., and Schaeffer, H. J., 1977, "Selectivity of action of an anti-herpetic agent, 9-(2-hydroxyethoxymethyl)guanine", <u>Proc</u>. <u>Natl</u>. <u>Acad</u>. <u>Sci</u>., <u>USA</u>, **74**: 5716.

Fujita, T., Iwasa, J. and Hansch, C., 1964, "A New Substituent Constant, π, Derived from Partition Coefficients", <u>J</u>. <u>Am</u>. <u>Chem</u>. <u>Soc</u>., **86**: 5175.

Furman, P. A., Fyfe, J. A., Clair, M. H. S., Weinhold, K., Rideout, J. L., Freeman, G. A., Lehrman, S. N., Bolognesi, D.P., Broder, S., Mitsuya, H., and Barry D., 1986, " Phosphorylation of 3'-azido-3'-deoxythymidine and selective interaction of the 5'-triphosphate with human immunodeficiency virus reverse transcriptase", <u>Proc</u>. <u>Natl</u>. <u>Acad</u>. <u>Sci</u>. <u>USAP</u>, **83**: 8333.

Fyfe, J. A., Keller, P. M., Furman, P. A., Miller, R. L., and Elion, G. B., 1978, "Thymidine kinase from herpes simplex virus phosphates the new antiviral compound, 9-(2-hydroxyethoxymethyl)guanine", <u>J</u>. <u>Biol</u>. <u>Chem</u>., **253**: 8721.

Gonzalez, M. E., Martinez-Abarca, F. and Carrasco, L., 1990, "Flavonoids: potent inhibitors of poliovirus RNA synthesis", <u>Antiviral</u> <u>Chem</u>. <u>Chemotherapy</u>, **1**: 203.

Hansch, C. and Fujita, T., 1964, "$\rho - \sigma - \pi$ Analysis. A Method for the Correlation of Biological Activity and Chemical Structure", <u>J</u>. <u>Am</u>. <u>Chem</u>. <u>Soc</u>., **8**; 1616.

Kariyone, K., Harada, H., Kurita, M. and Takano, T., 1970, "Cephalosporin. III. cefazolin, a new semisynthetic antibiotic I. Synthesis and chemical properties of cafazolin", <u>J</u>. <u>Antibiot</u>., **23**: 131.

Koehn, F. E., Gunasekere, S. P., Niel, D. N. and Cross, S. S., 1991, "Halitunal, unusual diterpene aldehyde from the marine alga <u>Halimeda</u> <u>tuna</u>", <u>Tetrahedron</u> <u>Lett</u>., **32**: 169.

Konoshima, T., Kozuka, M., Koyama, J., Okatani, T., Tagahara, K. and Tokuda, H., 1989, "Studies on inhibitors of skin tumor promotion. VI. inhibitory effects of quinones on Epstein-Barr virus activition", <u>J</u>. <u>Nat</u>. <u>Prod</u>., **52**: 987.

MacRae, W. D., Hudson, J. B. and Towers, G. H. N., 1989, "The antiviral action of lignans", <u>Planta</u> <u>Med</u>., **55**: 531.

Mitchell, J. W., Mandava, N., Worley, J. F., Plimmer, J. R. and Smith,M.V., 1970, "Brassins; a new family of plant hormones from rape pollen", <u>Nature</u>, **225**: 1065.

Mitsuya, H., Weinhold, K. J., Furman, P. A., Clair, M. H. S., Lehrman, S. N., Gallo, R. C., Bolognesi, D., Barry, D. W., and Broder, S., 1985, "3'-Azido-3'-deoxythymidine(BW A509U): An antiviral agent that inhibits the infectivity and cytopathic effect of human T-lymphotropic virus type III/lymphadenopathy-associated virus <u>in</u> <u>vitro</u>", <u>Proc</u>. <u>Natl</u>. <u>Acad</u>. <u>Sci</u>. <u>USA</u>, **82**: 7096.

Nagai, M., and Tada ,M., 1987, " Antimicrobial compounds, Chinesin I and II from Flowers of <u>Hypericum</u> <u>chinense</u> L.", <u>Chem</u>. <u>Lett</u>., **1987**: 1337.

Nagai, T., Miyaichi, Y., Tomimori, T., Suzuki, Y. and Yamada, H., 1990, "Inhibition of influenza virus sialidase and anti-influenza virus activity by plant flavonoids", <u>Chem</u>. <u>Pharm</u>. <u>Bull</u>., **38**: 1329.

Nakane, H., Arisawa, M., Fujita, A., Koshimura, S. and Ono, K., 1991, "Inhibition of HIV-reverse transcriptase activity by some phloroglucinol derivatives", <u>FEBS</u> <u>Lett</u>., **1991**: 83.

Nishizawa,M.; Yamada,H.; Sano,J.; Ito,S.; Hayashi,Y.; Ikeda,H.; Chairul, Shiro,M.; Tokuda,H., 1991, "Structure of syzygiol: A skin-tumor promotion inhibitor", <u>Tetrahedron</u> <u>Lett</u>., **32**: 211.

Nonaka, G., Nishioka, I., Nishizawa, M., Yamagishi, T., Kashiwada, Y., Dutschman, G. E., Bodner, A., Kilkuskie, R. E., Cheng, Y.and Lee, K., 1990, "Anti-AIDS agents,2: Inhibitory effects of tannins on HIV reverse transcriptase and HIV replication in H8 lymphocyte cells", J. Nat. Prod., 53: 587.

Parker, W. L. and Johnson, F., 1968, "The structure determination of antibiotic compounds from Hypericum uliginocum", J. Am. Chem. Soc., 90: 4716 and 4724.

Reusser, F., Tarpley, W. G., Dolak, L. and Althaus, I. W., 1989, "Coumarins to inhibit reverse transcriptase in humans for treatment of human immunodeficiency virus infection", Pct. Int. Appl., 11pp.

Riedl, W. and Mitteldorf, R., 1956, "Constituents of dilix mas and analogs. V. Synthesis of pseudoaspidinol, Chem. Ber., 89: 2595.

Sakagami, H., Ohhara, T., Kaiya, T., Kawazoe, Y., Nonoyama, M. and Konno, K., 1989, "Molecular species of antitumor and antiviral fraction from pine cone extract", Anticancer Res., 9: 1593.

Schönbeck, F., 1967, "Untersuchungen uber Bluteninfektionen", Phytopathol. Z., 59: 205.

Tada, M., Chiba, K., Yamada, H., and Maruyama, H., 1991, "Phloroglucinol Derivatives as Competitive Inhibitors Against Thromboxane A_2 and Leukotoriene D_4", Phytochemistry, 30: 2559. A misprint has been found in the structure of otogirone in Fig.2 of the literature.

Tada, M., Chiba, K., Takakuwa, T. and Kojima, E., 1992 "Analogues of Natural Phloroglucinols as Antagonists Against both Thromboxane A_2 and Leukotriene D_4", J. Med. Chem., 35: 1209.

Tada, M., Takakuwa, T., Nagai, M. and Yoshii, T., 1990, "Antiviral and Antimicrobial Activity of 2,4-Diacylphloroglucinols, 2-Acylcyclohexane-1,3-diones and 2-Carboxamidocyclohexane-1,3-diones", Agric. Biol. Chem., 54: 3061.

Takasaki, M., Konoshima, T., Fujitani, K., Yoshida, S., Nishimura, H., Tokuda, H., Nishino, H., Iwashima, A. and Kozuka, M., 1990a, "Inhibition of Skin-tumor Promotion. VIII. Inhibitory effects of Euglobals and Their Related Compounds on Epstein-Barr Virus Activation", Chem. Pharm. Bull., 38: 2737.

Takasaki, M., Konoshima, T., Shingu, T., Tokuda, H., Nishino, H., Iwashima, A. and Kozuka, M., 1990, "Structure of Euglobal-G1, -G2, and -G3 from Eucalyptus grandis, Three New Inhibitors of Epstein Barr Virus Activation", Chem. Pharm. Bull., 38: 1444.

Van Der Schye, C. J., Dekker, T. G., Fourie, T. G. and Snyckers, F. O., 1986, "Synthesis and antimicrobial activity of a series of caespitin derivatives", Antimicrobial Agents and Chemotherapy, 30: 375.

Van Hoof, L., Totte, J., Corthout, J., Pieters, L., Mertens, F., Berghe, D. A. V. and Vlietinck, A. J., 1989, "Plant antiviral agent, VI. Isolation of Antiviral Phenolic Glucosides from Populus cultivar Beaupre by droplet counter-current chromatography", J. Nat. Prod., 52: 875.

Yoneyama, K., Asami, T., Crow, W. C., Takahashi, N. and Yoshida, S., 1989, "Photosynthetic Electron Transport Inhibition by Phlorophenone Derivatives", Agric. Biol. Chem., 53: 471.

Yoshida, S., Asami, T., Kawano, T., Yoneyama, K., Crow, W. D., Dugald, M., Paton and Takahashi, N., 1988, "Photosynthetic Inhibitors in Eucalyptus grandis" Phytochemistry, 27: 1943.

Yoshii, T. and Kono, R., 1978, "Differential inhibitory effects of 5-bromodeoxyuridine on vaccinia and monkeypox viruses", J. Hyg., Camb. 81: 171.

Yoshii, T., Natori, K. and Kono, R., 1977, "Replication of Enterovirus 70 in Non-Primate Cell Cultures", J. Gen. Virol., 36: 377.

THE INDOLE ALKALOID TUMOR PROMOTER TELEOCIDINS AS EPSTEIN-BARR VIRUS

INDUCERS: STRUCTURE, BIOSYNTHESIS AND STRUCTURE-ACTIVITY RELATIONSHIP

Kazuhiro Irie and Koichi Koshimizu

Department of Food Science and Technology
Faculty of Agriculture, Kyoto University
Kyoto 606, Japan

INTRODUCTION

Epstein-Barr virus (EBV) belongs to the herpes virus group and is thought to be the cause of African Burkitt's lymphoma and anaplastic naso-pharyngeal carcinoma.[1] 12-O-Tetradecanoylphorbol-13-acetate (TPA), teleo-cidins and aplysiatoxin, which are potent skin tumor promoters in mouse skin[2-4] and have quite similar biological activities such as activation of protein kinase C,[5] efficiently induce the viral cycle in latently infected EBV genome carrying cells at nanogram order.[6-8] Since such EBV induction is specifically restricted only to these natural products, they are particularly useful as tools for studying the mechanism of induction of EBV antigens. This is especially true for teleocidins because of their smaller molecular weight and higher stability than TPA and aplysiatoxin (Fig. 1).

In 1984, in collaboration with Dr. Yohei Ito (Kyoto University) and Dr. Sawao Murao (Osaka Prefectural University) our group found one strain, Streptoverticillium blastmyceticum NA34-17, showing potent EBV induction.[9] This strain produced a new microbial metabolite, (-)-indolactam-V (21), the core structure of teleocidins, in large quantities, a compound first synthe-sized by Shudo et al.[10] (-)-Indolactam-V (21) is a key compound for investigating the relationship between the structure of teleocidins and EBV-inducing activity because it has the fundamental structure of teleocidins as well as being a potent inducer of the Epstein-Barr virus early antigen (EBV-EA).[9] Structure-activity studies using a variety of indolactam derivatives are especially important because they contribute to the elucidation of the structural requirement for EBV-inducing activity and provide a basis for

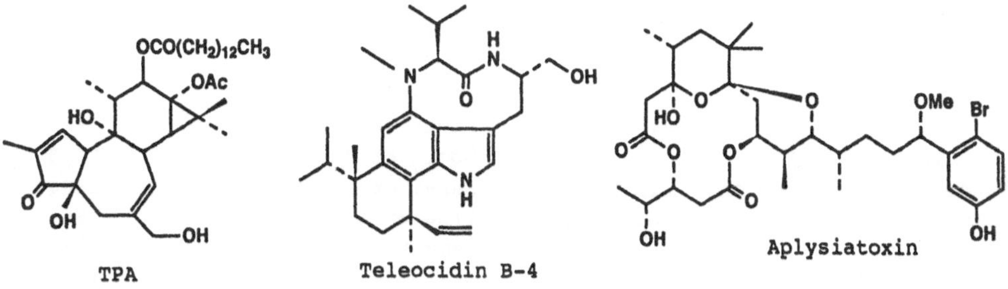

Fig. 1.   Structure of TPA, teleocidin B-4 (1) and aplysiatoxin.

Natural Products as Antiviral Agents, Edited by C.K. Chu
and H.G. Cutler, Plenum Press, New York, 1992

designing antiviral agents such as a TPA antagonist. Moreover, new probes for the receptor analysis of the EBV inducers can be developed on the basis of these results.

For several years, we have been isolating several new teleocidin-related microbial metabolites, examined their biosynthesis and metabolism, investigated the structure-activity relationships and synthesized new probes for the receptor analysis of EBV inducers. This chapter summarizes these studies.

STRUCTURE OF TELEOCIDINS AND THEIR-RELATED COMPOUNDS

Teleocidins were first isolated from Streptomyces mediocidicus as highly inflammatory substances by Takashima and Sakai.[11] The name "teleocidin" originates from its piscicidal activity to teleosts. Takashima and Sakai found two types of compounds, teleocidin A and B, whose molecular weights were 437 and 451, respectively. The structure of the latter was determined by X-ray crystallographic analysis of the monobromoacetate of its dihydro derivative.[12] After that, teleocidin A was proved to consist of two isomers, teleocidin A-1 (13) and A-2 (14)[13]; teleocidin B of four isomers, teleocidin B-1 - B-4 (1-4).[14] Teleocidin B, which Hirata et al. used for X-ray analysis, was identical with teleocidin B-4 (1).[14] Lyngbyatoxin A, isolated from the blue-green algae Lyngbya majuscula, found in Hawaii, as an inflammatory substance,[15] was proved identical with teleocidin A-1 (13).[13]

Hitherto, several teleocidin-producing organisms (actinomycetes and blue-green algae) have been reported (Table I).[9,11,15-20] Among these organisms, our strain Streptoverticillium blastmyceticum NA34-17 has a characteristic feature of producing (-)-indolactam-V (21) in quantity (40-50mg/l).[21] This characteristic was advantageous in isolating a variety of teleocidin-related compounds including the biosynthetic teleocidin intermediates. We scrutinized, therefore, the culture broth and mycelia of this microorganism and isolated twelve new teleocidin-related compounds, blastmycetin A (24),[22] B (25),[22] C (28),[22] D (17),[23] E (30),[24] des-O-methylolivoretin E (6),[24] 14-O-methylteleocidin A-1 (18),[25] des-N-methylteleocidin A-1 (19),[26] (-)-7-geranylindolactam-V (20) (manuscript in preparation), (-)-3-R-2-oxyindolactam-V (26),[25] (-)-3-S-2-oxyindolactam-V (29)[25] and (-)-indolactam-I (31).[27] (-)-Indolactam-V (21),[9] (-)-14-O-acetylindolactam-V (22)[9] and (-)-des-N-methylindolactam-V (23)[25] synthesized by Endo et al.[28] were also first isolated as naturally-occurring compounds. These metabolites contributed to the research of teleocidin chemistry and biology. Blastmycetin A (24) was valuable for structure-activity studies, blastmycetin B (25) and C (28), the 2-oxyindolactams (26,29) and des-N-methyl derivatives (11,19,23) for the metabolism of teleocidins, blastmycetin D (17) and E (30) for the biosynthesis of the monoterpenoid moiety of the teleocidins.

Table 1.   Teleocidin-producing organisms

Streptomyces mediocidicus	Takashima et al.[11]	1960
Streptoverticillium krissii	Knoev et al.[16]	1975
Lyngbya majuscula	Moore et al.[15]	1979
Streptoverticillium olivoreticuli	Sakai et al.[17]	1984
Streptoverticillium blastmyceticum	Irie et al.[9]	1984
Nocardiopsis sp.	Yamashita et al.[18]	1988
Streptoverticillium eurocidicum	Oka et al.[19]	1989
	Kumagai et al.[20]	1991

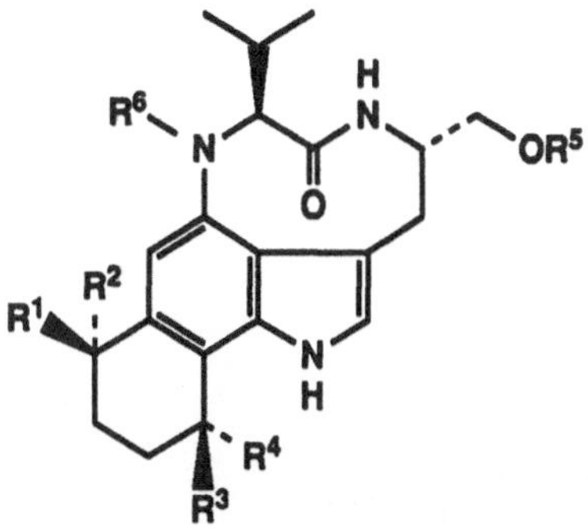

R¹	R²	R³	R⁴	R⁵	R⁶	
Me	i-Pr	Vinyl	Me	H	Me	Teleocidin B-4 (1)
i-Pr	Me	Vinyl	Me	H	Me	Teleocidin B-1 (2)
Me	i-Pr	Me	Vinyl	H	Me	Teleocidin B-2 (3)
i-Pr	Me	Me	Vinyl	H	Me	Teleocidin B-3 (4)
Me	Vinyl	i-Pr	Me	H	Me	Des-O-methyl-olivoretin C (5)
Me	Vinyl	t-Bu	H	H	Me	Des-O-methyl-olivoretin E (6)
Me	i-Pr	Vinyl	Me	Me	Me	Olivoretin A (7)
i-Pr	Me	Vinyl	Me	Me	Me	Olivoretin B (8)
Me	Vinyl	i-Pr	Me	Me	Me	Olivoretin C (9)
Me	Vinyl	t-Bu	H	Me	Me	Olivoretin E (10)
Me	i-Pr	Vinyl	Me	H	H	Des-N-methyl-teleocidin B-4 (11)

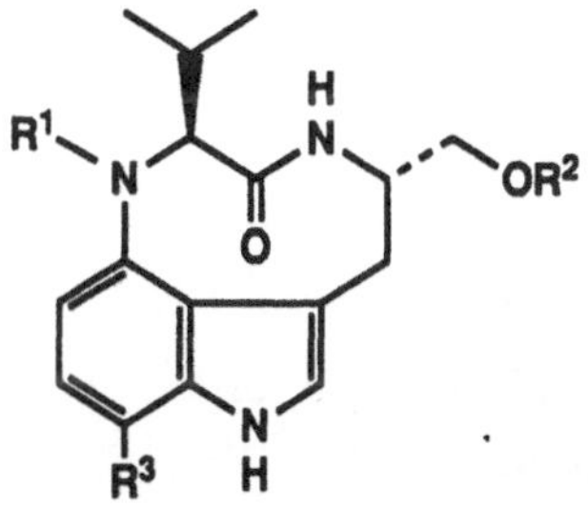

R¹	R²	R³	
Me	H		Pendolmycin (12)
Me	H	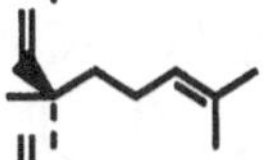	Teleocidin A-1 (13)
Me	H	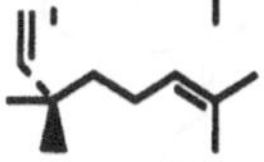	Teleocidin A-2 (14)
Me	H	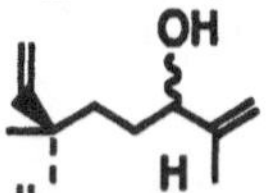	Lyngbyatoxin B (15)
Me	H	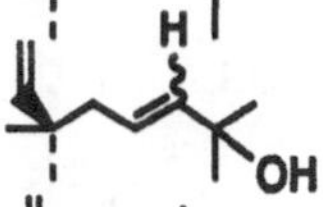	Lyngbyatoxin C (16)
Me	Me	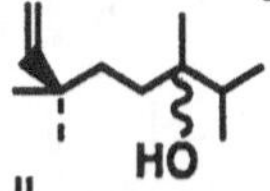	Blastmycetin D (17)
Me	Me	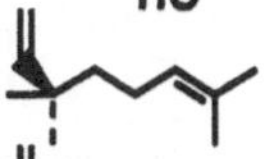	14-O-Methyl-teleocidin A-1 (18)
H	H	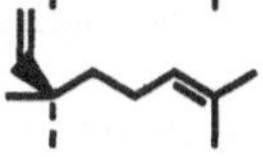	Des-N-methyl-teleocidin A-1 (19)
Me	H	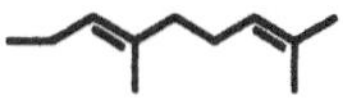	(−)-7-Geranyl-indolactam-V (20)
Me	H	H	(−)-Indolactam-V (21)
Me	Ac	H	(−)-14-O-Acetyl-indolactam-V (22)
H	H	H	(−)-Des-N-methyl-indolactam-V (23)

Fig. 2. Structure of naturally-occurring teleocidin-related compounds.

Structure elucidation of new teleocidin-related compounds has also been intensively carried out by Sakai et al. They isolated olivoretin A (7),[17] B (8),[29] C (9),[29] E (10),[30] des-O-methylolivoretin C (5),[29] des-N-methyl-teleocidin B-4 (11)[30] and N-methyl-L-valyl-L-tryptophanol (27)[31] from Streptoverticillium olivoreticuli. Moreover, they have recently reexamined the Lyngbya majuscula in Hawaii and isolated the new metabolites lyngbyatoxin B (15) and C (16).[32]

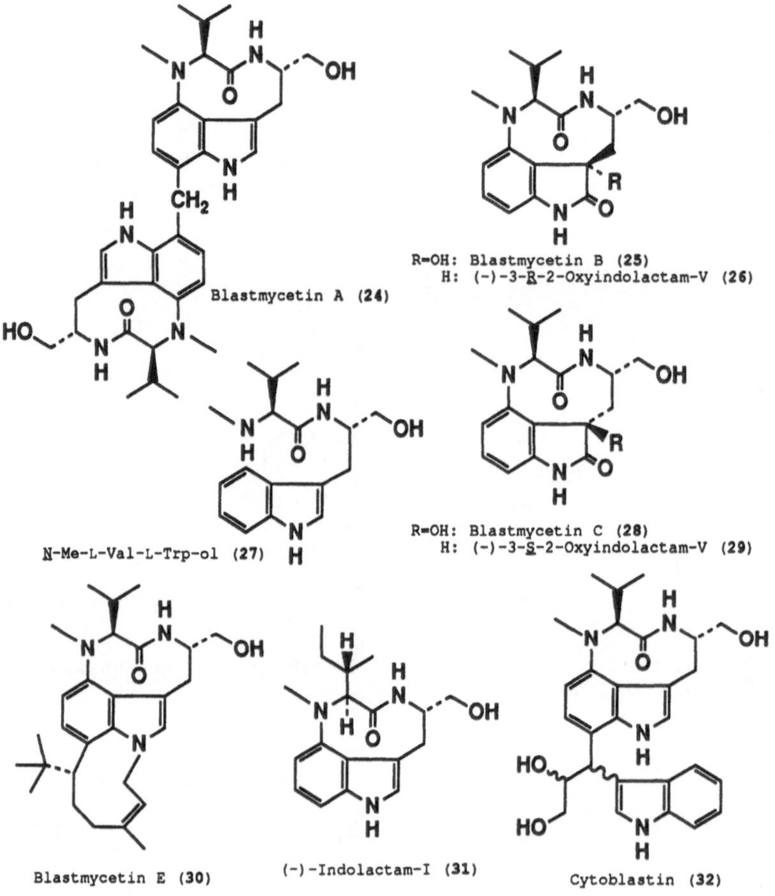

Fig. 2-cont. Structure of naturally-occurring teleocidin-related compounds.

Yamashita et al. isolated pendolmycin (12) from Nocardiopsis sp. in the course of screening for inhibitors of epidermal growth factor-induced phosphatidylinositol turnover.[18] More recently, cytoblastin (32) was found in the culture broth of Streptoverticillium eurocidicum as a T-cell growth stimulant.[20] At present, the total number of the teleocidin-related metabolites is 32, as shown in Fig. 2.

The teleocidins attract much attention in the area of organic chemistry because of their unique structure, especially because of the nine-membered lactam ring from position 3 to position 4 on the indole ring and the complex monoterpenoid side chains at positions 6 and 7 (Fig. 2). The last several years have seen intensive studies on the total syntheses of teleocidins and their related metabolites.[10,28,33-50] However, there are few reports on the biosynthesis of these fascinating indole alkaloids, whose elucidation might ultimately be applicable to the efficient syntheses of new teleocidin-related compounds. Among the teleocidin-producing organisms (Table 1), our strain of S. blastmyceticum NA34-17 has characteristic feature of producing (-)-indolactam-V (21) along with teleocidin B-4 (1) in quantity.[21] We exploited this characteristic to investigate the biosynthesis of teleocidins.

Biosynthesis of (-)-indolactam-V (21)

To elucidate the biosynthetic block of 21, feeding experiments with three logical precursors, DL-[1-^{13}C]-tryptophan, L-[1-^{13}C]-valine and L-[CD$_3$]-methionine, clearly showed that 21 was biosynthesized from L-tryptophan, L-valine and L-methionine.[51] Next, feeding experiments using ^{2}H or ^{13}C-labeled possible biosynthetic precursors were carried out to establish the biosynthetic pathway. Of the seven possible precursors examined (DL-tryptophanol, N-acetyl-L-tryptophanol, N-methyl-L-valine, L-valyl-L-tryptophan, L-valyl-L-tryptophanol, N-methyl-L-valyl-L-tryptophan and N-methyl-L-valyl-L-trypto-phanol), only the seco compound N-methyl-L-valyl-L-tryptophanol (27) was efficiently incorporated into (-)-indolactam-V (21),[51] suggesting that the first step including reduction, condensation and methylation, occurs successively through some multifunctional macromolecular enzyme. The occurrence of 27 in the culture broth and mycelia of this microorganism was also confirmed. From these results, a possible biosynthetic pathway for 21 was proposed as shown in Fig. 3.

Recently, total syntheses of 21 have been intensively investi-gated.[10,28,33,34,38,41,44] More convenient syntheses are, however, desirable from the view point of synthesizing various indolactam congeners for structure-activity studies. The most troublesome problem in the total synthesis of 21 is making the carbon-nitrogen bond at position 4 of the indole ring. Since N-methyl-L-valyl-L-tryptophanol can be chemically synthesized without difficulty and was efficiently converted into 21,[51] utilization of the microbial cyclization enzyme would be a convenient synthetic method for various indolactam analogues. We have recently shown that ten indolactam analogues, (-)-indolactam-Ala, Abu, γ,δ-Δ-Nva, Nva, Nle, tert-Leu, Leu, Ile, allo-Ile, Phg, can be efficiently synthesized from each seco compound.[52] This enzyme could also cyclize N-ethyl-L-norvalyl-L-tryptophanol (unpublished results). Since teleocidin-producing microorganisms are very common (Table 1), this microbial conversion might be a simple and convenient method to prepare various indolactam analogues. Kogan et al.[44] have recently tried to

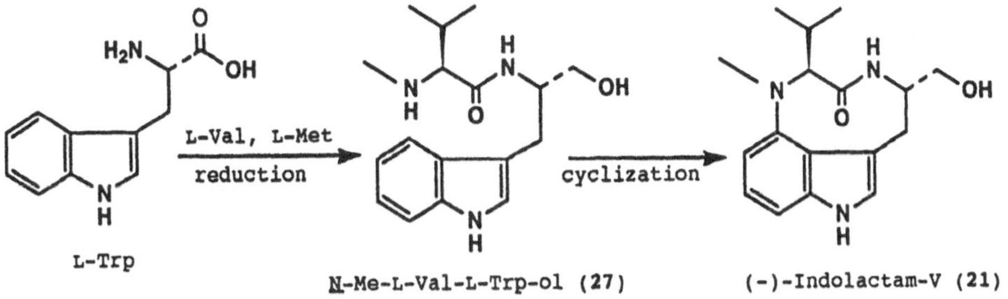

Fig. 3. Possible biosynthetic pathway of (-)-indolactam-V (21).

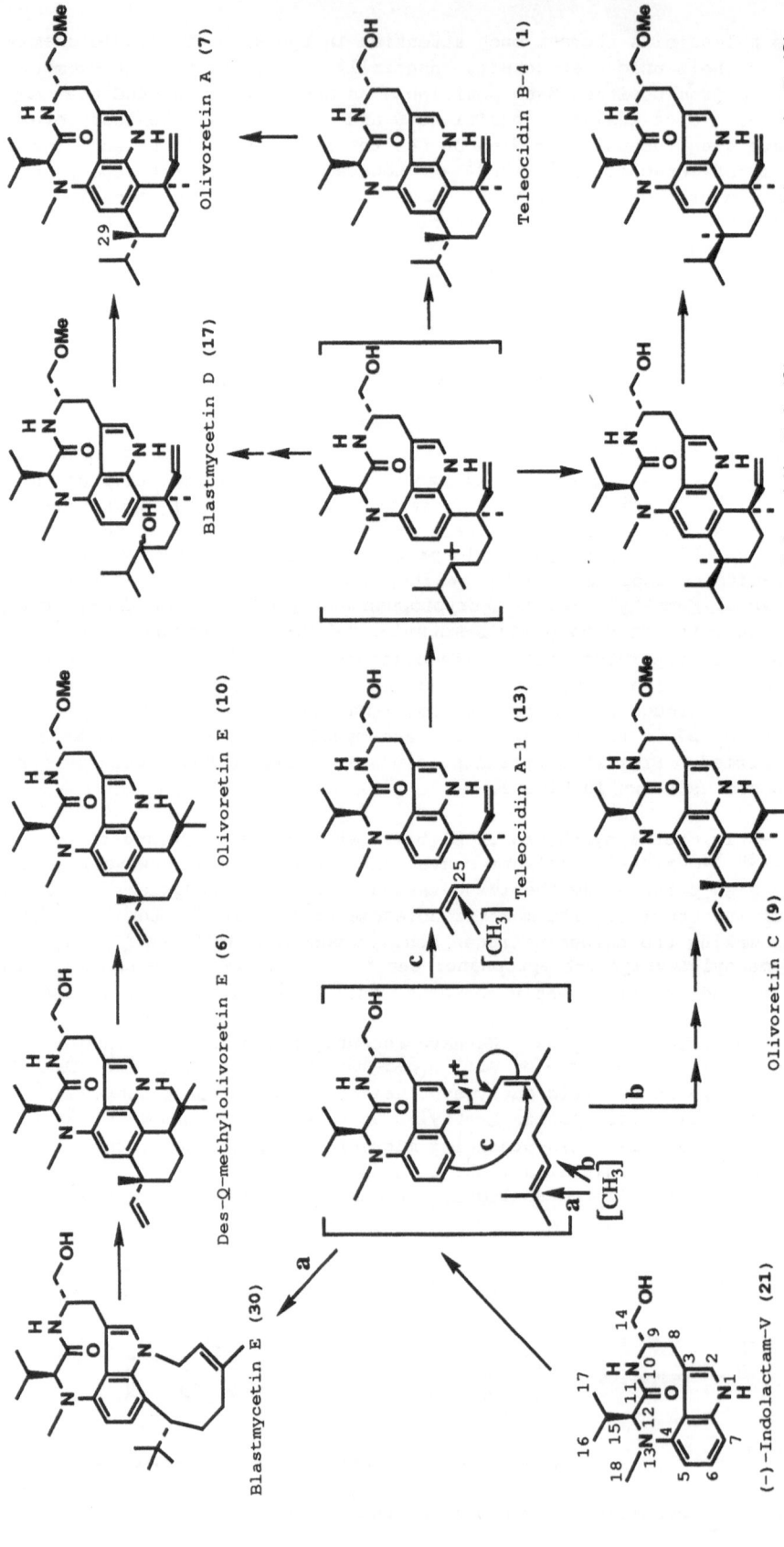

Fig. 4. Possible biosynthetic pathway of the monoterpenoid moieties of teleocidins.

cyclize 14-_O_-acetyl-8-oxo-_N_-methyl-L-valyl-L-tryptophanol by mean of regio-specific thallation in an analogy to the biosynthesis of **21**. However, this attempt was disappointing.

Biosynthetic pathway of the monoterpenoid moieties of teleocidins

The monoterpenoid moieties of teleocidins are classified into two classes: teleocidin As with a linalyl group at position 7, and teleocidin B's and olivoretins with a C_{11}-cyclic side chain at positions 6 and 7 (Fig. 2). The origin and insertion mechanism of the additional one carbon of the C_{11}-cyclic side chain of teleocidin B's and olivoretins are noteworthy. All monoterpenoid moieties of teleocidins except (-)-7-geranylindolactam-V (**20**) have a 1,1-dialkylallyl structure, which is thought to be biosynthesized through some Claisen type rearrangement. Moreover, the vinyl group is located on the 7 position side in teleocidin B's (**1-4**), while on the 6 position side in olivoretin C (**9**) and E (**10**). Olivoretin E (**10**) has a rare structure of the _tert_-butyl group. To make clear the biosynthetic pathway of these complex monoterpenoid moieties of teleocidins, we searched for biosynthetic intermediates between (-)-indolactam-V (**21**) and teleocidins in the mycelia of _S. blastmyceticum_ NA34-17, and found two new metabolites named blastmycetin D (**17**)[23] and E (**30**)[24] which gave invaluable information on the biosynthesis of the monoterpenoid moieties of teleocidins. The facts that blastmycetin D (**17**) gave olivoretin A (**7**) by treatment with phosphoric acid, and that the C-29 methyl group of **7** originated from L-methionine, indicated that the monoterpenoid moieties of teleocidin B type compounds were constructed by the methylation at position 25 of a teleocidin A-1 (**13**) type compound, and subsequent intramolecular cyclization as shown in Fig. 4. The involvement of the aza-Claisen rearrangement in the insertion of the monoterpenoid moieties on **21** was strongly suggested by the isolation of blastmycetin E (**30**), which was deduced to be formed through (-)-_N_[1]-nerylindolactam-V and converted into des-_O_-methylolivoretin E (**6**) by treatment with 1% acetic acid in methanol and water (1:1). This aza-Claisen rearrangement of **30** into **6** also provides knowledge concerning the biosynthetic pathway of olivoretin C (**9**) and E (**10**), whose vinyl groups of the cyclohexene ring are located on the 6 position side of the indole ring: namely, the C_{11} terpenoid moiety of **10** was deduced to be constructed by _N_[1]-nerylation of **21** followed by methylation at position **a** and intramolecular cyclization at position 7 like the case of blastmycetin D (**17**), and aza-Claisen rearrangement from position 1 to position 7 and subsequent 1,2-shift. The biosynthesis of olivoretin C (**9**) can be similarly explained by the difference of the position of methylation (position **b**). On the basis of these considerations, the possible biosynthetic pathway of the monoterpenoid moieties of teleocidins was proposed as shown in Fig. 4.[53]

STRUCTURE-ACTIVITY STUDIES IN THE INDUCTION OF EPSTEIN-BARR VIRUS BY TELEOCIDIN DERIVATIVES

The EBV-EA-inducing activity was measured in Raji cells by the method of Ito et al.[54] using n-butyric acid as a synergist. Figure 5 shows the EBV-EA-inducing activity of the typical compounds, (-)-indolactam-V (**21**) and teleocidin B-4 (**1**). The activity was expressed as the percentage of EA-positive cells, and evaluated in terms of this percentage and of the effective concentration required to induce EA-production. The effective concentration, EC_{50}, was defined as the concentration that induces half the percentage of EA-positive cells induced at the optimum concentration. A computer program (SAS)[55] of the probit procedure was used to calculate these concentrations. The percentage of EA-positive cells decreased due to cytotoxic effects at a superoptimum concentration of the drugs. Such a phenomenon was also reported by Eliasson et al.[8]

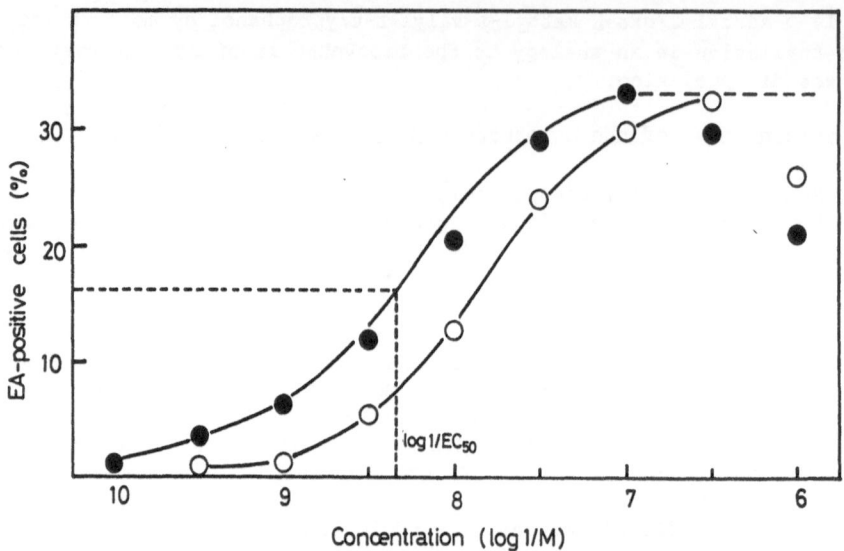

Fig. 5. Dose-response curves of (-)-indolactam-V (o) and teleocidin B-4
(●) in the induction of EBV-EA.

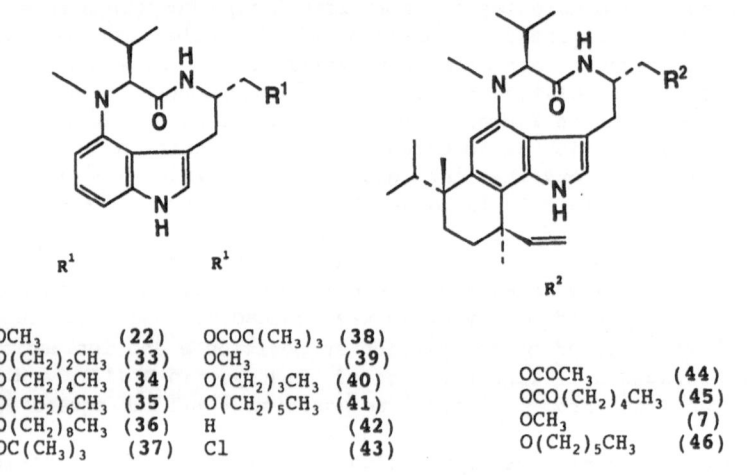

OCOCH$_3$	(22)	OCOC(CH$_3$)$_3$	(38)
OCO(CH$_2$)$_2$CH$_3$	(33)	OCH$_3$	(39)
OCO(CH$_2$)$_4$CH$_3$	(34)	O(CH$_2$)$_3$CH$_3$	(40)
OCO(CH$_2$)$_6$CH$_3$	(35)	O(CH$_2$)$_5$CH$_3$	(41)
OCO(CH$_2$)$_8$CH$_3$	(36)	H	(42)
OCOC(CH$_3$)$_3$	(37)	Cl	(43)

OCOCH$_3$	(44)
OCO(CH$_2$)$_4$CH$_3$	(45)
OCH$_3$	(7)
O(CH$_2$)$_5$CH$_3$	(46)

Fig. 6. Derivatives at position 14 of (-)-indolactam-V (21)
and teleocidin B-4 (1).

Modification of the hydroxyl group at position 14 of (-)-indolactam-V (21) and teleocidin B-4 (1)

Since teleocidins have both a hydrophilic region of the lactam moiety
and a hydrophobic region of the monoterpenoid side chains, the contribution
of the hydrophilic region to the EBV-EA-inducing activity was first examined.
We have synthesized a series of acyl and alkyl derivatives of 21 (33-41)
along with those of 1 (44-46), and examined their EBV-EA-inducing
activity.[56-58] As shown in Table 2, all 14-O-alkyl derivatives of (-)-
indolactam-V (39-41) were found to be far less active than (-)-indolactam-V
(21). Olivoretin A (14-O-methylteleocidin B-4) (7) showed weaker activity
than teleocidin B-4 (1), and 14-O-hexylteleocidin B-4 (46) was inactive.
Moreover, the activity of (-)-14-dehydroxyindolactam-V (42) and (-)-14-
chloroindolactam-V (43) was very weak. These results indicate that the free
hydroxyl group at position 14 plays an important role in the EBV-EA
induction.

Table 2. EBV-EA-inducing activity of teleocidin derivatives[a]

Compound number	Percentage of EA-positive cells at the indicated concentration (log 1/M)											log 1/EC$_{50}$ (log 1/M)	± SEM	n
	10.00	9.52	9.00	8.52	8.00	7.52	7.00	6.52	6.00	5.52	5.00			
1	1.1	3.4	6.1	11.8	20.4	28.9	32.8	29.5	21.0			8.35	0.04	8
7			0.1	0.1	2.4	2.9	12.0	22.5	28.4	17.4		6.98	0.01	2
11			0.1	5.6	10.0	16.2	22.0	31.8	25.0	22.0		7.61		1
19			0.1	2.1	9.2	14.2	21.8	29.5	20.4			7.57		1
21	0.1	0.2	1.1	5.4	12.6	24.0	29.9	32.2	27.0	22.1		7.90	0.04	8
22				0.1	0.1	9.1	17.5	27.0	29.9	14.7		7.16	0.01	2
23				0.1	0.3	1.0	1.8	4.6	12.9	14.8		not active		4
24	0.1	2.2	5.5	11.0	17.9	25.2	28.4	22.4	11.3			8.36	0.05	2
25						0.1	0.1	0.1	0.1	0.1	1.2	not active		1
26				0.1	2.6	4.2	5.3	11.9	9.7	9.2		not active		2
28						0.1	0.1	0.1	0.1	0.1	0.1	not active		1
29				0.1	0.1	0.1	6.1	11.9	7.4	6.1		not active		2
30				5.3	12.4	17.3	23.9	30.1	26.6			7.78	0.09	2
33				7.3	9.1	15.5	26.7	30.3	27.6	20.7		7.74		2
34				4.9	9.1	10.5	23.9	32.4	27.3	21.4		7.51	0.06	2
35				4.2	5.5	15.5	27.8	31.8	30.5	23.4		7.59		2
36				3.5	4.1	7.7	17.0	31.8	26.5	22.6		7.26		2
37				6.0	8.4	13.1	19.1	19.1	24.6	15.0		7.64		2
38				3.1	5.3	9.2	19.1	23.0	24.1	20.6		7.50		2
39				0.1	0.1	0.1	0.1	0.1	0.1	1.6		not active		2
40				0.1	0.1	0.1	0.1	0.1	0.1	0.1		not active		2
41				0.1	0.1	0.1	0.1	0.1	0.1	0.1		not active		2
42						7.2	10.2	13.7	18.9	21.4	15.7	7.02	0.21	2
43					4.3	5.2	9.2	14.2	20.9	29.4	22.3	6.69	0.15	2
44			0.1	3.8	7.5	13.5	23.8	31.2	27.8	19.9		7.55	0.00	2
45			0.1	3.8	11.1	18.4	25.1	31.2	24.4	19.0		7.70	0.02	2
46							0.1	0.1	2.8	11.5		not active		2
47			2.8	4.0	9.7	11.8	16.1	20.5	24.9	31.0	29.4	7.22	0.05	2
48			1.4	3.8	10.7	12.2	14.9	22.4	24.7	29.6	27.1	7.29	0.01	2
49			3.1	5.3	9.5	12.3	15.2	22.2	30.5	22.2	19.2	7.37	0.08	2
50						2.1	4.5	10.0	15.1	27.5	12.7	6.34	0.02	2
51						4.7	8.4	12.5	15.5	27.4	18.6	6.53	0.04	2
52				2.5	6.7	13.4	16.3	17.0	15.8			not active		1
53			0.1	8.3	16.4	27.3	32.8	25.7	22.9	11.6		8.05	0.03	2
54				0.1	0.1	0.1	0.1	0.1	3.0	5.7		not active		2
55				0.1	0.1	0.6	2.0	2.4	5.1	5.9		not active		2
56				0.1	1.8	4.3	12.2	21.9	30.9	20.3		6.92	0.01	2
57				1.2	4.2	9.9	12.5	15.2	23.5	14.2		7.17		1
58			0.1	2.8	13.3	18.5	25.4	30.5	24.4	18.1		7.74	0.03	2
59				0.1	0.1	0.1	0.1	0.1	0.1	0.6		not active		2
60				0.1	0.1	0.1	0.1	0.1	1.0	4.4		not active		2
61	1.0	2.1	5.8	10.4	21.0	28.4	32.7	28.9	22.2	15.6		8.33	0.06	2
62	0.1	1.1	3.7	7.4	21.2	25.2	28.7	31.8	24.1	20.9		8.13	0.03	2
63			0.1	1.8	12.7	25.1	30.7	27.8	24.4	21.0		7.91	0.06	2
64	0.1	0.3	2.8	7.2	17.2	26.3	30.3	26.1	17.2	8.6		8.16	0.03	5
65	1.9	6.6	7.9	11.7	19.7	25.0	30.6	26.9	15.5	7.2		8.46	0.02	2
66	1.6	2.5	7.7	10.6	17.0	26.8	29.4	21.9	13.7	7.4		8.40	0.00	2
67	1.1	2.9	7.6	9.8	17.3	25.9	28.3	21.6	12.6	8.4		8.42	0.05	3
68	1.1	2.1	3.8	12.2	15.8	21.4	26.7	19.4	12.0	5.9		8.32	0.05	2
69	2.2	5.2	13.4	16.1	20.6	25.7	30.3	24.3	19.8	17.8	11.7	8.64	0.07	2
70		0.1	4.6	9.0	11.8	20.7	26.2	28.1	30.0	29.0	25.7	8.01	0.01	2
71	0.1	0.1	6.4	7.9	18.0	24.0	29.3	26.6	18.7	8.0		8.23	0.04	2
72	1.2	7.1	8.7	13.0	20.5	27.7	28.9	24.5	12.7	8.0		8.63	0.01	2
73	2.1	7.1	14.2	16.9	23.0	27.0	31.1	24.9	22.4	19.2	11.8	8.72	0.07	2
74				1.2	4.5	10.1	15.8	23.8	31.1	31.0	27.1	7.66		1
75				0.1	2.5	10.3	16.7	25.8	27.0	20.5		7.79		1
76			6.1	10.5	21.6	28.2	33.0	16.6	15.6			8.29		1
77			2.0	7.3	13.5	20.4	25.6	31.0	15.2	12.0		7.88	0.08	2
78				1.2	5.5	11.8	18.1	23.7	19.6			7.55	0.09	2
79			4.0	11.0	17.6	26.4	29.1	31.1	21.1			8.19	0.09	2

[a] Sodium n-butyrate was added to all samples to enhance the sensitivity of Raji cells. Only 0.1% EA-induction was demonstrated at 4mM. The viability of the cells exceeded 60% in each experiment.

Table 3. Effect of DEF on the EBV-EA-inducing activity of teleocidin B-4
(1), (-)-indolactam-V (21) and (-)-14-O-hexanoylindolactam-V
(34)[a]

Compound	Percentage of EA-positive cells when incubated with DEF at the indicated concentrations (μM)			
	0	1	10	100
Teleocidin B-4 (1)	31.6	30.3 (4.1)[b]	30.1 (4.7)	31.7 (0.0)
(-)-indolactam-V (21)	30.1	30.0 (0.0)	29.6 (1.7)	27.1 (10.0)
(-)-14-O-Hexanoyl- indolactam-V (34)	23.4	22.1 (5.6)	12.6 (46.2)	9.6 (59.0)

[a]Raji cells were incubated with sodium n-butyrate (4mM), each test compound
(0.1μM) and DEF at concentrations of 1, 10 or 100 μM. The values represent
the average of 2 experiments. DEF showed only 0.1% EA-positive cells at all
concentrations examined. Cell viability exceeded 60% in each experiment.
[b]Values in parentheses are percentage of inhibition.

The series of acyl derivatives (22,33-38,44,45) showed, however,
potent EBV-EA-inducing activity. Large differences in activity between the
14-O-acyl derivatives and the 14-O-alkyl derivatives suggests that the
activity of the 14-O-acyl derivatives is attributable to (-)-indolactam-V
(21) or teleocidin B-4 (1) formed through hydrolysis of the ester groups. To
prove this mechanism, Raji cells were incubated with (-)-14-O-
hexanoylindolactam-V (34) and the esterase inhibitor, S,S,S-tributyl
phosphorotrithioate (DEF), which had been found to be a potent potentiator of
the insecticide [(dimethoxyphosphinothioyl)thio]butanedioic acid diethyl
ester (malathion).[59] As shown in Table 3, the activity of 34 fell
drastically with increasing inhibitor concentration, in contrast to those of
21 and 1.[58] In addition, the existence of an esterase in Raji cells was
confirmed by a conventional biochemical method (unpublished results). These
results strongly suggest that at least some part of the activity of the 14-O-
acyl derivatives arise through hydrolysis of the ester groups.

Introduction of substituents into the indole ring of (-)-indolactam-V (21)

The EBV-EA-inducing activity of (-)-indolactam-V (21) was 2.8 times
weaker than that of teleocidin B-4 (1), indicating that the monoterpenoid
side chain of 1 acted to amplify the activity. To examine the role of this
moiety, a series of N[1]-alkyl derivatives of 21 (47-51) was first prepared.
As shown in Table 2, the activity of (-)-N[1]-methyl, ethyl- and butyl-
indolactam-V (47-49) was about 5 times weaker than that of 21, and (-)-N[1]-
hexyl- and octylindolactam-V (50, 51) were even less active. Moreover,
blastmycetin E (30), whose monoterpenoid side chain cyclized from position 1
to position 7, was less active than teleocidin B-4 (1). These results
revealed that the free imino group at position 1 played a significant role in
the activity, and that a substituent should attach to a position other than
position 1 of the indole ring.[58]

We examined, therefore, substitution reactions on the indole ring of 21
and obtained the 25 derivatives shown in Fig. 7 (52-76), in which the hydro-

gen at position 2, 5 or 7 of **21** was replaced by various substituents.[60-62] The EBV-EA-inducing activity of these derivatives is also summarized in Table 2. (-)-2-Prenylindolactam-V (**52**) and (-)-2,7-dibromoindolactam-V (**53**) had lower activity than (-)-7-prenylindolactam-V (**76**) and (-)-7-bromoindolactam-V (**61**), respectively. Moreover, (-)-2,7-diacylindolactam-V's (**54-56**) were inactive or only slightly active. As to the substituents at position 5, the activity of (-)-5-prenylindolactam-V (**57**) and (-)-5-chloroindolactam-V (**58**) was weaker than (-)-indolactam-V (**1**), and (-)-5-acetylindolactam-V (**59**) and (-)-5,7-diacetylindolactam-V (**60**) were inactive. Thus, large substituents at positions 2 and 5 of the indole ring of **21** remarkably lowered the activity.[63,64]

On the other hand, all 7-substituted derivatives (**61-76**) showed high EBV-EA-inducing activity.[63] The activity of a series of (-)-7-acylindolactam-V's (**64-68,74**) increased as the acyl chain extended, and reached the maximum at (-)-7-butanoylindolactam-V (**65**), whose activity was comparable to that of teleocidin B-4 (**1**). However, further elongation of the acyl chain resulted in a slight decrease in activity, suggesting the existence of an optimum hydrophobicity of the substituents for activity. Similar activity of

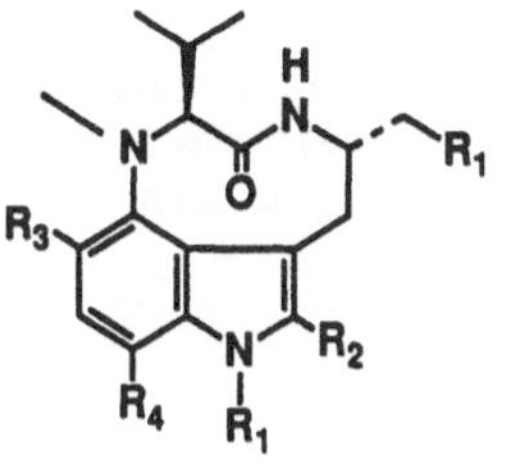

R_1	R_2	R_3	R_4	
CH_3	H	H	H	(47)
CH_2CH_3	H	H	H	(48)
$CH_2(CH_2)_2CH_3$	H	H	H	(49)
$CH_2(CH_2)_4CH_3$	H	H	H	(50)
$CH_2(CH_2)_6CH_3$	H	H	H	(51)
H	$CH_2CH=CH(CH_3)_2$	H	H	(52)
H	Br	H	Br	(53)
H	$COCH_3$	H	$COCH_3$	(54)
H	$CO(CH_2)_2CH_3$	H	$CO(CH_2)_2CH_3$	(55)
H	$CO(CH_2)_8CH_3$	H	$CO(CH_2)_8CH_3$	(56)
H	H	$CH_2CH=CH(CH_3)_2$	H	(57)
H	H	Cl	H	(58)
H	H	$COCH_3$	H	(59)
H	H	$COCH_3$	$COCH_3$	(60)
H	H	H	Br	(61)
H	H	H	I	(62)
H	H	H	SCH_3	(63)
H	H	H	$COCH_3$	(64)
H	H	H	$CO(CH_2)_2CH_3$	(65)
H	H	H	$CO(CH_2)_4CH_3$	(66)
H	H	H	$CO(CH_2)_6CH_3$	(67)
H	H	H	$CO(CH_2)_{10}CH_3$	(68)
H	H	H	COC_6H_5	(69)
H	H	H	$CO(CH_2)_2COOCH_3$	(70)
H	H	H	CH_2CH_3	(71)
H	H	H	$CH_2(CH_2)_6CH_3$	(72)
H	H	H	$CH_2C_6H_5$	(73)
H	H	H	CHO	(74)
H	H	H	$CH=CHNO_2$	(75)
H	H	H	$CH_2CH=CH(CH_3)_2$	(76)

Fig. 7. Structures of 1, 2, 5 and/or 7 substituted indolactams.

the derivatives with electron-withdrawing groups (64,67,69) and those with electron-donating groups (71-73) suggested that the electronic factor was not significant for the activity. The activity of (-)-7-benzoyl- and benzyl-indolactam-V (69,73) with bulky substituents was almost equal to that of (-)-7-acyl- and alkylindolactam-V, and even blastmycetin A (24), a dimer of 21, was proved to be a potent EBV-EA inducer. From these results, the low structural requirement at position 7 of (-)-indolactam-V (21) was revealed. However, it is nearly impossible to determine what are the physicochemical parameters of the substituents at position 7, for example, steric, hydrophobic and electronic, significant in EBV-EA induction because the difference in the activity between the weakest derivative (74) and the strongest one (73) was very small (at the outside, 10-fold). This means that the structure of (-)-indolactam-V (21) already has a requisite condition for EBV-EA induction.

The contribution of the isopropyl group at position 12 of 21 to the EBV-EA-inducing activity has not yet been examined. However, our latest studies[65] on the new indolactam congeners obtained by the above-mentioned microbial conversion,[52] and the earlier work of Shudo et al.[66] indicated that the isopropyl group of 21 was involved in the hydrophobic interaction on the receptor site. Since there was a good correlation between the EBV-EA-inducing activity and several in vitro biological activities related to tumor promotion, it would be safe to say that hydrophobic substituents at position 12 of 21, such as (-)-indolactam-I (31), would enhance EBV-EA-inducing activity. On the basis of these considerations, the structural factors required for EBV-EA-inducing activity is summarized in Fig. 8.

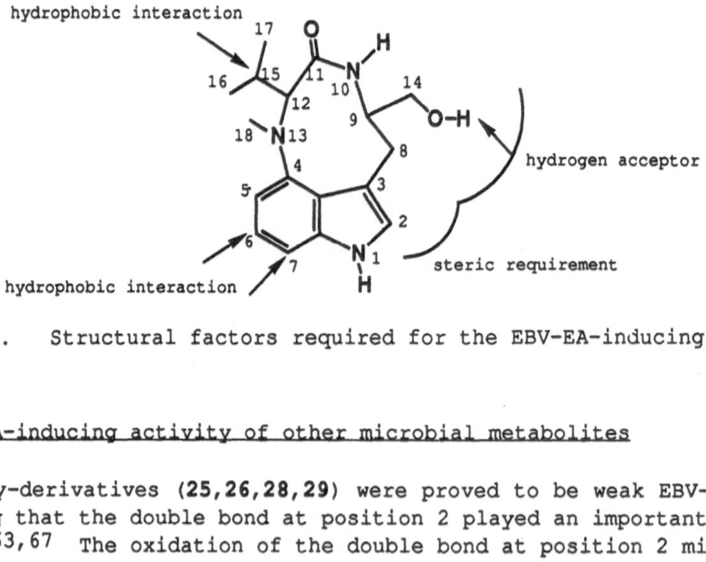

Fig. 8.   Structural factors required for the EBV-EA-inducing activity.

The EBV-EA-inducing activity of other microbial metabolites

2-Oxy-derivatives (25,26,28,29) were proved to be weak EBV-EA inducers, suggesting that the double bond at position 2 played an important role in the activity.[63,67] The oxidation of the double bond at position 2 might be thought to lower the activity by changing the conformation of the nine-membered lactam ring. The activities of des-N-methylteleocidin B-4 (11) and des-N-methylteleocidin A-1 (19) were lower than that of teleocidin B-4 (1), while (-)-des-N-methylindolactam-V (23) was almost inactive. These results indicate that the N^{13}-methyl group is significant in EBV-EA-inducing activity.[26] Our recent study on the metabolism of (-)-indolactam-V (21) and teleocidin B-4 (1)[26,67] revealed that the major metabolites by rat liver microsomes were (-)-des-N-methylindolactam-V (23), (-)-2-oxyindolactam-V's (26,29) and des-N-methylteleocidin B-4 (11). The weak activities of these metabolites thus demonstrated that the metabolism of 21 and 1 resulted in detoxification and that they themselves were the active principles.

Concluding remarks

These findings on structure-activity relationship and metabolism could provide important information for synthesizing stable radioactive, fluorescent or photolabile teleocidin derivatives as new probes for the receptor analysis of EBV-EA-inducers. We have recently synthesized several biologically active fluorescent or photolabile teleocidin derivatives as shown in Fig. 9.[61,62,68] Fluorescent or photolabile groups were introduced at position 7 of the indole ring on the basis of our structure-activity studies. As shown in Table 2, (-)-7-(2'-N-dansylaminoethyl)-indolactam-V (77), (-)-7-[2'-N-(2,6-mansyl)aminoethyl]-indolactam-V (78) and (-)-7-[2'-N-(4"-azidobenzoyl)aminoethyl]-indolactam-V (79) induced EA in up to 30% of total cells at a concentration of 0.3μM.[64,68,69] The activity of 80 and 81 has not yet been tested. Furthermore, ^{3}H-labeled photoaffinity probes with a high specific radioactivity of more than 50Ci/mmol and high radiochemical purity of more than 96% were achieved using commercially available ^{3}H-labeled succinimidyl-4-azidobenzoate (Du Pont/NEN Research Products). These compounds could give clues to the identification of the target sites of teleocidins and the elucidation of the mechanism of EBV-EA induction. These results might ultimately lead to the development of new anti-viral agents.

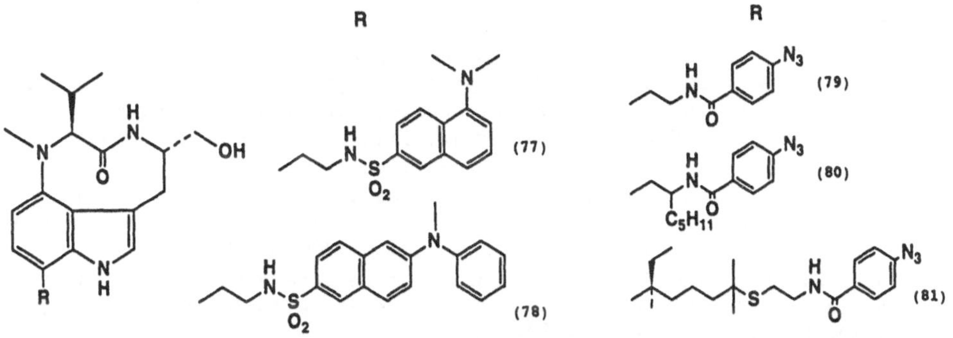

Fig. 9.   Structure of fluorescent and photolabile indolactam derivatives.

ACKNOWLEDGEMENTS

This paper is the summary of the collaborative work of Drs M. Hirota, N. Hagiwara, A. Funaki, S. Okuno and S. Kajiyama of the Department of Food Science and Technology, Kyoto University; Drs. H. Hayashi, M. Arai and S. Murao of the Department of Agricultural Chemistry, University of Osaka Prefecture; Drs. H. Nishino, H. Tokuda and A. Iwashima of the Department of Biochemistry, Kyoto Prefectural University of Medicine; Dr. T. Osato of the Faculty of medicine, Hokkaido University; and the late Dr. Y. Ito of the Faculty of Medicine, Kyoto University.

The authors wish to express their thanks to Dr. H. Fujiki of the National Cancer Research Institute for valuable discussion, to Dr. K. Shudo of the Faculty of Pharmaceutical Sciences at University of Tokyo and Dr. S. Sakai of the Faculty of Pharmaceutical Sciences at Chiba University, for the spectral data and authentic samples of several teleocidin-related compounds.

REFERENCES

1. M. A. Epstein and B. G. Achong, "Epstein-Barr Virus," Springer-Verlag, Berlin (1979).
2. E. Hecker, Phorbol esters from croton oil: chemical nature and biological activities, *Naturwissenschaften*, **54**, 282 (1967).
3. H. Fujiki, M. Mori, M. Nakayasu, M. Terada, T. Sugimura and R. E. Moore, Indole alkaloids: dihydroteleocidin B, teleocidin, and lyngbyatoxin A as members of a new class of tumor promoters, *Proc. Natl. Acad. Sci. USA*, **78**, 3872 (1981).
4. H. Fujiki, M. Suganuma, M. Nakayasu, H. Hoshino, R. E. Moore and T. Sugimura, The third class of new tumor promoters, polyacetates (debromoaplysiatoxin and aplysiatoxin), can differentiate biological actions relevant to tumor promoters, *Jpn. J. Cancer Res. (Gann)*, **73**, 495 (1982).
5. H. Fujiki, Y. Tanaka, R. Miyake, U. Kikkawa, Y. Nishizuka and T. Sugimura, Activation of calcium-activated, phospholipid-dependent protein kinase (protein kinase C) by new classes of tumor promoters: teleocidin and debromoaplysiatoxin, *Biochem. Biophys. Res. Commun.*, **120**, 339 (1984).
6. H. zur Hausen, G. W. Bornkamm, R. Schmidt and E. Hecker, Tumor initiators and promoters in the induction of Epstein-Barr virus, *Proc. Natl. Acad. Sci. USA*, **76**, 782 (1979).
7. H. Yamamoto, T. Katsuki, Y. Hinuma, H. Hoshino, M. Miwa, H. Fujiki and T. Sugimura, Induction of Epstein-Barr virus by a new tumor promoter, teleocidin, compared to induction by TPA, *Int. J. Cancer*, **28**, 125 (1981).
8. L. Eliasson, B. Kallin, M. Patarroyo, G. Klein, H. Fujiki and T. Sugimura, Activation of the EBV-cycle and aggregation of human blood lymphocytes by the tumor promoters teleocidin, lyngbyatoxin A, aplysiatoxin and debromoaplysiatoxin, *Int. J. Cancer*, **31**, 7 (1983).
9. K. Irie, M. Hirota, N. Hagiwara, K. Koshimizu, H. Hayashi, S. Murao, H. Tokuda and Y. Ito, The Epstein-Barr virus early antigen inducing indole alkaloids, (-)-indolactam V and its related compounds, produced by Actinomycetes, *Agric. Biol. Chem.*, **48**, 1269 (1984).
10. Y. Endo, K. Shudo and T. Okamoto, Molecular requirements for epigenetic modulators. Synthesis of active fragments of teleocidins and lyngbyatoxin, *Chem. Pharm. Bull.*, **30**, 3457 (1982).
11. M. Takashima and H. Sakai, A new toxic substance, teleocidin, produced by Streptomyces. Part I. Production, isolation and chemical studies, *Bull. Agric. Chem. Soc. Japan*, **24**, 647 (1960).
12. H. Nakata, H. Harada and Y. Hirata, X-Ray structure determination of dihydroteleocidin B monobromoacetate, *Tetrahedron Lett.*, 2523 (1966).
13. S. Sakai, Y. Hitotsuyanagi, N. Aimi, H. Fujiki, M. Suganuma, T. Sugimura, Y. Endo and K. Shudo, Absolute configuration of lyngbyatoxin A (teleocidin A-1) and teleocidin A-2, *Tetrahedron Lett.*, **27**, 5219 (1986).
14. Y. Hitotsuyanagi, H. Fujiki, M. Suganuma, N. Aimi, S. Sakai, Y. Endo, K. Shudo and T. Sugimura, Isolation and structure elucidation of teleocidin B-1, B-2, B-3, and B-4, *Chem. Pharm. Bull.*, **32**, 4233 (1984).
15. J. H. Cardellina II, F. -J. Marner and R. E. Moore, Seaweed dermatitis: structure of lyngbyatoxin A, *Science*, **204**, 193 (1979).
16. Y. E. Konev, Y. D. Shenin, V. V. Venediktov and V. P. Sergutina, *Streptoverticillium krissii* species nova producing teleocidins A and B, *Mikrobiologiya*, **44**, 248 (1975).
17. S. Sakai, N. Aimi, K. Yamaguchi, Y. Hitotsuyanagi, C. Watanabe, K. Yokose, Y. Koyama, K. Shudo and A. Itai, Elucidation of the structure of olivoretin A and D (teleocidin B), *Chem. Pharm. Bull.*, **32**, 354 (1984).

18. T. Yamashita, M. Imoto, K. Isshiki, T. Sawa, H. Naganawa, S. Kurasawa, B.-Q. Zhu and K. Umezawa, Isolation of a new indole alkaloid, pendolmycin, from Nocardiopsis, J. Natl. Prod., **51**, 1184 (1988).

19. S. Oka, T. Owa, M. Sugie and H. Tanaka, Actions of teleocidins on in vitro CFU-GM and CFU-E colony formation, Agric. Biol. Chem., **53**, 2261 (1989).

20. H. Kumagai, M. Iijima, M. Osono, T. Masuda, K. Dobashi, H. Naganawa, M. Hamada, T. Sawa, M. Ishizuka and T. Takeuchi, Studies on the new immuno-regulating compound, cytoblastin, produced by Streptoverticillium eurocidicum MI43-37, Nippon Nogeikagaku Kaishi, **65**, 190 (1991).

21. K. Irie, N. Hagiwara, K. Koshimizu, H. Hayashi, S. Murao and Y. Ito, Identification of the microorganism which produces Epstein-Barr virus early antigen-inducing indole alkaloids, Agric. Biol. Chem., **49**, 845 (1985).

22. K. Irie, N. Hagiwara, T. Kurome, H. Hayashi, M. Arai and K. Koshimizu, New teleocidin-related metabolites from Streptoverticillium blastmyceticum producing tumor-promoting indole alkaloids, Agric. Biol. Chem., **51**, 285 (1987).

23. K. Irie, N. Hagiwara, A. Funaki, H. Hayashi, M. Arai and K. Koshimizu, Isolation of blastmycetin D, a possible precursor of teleocidins, from Streptoverticillium blastmyceticum, Agric. Biol. Chem., **51**, 1733 (1987).

24. K. Irie, A. Funaki, K. Koshimizu, H. Hayashi and M. Arai, Structure of blastmycetin E, a new teleocidin-related compound, from Streptoverticillium blastmyceticum, Tetrahedron Lett., **30**, 2113 (1989).

25. N. Hagiwara, K. Irie, A. Funaki, H. Hayashi, M. Arai and K. Koshimizu, Structure and tumor-promoting activity of new teleocidin-related metabolites (blastmycetins) from Streptoverticillium blastmyceticum, Agric. Biol. Chem., **52**, 641 (1988).

26. K. Irie, N. Hagiwara, A. Funaki, H. Hayashi, M. Arai, H. Tokuda and K. Koshimizu, Isolation and biological activities of N^{13}-desmethyl analogues of tumor promoter teleocidin from Streptoverticillium blastmyceticum, Agric. Biol. Chem., **52**, 3193 (1988).

27. K. Irie, S. Kajiyama, K. Koshimizu, H. Hayashi and M. Arai, Isolation and biosynthesis of (-)-indolactam I, a new congener of indole alkaloid tumor promoter teleocidins, Tetrahedron Lett., **31**, 7337 (1990).

28. Y. Endo, K. Shudo, A. Itai, M. Hasegawa and S. Sakai, Synthesis and stereochemistry of indolactam-V, an active fragment of teleocidins. Structural requirements for tumor-promoting activity, Tetrahedron, **42**, 5905 (1986).

29. Y. Hitotsuyanagi, K. Yamaguchi, K. Ogata, N. Aimi, S. Sakai, Y. Koyama, Y. Endo, K. Shudo, A. Itai and Y. Iitaka, Elucidation of the structures of olivoretin B and C, Chem. Pharm. Bull., **32**, 3774 (1984).

30. S. Sakai, Y. Hitotsuyanagi, K. Yamaguchi, N. Aimi, K. Ogata, T. Kuramochi, H. Seki, R. Hara, H. Fujiki, M. Suganuma, T. Sugimura, Y. Endo, K. Shudo and Y. Koyama, The structures of additional teleocidin class tumor promoters, Chem. Pharm. Bull., **34**, 4883 (1986).

31. Y. Hitotsuyanagi, K. Yamaguchi, K. Ogata, N. Aimi, S. Sakai, H. Fujiki, M. Suganuma, T. Sugimura, Y. Endo, K. Shudo and Y. Koyama, Chemical studies on cancer promoter "teleocidins", in: 27th Symposium on the Chemistry of Natural Products, Hiroshima 1985, Symposium papers, pp. 624.

32. N. Aimi, H. Odaka, S. Sakai, H. Fujiki, M. Suganuma, R. E. Moore and G. M. L. Patterson, Lyngbyatoxin B and C, two new irritants from Lyngbya majuscula, J. Natl. Prod., **53**, 1593 (1990).

33. S. E. de Laszlo, S. V. Ley and R. A. Porter, Synthetic approaches to the teleocidin-related tumor promoters: a total synthesis of (±)-indolactam V, J. Chem. Soc., Chem. Commun., 344 (1986).

34. S. Nakatsuka, T. Masuda, K. Sakai and T. Goto, Synthetic studies on teleocidin IV. An efficient synthesis of (-)-indolactam V, Tetrahedron Lett., **27**, 5735 (1986).

35. H. Muratake and M. Natsume, Total synthesis of lyngbyatoxin A (teleocidin A-1) and teleocidin A-2, Tetrahedron Lett., **28**, 2265 (1987).

36. S. Nakatsuka, T. Masuda and T. Goto, Total syntheses of (±)-teleocidin B-3 and B-4, Tetrahedron Lett., **28**, 3671 (1987).

37. Y. Endo, M. Hasegawa, A. Itai and K. Shudo, Synthesis and stereochemistry of indolactam congeners. Conformational behavior of the nine-membered lactams, Tetrahedron, **43**, 3695 (1987).

38. M. Mascal and C. J. Moody, Synthesis of (-)-indolactam V, J. Chem. Soc., Chem. Commun., 589 (1988).

39. H. Muratake, K. Okabe and M. Natsume, Total synthesis of dihydroteleocidin B-4 (dihydroteleocidin B), Tetrahedron Lett., **29**, 6267 (1988).

40. K. Okabe, H. Muratake and M. Natsume, Synthesis of optically active teleocidins B-3 and B-4, Chem. Pharm. Bull., **37**, 563 (1989).

41. T. Masuda, S. Nakatsuka and T. Goto, Efficient syntheses and chemistry of indolactam-V and its analogs, Agric. Biol. Chem., **53**, 2257 (1989).

42. A. P. Kozikowski, K. Sato, A. Basu and J. S. Lazo, Synthesis and biological studies of simplified analogues of lyngbyatoxin A: Use of an isoxazoline-based indole synthesis. Quest for protein kinase C modulators, J. Am. Chem. Soc., **111**, 6228 (1989).

43. K. Okabe, H. Muratake and M. Natsume, Total synthesis of indole alkaloid pendolmycin, Tetrahedron, **46**, 5113 (1990).

44. T. P. Kogan, T. C. Somers and M. C. Venuti, A regio- and stereocontrolled total synthesis of (-)-indolactam-V, Tetrahedron, **46**, 6623 (1990).

45. R. R. Webb II, M. C. Venuti and C. Eigenbrot, Synthesis of a tetramethyl analogue of teleocidin, J. Org. Chem., **56**, 4706 (1991).

46. A. P. Kozikowski, P. W. Shum, A. Basu and J. S. Lazo, Synthesis of structural analogues of lyngbyatoxin A and their evaluation as activators of protein kinase C, J. Med. Chem., **34**, 2420 (1991).

47. K. Okabe and M. Natsume, The second generation synthesis of a tumor promoter pendolmycin, Tetrahedron, **47**, 7615 (1991).

48. H. Muratake and M. Natsume, Synthesis of teleocidins A, B and their congeners. Part 1. An efficient synthesis method of N-(7-alkyl-4-indolyl)-N-methyl-L-valine esters, essential intermediates for teleocidin synthesis, Tetrahedron, **47**, 8535 (1991).

49. H. Muratake, K. Okabe and M. Natsume, Synthesis of teleocidins A, B and their congeners. Part 2. Synthesis of lyngbyatoxin A (teleocidin A-1), teleocidin A-2, pendolmycin, and (R,E)- and (S,E)-7-(3,7,11-trimethyl-1,6,10-dodecatrien-3-yl)-(-)-indolactams V, Tetrahedron, **47**, 8545 (1991).

50. K. Okabe, H. Muratake and M. Natsume, Synthesis of teleocidins A, B and their congeners. Part 3. Synthesis of dihydroteleocidin B-4 (dihydroteleocidin B), teleocidin B-3 and teleocidin B-4, Tetrahedron, **47**, 8559 (1991).

51. K. Irie, S. Kajiyama, A. Funaki, K. Koshimizu, H. Hayashi and M. Arai, Biosynthesis of (-)-indolactam V, the basic ring-structure of tumor promoters teleocidins, Tetrahedron Lett., **31**, 101 (1990).

52. S. Kajiyama, K. Irie, T. Kido, K. Koshimizu, H. Hayashi and M. Arai, Synthesis of new indolactam analogues by microbial conversion, Tetrahedron, **47**, 5453 (1991).

53. K. Irie, S. Kajiyama, A. Funaki, K. Koshimizu, H. Hayashi and M. Arai, Biosynthesis of indole alkaloid tumor promoters teleocidins (I). Possible biosynthetic pathway of the monoterpenoid moieties of teleocidins, Tetrahedron, **46**, 2773 (1990).

54. Y. Ito, S. Yanase, J. Fujita, T. Harayama, M. Takashima and H. Imanaka, A short-term in vitro assay for promoter substances using human lymphoblastoid cells latently infected with Epstein-Barr virus, Cancer Lett., **13**, 29 (1981).

55. H. G. James, Probit procedure, in: "SAS User's Guide", T. H. Jane and A. C. Kathryn, ed., SAS Institute, Cary, NC, pp. 357 (1979).

56. K. Irie, N. Hagiwara, K. Koshimizu, H. Hayashi, S. Murao, H. Tokuda and Y. Ito, Isolation and the Epstein-Barr virus early antigen inducing activity of olivoretins from Streptoverticillium blastmyceticum, Agric. Biol. Chem., **49**, 221 (1985).

57. K. Irie, N. Hagiwara, K. Koshimizu, H. Tokuda, Y. Ito, H. Hayashi and S. Murao, Epstein-Barr virus early antigen inducing activity of 14-O-derivatives of (-)-indolactam V, Agric. Biol. Chem., **49**, 1441 (1985).

58. K. Irie, H. Tokuda, N. Hagiwara, K. Koshimizu, H. Hayashi, S. Murao and Y. Ito, Structure-activity relationship in the induction of Epstein-Barr virus by teleocidin derivatives, Int. J. Cancer, **36**, 485 (1985).

59. J. E. Casida, Mixed-function oxidase involvement in the biochemistry of insecticide synergists, J. Agr. Food. Chem., **18**, 753 (1970).

60. K. Irie, H. Hayashi, M. Arai and K. Koshimizu, Substitution reaction on the indole ring of (-)-indolactam V, the fundamental structure of teleocidins, Agric. Biol. Chem., **50**, 2679 (1986).

61. K. Irie, N. Hagiwara and K. Koshimizu, Synthesis of a biologically active fluorescent indolactam derivative; a method preparing new probes for receptor analysis of tumor promoters, Tetrahedron Lett., **28**, 3267 (1987).

62. K. Irie, N. Hagiwara and K. Koshimizu, New probes for receptor analysis of tumor promoters; synthesis of fluorescent derivatives of (-)-indolactam V, the basic ring-structure of teleocidins, Tetrahedron, **43**, 5251 (1987).

63. K. Irie, N. Hagiwara, H. Tokuda and K. Koshimizu, Structure-activity studies of the indole alkaloid tumor promoter teleocidins, Carcinogenesis, **8**, 547 (1987).

64. K. Irie and K. Koshimizu, Structure-activity studies of indole alkaloid tumor promoters, Mem. Coll. Agric., Kyoto Univ., **132**, 1 (1988).

65. K. Irie, S. Okuno, S. Kajiyama, K. Koshimizu, H. Nishino and A. Iwashima, Quantitative structure-activity studies on indole alkaloid tumor promoter indolactam congeners, Carcinogenesis, **12**, 1883 (1991).

66. M. Hirota, M. Suganuma, S. Yoshizawa, T. Horiuchi, M. Nakagawa, M. Hasegawa, Y. Endo, K. Shudo and H. Fujiki, Synthetic analogues (indolactams) of (-)-indolactam-V are new congeners of the teleocidin class of tumor promoters, Jpn. J. Cancer Res. (Gann), **78**, 577 (1987).

67. N. Hagiwara, K. Irie, H. Tokuda and K. Koshimizu, The metabolism of indole alkaloid tumor promoter, (-)-indolactam V, which has the fundamental structure of teleocidins, by rat liver microsomes, Carcinogenesis, **8**, 963 (1987).

68. S. Okuno, K. Irie, H. Nishino, A. Iwashima and K. Koshimizu, Syntheses and biological activities of photolabile indolactam derivatives; new probes for the receptor analysis of tumor promoters, Agric. Biol. Chem., **54**, 1885 (1990).

69. K. Irie, S. Okuno, K. Koshimizu, H. Tokuda, H. Nishino and A. Iwashima, Biological activities and cellular uptake studies of fluorescent derivatives of indole alkaloid tumor promoter teleocidin, Int. J. Cancer, **43**, 513 (1989).

INDEX